Abner Shimony is an eminent philosopher and theoretical physicist, best known for contributions to experiments on foundations of quantum mechanics, notably the polarization correlation test of Bell's Inequality and thereby of the family of local hidden-variables theories. *Search for a Naturalistic World View* consists of essays written over a period of four decades and aims at linking the natural sciences, especially physics, biology, and psychology, with epistemology and metaphysics. It will serve professionals and students in all of these disciplines.

Volume I, *Scientific Method and Epistemology,* advocates an "integral epistemology" combining conceptual analysis with results of empirical science. It proposes a version of scientific realism that emphasizes causal relations between physical and mental events and rejects a physicalist account of mentality. It offers a "tempered personalist" version of scientific methodology, which supplements Bayesianism with *a posteriori* principles distilled from exemplary cognitive achievements. It defends the general reliability, corrigibility, and progressiveness of empirical knowledge against relativism and skepticism.

Volume II, *Natural Science and Metaphysics,* widely illustrates "experimental metaphysics." Quantum-mechanical studies argue that potentiality, chance, probability, entanglement, and nonlocality are objective features of the physical world. The variety of relations between wholes and parts is explored in complex systems. One essay proposes that in spite of abundant phenomena of natural selection, there exists no principle of natural selection. A defense is given of the reality and objectivity of transiency. A final section consists of historical, speculative, and experimental studies of the mind–body problem.

"Abner Shimony's work makes signal contributions to every subject his hand has touched in general epistemology and philosophy of science: probability theory, foundations of quantum mechanics, scientific inference, measurement, holism, and many more. He was a pioneer in the transformation of Bayesian statistics into a core resource for philosophical reflection of scientific methodology, as well as in the experimental and theoretical exploration of Bell's inequalities for the possibility of hidden variables in quantum mechanics. His writings are invaluable as both challenge and resource for current research."

Bas van Fraassen, Princeton University

Search for a naturalistic world view
Natural science and metaphysics

Search for a naturalistic world view

VOLUME II
Natural science and metaphysics

ABNER SHIMONY

Professor of Philosophy and Physics
Boston University

CAMBRIDGE
UNIVERSITY PRESS

Published by the Press Syndicate of the University of Cambridge
The Pitt Building, Trumpington Street, Cambridge CB2 1RP
40 West 20th Street, New York, NY 10011-4211, USA
10 Stamford Road, Oakleigh, Victoria 3166, Australia

First published 1993

Printed in the United States of America

Library of Congress Cataloging-in-Publication Data
Shimony, Abner.
Search for a naturalistic world view / Abner Shimony.
p. cm.
Includes index.
Contents: v. 1, Scientific method and epistemology –
v. 2, Natural science and metaphysics.
ISBN 0-521-37352-2 (v. 1), – ISBN 0-521-37744-7 (v. 1 : pbk.)
ISBN 0-521-37353-0 (v. 2), – ISBN 0-521-37745-5 (v. 2 : pbk.)
1. Science – Methodology. 2. Science – Philosophy. 3. Knowledge,
Theory of. 4. Metaphysics. I. Title.
Q175.S53483   1993
501 – dc20   92-38168
CIP

A catalog record for this book is available from the British Library.

Volume 1
ISBN 0-521-37352-2 hardback
ISBN 0-521-37744-7 paperback

Volume 2
ISBN 0-521-37353-0 hardback
ISBN 0-521-37745-5 paperback

To Ammemie,
la vita nuova

but until now I never knew
That fluttering things have so distinct a shade
— Wallace Stevens, "Le Monocle de Mon Oncle"

Contents

Contents

PART A

Measurement in quantum mechanics

1

Role of the observer in quantum theory

In quantum theory as it is currently formulated the measurement of an observable quantity of a physical system is the occasion for a change of state of the system, except when the state prior to the measurement is an eigenstate of the observable. Two proposals for interpreting this kind of change are examined in detail, and several variant proposals are considered briefly. According to the interpretation proposed by von Neumann, and by London and Bauer the change of state is completed only when the result of the observation is registered in the observer's consciousness. Although this interpretation appears to be free from inconsistencies, it is not supported by psychological evidence and it is difficult to reconcile with the intersubjective agreement of several independent observers. According to the interpretation proposed by Bohr, the change of state is a consequence of the fundamental assumption that the description of any physical phenomenon requires reference to the experimental arrangement. Bohr's proposals are valuable as practical maxims in scientific theory, but they are shown to involve the renunciation of any ontological framework in which all types of events – physical and mental, microscopic and macroscopic – can be located. It is concluded that a satisfactory account of the observation of microphysical quantities is unlikely if the present formulation of quantum theory is rigorously maintained.

I. INTRODUCTION

There are two distinct problems concerning the relationship between physical objects and consciousness. One is the ontological problem of accounting for the fact that two such diverse kinds of entities occur in nature and interact with each other. The other is the epistemological problem of justifying physical theories by reference to human experience. A complete solution to either of these problems would surely require a solution to the other as well. In particular, it seems that the epistemological problem cannot be completely solved without understanding how the effects of physical entities can be registered upon consciousness, since performing observations and formulating theories constitute a series of acts of consciousness. It is a remarkable fact about classical physical theory that considerable progress was made on the epistemological problem, at least

This work originally appeared in *American Journal of Physics* 31 (1963), pp. 755–73. Reprinted by permission of the publisher.

on that part of the problem which has been demarcated as "scientific method," while the ontological problem remained obscure. Classical physical theory was consistently "mechanical" in the sense that the fundamental physical entities were considered devoid of sensuous qualities; and it was "empirical" in the sense that most of the classical masters recognized that the truth of physical theory is tested by its predictions regarding the observable behavior of things. However, the apparent discrepancy between the "mechanical" and the "empirical" aspects of classical physics did not seem to impede the development of the science. It was possible to relate fundamental physical concepts to common characteristics of the objects encountered in daily life and in the laboratory, and these common characteristics could somehow be directly recognized by an observer.[1] (This was already understood, in a matter-of-fact way, by Galileo and Newton, who were, after all, experimenters as well as theoreticians.) Thus, a relation between the physical *Weltbild* and experience could be established, even though the process whereby the observer performed his act of recognition was very obscure. Indeed, classical physics was indifferent to attempts to explain this relation ontologically – e.g., by means of a causal connection between mental and physical events. The network of more or less tacit rules for applying theoretical physics to ordinary objects and of normal procedures for dealing with the common characteristics of these objects allowed classical physics to by-pass the fundamental ontological problem. The "bifurcation of nature" – as Whitehead named the extreme separation of physical and mental entities in nature – may have been a scandal from the standpoint of metaphysics, but it became a convenient working arrangement from the standpoint of physical theory.

The replacement of classical physics by quantum theory was, of course, the result of puzzling sets of physical phenomena, and not of philosophical reconsideration. Nevertheless, there are several philosophically significant respects in which quantum physics differs from classical physics, the most important being that the concept of an "observation" plays a central role in the quantum physical picture of the world. The relation of elements of the physical theory to experience no longer seems to be extraneous to physics, but seems to be an intrinsic part of physical theory itself. Whether this is a correct characterization of quantum theory depends, to

---

1. It is not important, for the purposes of this paper, to distinguish the various common characteristics of objects, although they range from attributes which apparently are purely perceptual, such as "green," to those which involve theory in an essential way, such as "being an ammeter." The essential point is not whether there is a stage of perception which is independent of memory and of theoretical judgment, but that there exists some procedure for relating to experience a picture of physical reality in which consciousness has no part.

be sure, upon the exact meaning of 'observation,' and this has to be examined in detail. But *prima facie* quantum theory does differ from classical physics regarding the relationship between physical entities and consciousness. The founders of quantum theory, particularly Heisenberg[2] and Bohr,[3] discussed with much subtlety the epistemological problem of relating atomic physics to human experience; but except for a few remarks, usually tentative and oblique, they do not consider the ontological problem of mind and physical reality. It is reasonable to inquire, therefore, whether the ontological problem can be by-passed by quantum theory in much the same way as was done by classical physics. And if not, can an account of the relationship of mind to physical reality be given which is consistent with both our knowledge of psychology and the present formulation of quantum theory? In this paper I contend that the answer to both questions is negative.

II. RESUMÉ OF THE QUANTUM THEORY OF OBSERVATION

According to quantum theory, a physical system is in a definite *state* at every moment. Intuitively, the state of the system is the totality of its observable properties, but the relation of this "totality" to individual observables is peculiar to quantum mechanics. Furthermore, the states have the peculiar characteristic of obeying the *superposition principle,* and this permits us to represent the states of the system by vectors ψ of a linear vector space. An inner product (ψ, ψ') is, also, defined which is an index of the extent to which two states "overlap."

A quantity which can be assigned to the system as the result of a measurement is an *observable.* A well-performed measurement of an observable F yields a definite number belonging to a set of possible values $f_1, f_2,$ ... (where, for simplicity, the possibility of a continuum of values of the observable has been neglected). A state which is unchanged when a sufficiently careful measurement of F is performed is an *eigenstate* of F. Quantum theory supposes that the eigenstates of F are *complete,* in the sense that a set of vectors $u_1, u_2,$ ... can be selected, each representing an eigenstate, such that an arbitrary vector can be expressed in the form

$$(1) \qquad \psi = \psi_1 u_1 + \psi_2 u_2 + \cdots,$$

where the ψ_i are complex numbers. The u_i can be chosen orthonormal, i.e. the inner product $(u_i, u_j) = 1$ if $i = j$ and $= 0$ if $i \neq j$. In the following

2. W. Heisenberg, *The Physical Principles of the Quantum Theory* (University of Chicago Press, Chicago, Illinois, 1930).
3. N. Bohr: (a) *Atomic Theory and the Description of Nature* (Cambridge University Press, Cambridge, England, 1934); (b) Phys. Rev. **48**, 696 (1935); (c) *Atomic Physics and Human Knowledge* (Science Editions, Inc., New York, 1961).

discussion, each value of F is assumed to be associated with only one of the u_i, since no essential question of interpretation is affected by this simplification. The representation of F by a *linear operator* α is implicit in (1), for it suffices to specify that

(2) $$\alpha u_i = f_i u_i$$

and to require *linearity* of α, i.e., $\alpha(a\phi + b\psi) = a\alpha\phi + b\alpha\psi$. Because the eigenvalues f_i in (2) are real numbers, the operator α is *hermitian*: $(\alpha\phi, \psi) = (\phi, \alpha\psi)$ for any vectors ϕ and ψ. It is assumed that when F is measured in a state which is not an eigenstate, the probability that the value f_i will be found is

(3) $$\mathrm{Prob}_\psi(f_i) = |\psi_i|^2,$$

where ψ_i is the coefficient in the expansion (1). The fundamental Eq. (3) cannot be interpreted as asserting that ψ represents a body of partial information, to the effect that there is a probability $|\psi_i|^2$ that the actual, but unknown, value of the observable F is f_i. Such an interpretation can be shown to be contradictory by considering a different observable F' with eigenstates differing from the eigenstates of F. Predictions regarding observables other than F depend, in general, not only on the absolute values of the complex coefficients ψ_i, but on their phases as well; and these phases are neglected in the interpretation of a superposition as representing partial information. When ψ is known, this is maximal information about the system, and, therefore, (3) must be interpreted as asserting that in an objective sense F is indefinite in the state represented by ψ.

There are two ways in which the state of a system may change: (1) If the system in the state represented by ψ is subjected to a measurement of F, then a transition $\psi \rightarrow u_i$ occurs, for some i, and the probability of a particular transition is $|\psi_i|^2$. (2) If the system is undisturbed during a time interval, then in that interval the "equation of motion" is the Schrödinger equation

(4) $$i\hbar(\partial\psi/\partial t) = H\psi,$$

where H is the Hamiltonian operator, i.e., the operator corresponding to the energy of the system. It follows from Eq. (4) that if the system is undisturbed, its states at two times t and t_0 are related by the equation

(5) $$\psi(t) = U(t - t_0)\psi(t_0),$$

where

$$U(t - t_0) = \exp[-(i/\hbar)H(t - t_0)].$$

U is a linear operator, and furthermore, since the Hamiltonian operator H is hermitian, U is *unitary*, i.e., $(U\phi, U\psi) = (\phi, \psi)$, for arbitrary vectors ϕ and ψ.

The most systematic theory of observation in quantum mechanics was proposed by von Neumann[4] and later presented more simply (and in some ways more deeply) by London and Bauer.[5] They consider transitions of type 1, discontinuous transitions due to the performance of a measurement, to be an uneliminable aspect of quantum theory; and they explicitly understand by 'measurement' the registration of the result in a consciousness.

That a transition of type 1 cannot be due simply to interaction of the system with a measuring apparatus may be seen as follows: The apparatus is itself a physical system (system y) and, therefore, its states and observables are describable by quantum theory. For the apparatus to be usable for the purpose of measuring F in the original system of interest (system x) it must be possible to prepare the apparatus in an initial state $\phi_0(y)$ which is, in a certain sense, sensitive to states of x. That is, if a composite system $x + y$ is formed by placing x and y into contact, and if the initial state of $x + y$ is represented by

(6) $$\Psi(t_0) = u_i(x)\phi_0(y),$$

where $u_i(x)$ represents one of the eigenstates of F. Then, in time t_1, system $x + y$ evolves to

(7) $$\Psi(t_0 + t_1) = u_i(x)v_i(y),$$

where $v_i(y)$ represents an eigenstate of some observable G of y such that $g_i \neq g_j$ if $i \neq j$. The transition of the state from $\Psi(t_0)$ to $\Psi(t_0 + t_1)$ is continuous and is governed by the Schrödinger equation

(8) $$i\hbar[\partial\Psi(t)/\partial t] = H_{x+y}\Psi(t),$$

where H_{x+y} is the Hamiltonian operator of the composite system. Eq. (7) implies that a determination of the value of G in the apparatus indirectly but unequivocally determines the value of F in the original system. Suppose, however, that the initial state of x is not an eigenstate of F, but is represented by the vector ψ which can be expanded as in Eq. (1). Then

(9) $$\Psi(t_0) = \psi_2 u_1(x)\phi_0(y) + \psi_2 u_2(x)\phi_0(y) + \cdots.$$

Since H_{x+y} is a linear operator, it follows from (6)–(9) that

(10) $$\Psi(t_0 + t_1) = \psi_1 u_1(x)v_1(y) + \psi_2 u_2(x)v_2(y) + \cdots.$$

4. J. von Neumann, *Mathematical Foundations of Quantum Mechanics* (Princeton University Press, Princeton, New Jersey, 1955).

5. F. London and E. Bauer, *La Théorie de l'observation en mécanique quantique* (Hermann & Cie, Paris, 1939). An English translation is included in *Quantum Theory and Measurement*, ed. by J. A. Wheeler and W. H. Zurek (Princeton University Press, Princeton, New Jersey, 1983).

The fact that ψ is a superposition of different u_i is reflected in the fact that the final state vector of $x+y$ is a superposition of different $u_i(x)v_i(y)$. In other words, the initial indefiniteness of F in ψ implies an indefiniteness of G in the state represented by $\Psi(t_0+t_1)$. The latter indefiniteness can, of course, be resolved by a measurement of G, which would produce the transition

$$(11) \qquad \Psi(t_0+t_1) \rightarrow u_n(x)v_n(y)$$

for some value of n. But this transition is of type 1 and requires the registration of the result of the measurement in a consciousness. If, instead of such a registration, an apparatus y' sensitive to the states of $x+y$ is used, then in exact analogy to (10) the state of $x+y+y'$ at some time $t_0+t_1+t_2$ will be of the form

$$(12) \qquad \Xi(t_0+t_1+t_2) = \psi_1 u_1(x)v_1(y)v_1'(y') + \cdots.$$

This is a further stage in an infinite regress which seems to be terminated only by conscious awareness of the result of a measurement.

One further feature of the quantum theory of interacting systems should be mentioned: that a composite system $x+y$ may be in a definite quantum state without either x or y being in definite states. In other words, it may be impossible to express the vector Ψ, which represents the state of the composite system, in the form $\phi(x)\psi(y)$, where $\phi(x)$ and $\psi(y)$ represent states of x and y, respectively. In general, one can only describe system x as having probabilities $p_1, p_2, \ldots$ of being in appropriate states represented by $\phi_1, \phi_2, \ldots$. Such a description is called a 'mixture,' and it differs from a superposition by the absence of any specification of phase relations among the vectors ϕ_i, thereby eliminating the possibility that a mixture is really a state in disguise. A similar description by means of a mixture is, of course, also possible for system y. This situation is counter-intuitive and in sharp contrast to that of classical physics, where the specification of the state of a composite system implies the specification of states of all subsystems.

III. THE ABILITY OF THE MIND TO REDUCE SUPERPOSITIONS

The foregoing account of the proposals of von Neumann and of London and Bauer is incomplete, since their treatment of the problem of agreement among different observers has not been summarized. Before turning to this problem, however, I believe it enlightening to examine in detail the operation whereby a single observer effects a reduction of a superposition.

Von Neumann says almost nothing about the consciousness of the observer, except that "the intellectual inner life of the individual . . . is extra-observational by its very nature (since it must be taken for granted by any conceivable observation of experiment)." (Ref. 4, p. 418.) London and Bauer say somewhat more:

. . . let us consider the set of three systems, (object x) (apparatus y) (observer z), as a single composite system. We shall describe it by a total wave function . . .

$$\Psi(x, y, z) = \sum \psi_k u_k(x) v_k(y) w_k(z),$$

where the w_k represent the different states of the observer.

'Objectively' – that is to say, *for us* who consider as 'object' the composite system x, y, z – the situation seems little changed in comparison with what we have encountered before, when we only considered the apparatus and the object. We now have three mixtures, one for each system, with the statistical correlations among them bound to a pure case for the total system. Indeed, the function $\Psi(x, y, z)$ represents a maximum description of the composite 'object,' consisting of x, which is the object in the strict sense, the apparatus y, and the observer z. Nevertheless, we do not know in what state the object x is found.

The observer has an entirely different point of view. For him it is only the object x and the apparatus y which belong to the external world – to that which he calls 'objective.' By contrast, he has *with himself* some relations of a completely special character: He has at his disposal a characteristic and quite familiar faculty, which we may call the 'faculty of introspection.' He can thus give an account of his own state in an immediate manner. It is in virtue of this 'immanent knowledge' that he claims the right to create for himself his own objectivity, that is to say, to cut the chain of statistical coordinations expressed by $\sum \psi_k(x) v_k(y) w_k(z)$ by certifying: 'I am in the state w_k' or more simply 'I see $G = g_k$' or even directly '$F = f_k$.'

It is therefore not a mysterious interaction between the apparatus and the object which brings about a new ψ during the measurement. It is only the consciousness of an '*I*' which can separate itself from the old function $\Psi(x, y, z)$ by henceforth attributing to the object a new function $\psi(x) = u_k(x)$. (Translated from Ref. 5, pp. 42–3.)

In this passage London and Bauer seem to be stating some important, though incompletely developed, propositions regarding the place of the mind in nature. (i) The use of the product formalism $u_k(x) v_k(y) w_k(z)$ indicates that the observer z is ontologically on the same level as the microscopic system x and the apparatus y. The observer z is one system among many, and the fact that the state $w_k(z)$ can be correlated with the states $u_k(x)$ and $v_k(y)$ implies that the physical systems x and y can interact with the observer, even though the details of the interaction are unknown. In particular, London and Bauer do not seem to be attributing a transcendental position to the observer, such that the physical systems x and y somehow derive their existence from the observer. (ii) The assertion that the observer knows his own state by direct introspection implies that London and Bauer understand z to include the mind of the observer, possibly together with all or part of his body (though it is clearly also possible to consider parts of the body as physical apparatus, as von Neumann explicitly suggests). (iii) At least some of the usual principles of quantum theory are implicitly asserted to apply to states $w_k(z)$ of the observer. In particular, the reference to the sum $\sum \psi_k u_k(x) v_k(y) w_k(z)$ requires that the states of the observer be superposable and that phase relations among them be meaningful. (iv) The dynamical laws governing the evolution of states of the observer are such that the transition $\sum \psi_k u_k(x) v_k(y) w_k(z) \rightarrow u_n(x) v_k(y) w_k(z)$ occurs without any outside disturbance of the composite system $x + y + z$. This transition is possible because the observer has a property which is not shared by any other system in nature, the faculty of introspection, whereby the observer can "cut the chain of statistical coordinations." (Translated from Ref. 5, pp. 41–2.)

Propositions (i) and (ii) are not novel: they merely state without elaboration an ontology, like that of Aristotle or Locke, in which both mental and material systems occur in nature and interact with each other. Proposition (iii) is a remarkable extrapolation of ordinary quantum theoretical characteristics to states of mind. Proposition (iv) is essentially a qualification of (iii), for it asserts that at least one of the fundamental principles of quantum theory fails to hold of states of the observer. The transition $\sum \psi_k u_k(x) v_k(y) w_k(z) \rightarrow u_n(x) v_n(y) w_n(z)$ is a nonlinear transition and is stochastic, whereas ordinary quantum theory asserts that all transitions of isolated systems are linear and nonstochastic. It follows, of course, that the temporal evolution of a state Ψ of a composite system of which a mind is a subsystem cannot be governed by a Schrödinger equation, since we have seen in Eq. (5) of Sec. II that the Schrödinger equation implies a deterministic and linear relation between the states of the system at different times. Two considerations make this conclusion seem reasonable. First, it is doubtful that there exists a Hamiltonian operator H for a system containing a mind (for this would require that energy could be expressed partially in terms of psychological variables), and without a Hamiltonian operator the Schrödinger equation cannot even be formulated. The second consideration is based on the fact, previously mentioned, that a composite system can be in a definite quantum state and yet its components need not be in definite states. In particular, if the composite system $x + y + z$ is in the state represented by $\sum \psi_k u_k(x) v_k(y) w_k(z)$, then in general the observer z is not in a definite state but must be described by a mixture. It is the peculiar property of the observer, however, that by possessing the faculty of introspection he can attend to himself in abstraction from the physical systems with which he interacts. The result of his introspection is to establish himself in a definite state. There must be an element of chance in this process, since prior to introspection there were only various probabilities for the observer to be in various definite states.

In brief, then, London and Bauer seem to be proposing that states of the observer satisfy the vectorial relations required by ordinary quantum mechanics, but do not evolve temporally in the ordinary quantum mechanical manner. Although it is a strange proposal, consisting of a partial extension of quantum theory into the domain of psychology, it contains no obvious inconsistency. Whether it is factually correct, however, is another matter, and to judge this two psychological questions must be investigated: whether mental states satisfy a superposition principle, and whether there is a mental process of reducing a superposition.

Perhaps the most obvious mental phenomenon to investigate in connection with the superposition principle is the phenomenon of perceptual vagueness. It is often said that in a physical state ψ which is not an eigenstate of an operator α, the observable corresponding to α has an indefinite or "blurred" value; and this suggests that a mental state in which certain perceptions are blurred or indistinct is a superposition of states of clear perception. An obvious difficulty in such a proposal is the obscurity of the meaning of the phase relations in a superposition of mental states. A partial answer can perhaps be given in those cases where perceptual vagueness in one area is correlated with distinctness in another area, e.g., indistinctness of visual perception while concentrating on music; for such a case is reminiscent of the fact that the phase relations in a superposition $\psi = \sum c_i u_i$ ensure that some observable F' has a sharp value in ψ, while another observable F (complementary to F') which has different values f_i in the various u_i does not have a sharp value in ψ. This answer is unsatisfactory, however, because in general perceptual vagueness does not occur as a price paid for a distinct perception of a "complementary" quality. Vagueness may be due to a variety of factors, such as sleepiness, ill-health, or emotional turmoil, which dull all perceptions and do not sharpen one at the expense of others. An even more decisive consideration, however, is the following. Suppose the observer z examines the composite system consisting of a microscopic object x and a detecting apparatus y. Let the initial state of $x+y$ be $\sum \psi_k u_x(x) v_k(y)$, and suppose that some macroscopic observable G has distinct values g_k in the various $v_k(y)$. Then, if London and Bauer are correct, the initial state of $x+y+z$ is the superposition $\sum \psi_k u_k(x) v_k(y) w_k(z)$, but very quickly this state passes over into $u_n(x) v_n(y) w_n(z)$. If vagueness has any connection with superpositions, the observer should initially experience a vague perception regarding G, but should find that this vagueness is rapidly dispelled and a sharp perception is somehow crystallized. However, introspection does not seem to reveal any such psychological process. If the observer is healthy, alert, and undistracted, and if the light is good, etc., there is no initial state of perceptual vagueness; and if these conditions are not fulfilled, there may be an initial state of perceptual vagueness even if the physical system $x + y$ is not in a superposition relative to the macroscopic observable G. In short, there seems to be no correlation between the phenomenon of vagueness and the superposed character of the physical "input" into the observer.

Several other psychological phenomena could possibly be interpreted as instances of the superposition principle: e.g., indecision, conflict of loyalty, ambivalence. The crucial objection which was brought against such an interpretation of perceptual vagueness cannot be raised in these cases, because we do not know how to correlate such phenomena in an unambiguous manner with physical "input" into the observer; and therefore, we cannot prepare the "input" in a superposed state, make a prediction regarding the psychological effect, and test the prediction by introspection. However, even though a crucial negative test of the interpretation of decision, ambivalence, etc. in terms of superpositions does not seem to be forthcoming, there is not a trace of evidence in favor of this interpretation. In particular, in all such phenomena, the meaning of phase relations remains profoundly obscure.

Perhaps it is not too fanciful to carry the speculation one step further and suppose that when the observer z examines the system $x + y$ in a state $\sum \psi_k u_k(x) v_k(y)$, the initial state of the composite system $x + y + z$ including the observer is indeed the superposition $\sum \psi_k u_k(x) v_k(y) w_k(z)$, but that such a superposition of states of mind is not conscious. Freud's term, 'preconscious',[6] might appropriately describe such a superposition. There is psychological evidence, especially in dreams, that something which might be called a 'superposition principle' is operative in the preconscious – for instance, the image of a parent and that of a spouse may be superposed. One might suppose that when a physical input first influences the mind, the physical superposition is somehow reflected by a superposition of images in the preconscious, since the preconscious does indeed seem to be endowed with the capacity for performing superpositions. The reduction of a superposition would then occur at the threshold from the preconscious to the conscious. The foregoing piece of speculation is vitiated, however, by the reflection that all the evidence we have of combinations of images in the preconscious concerns memories with interlocking emotional associations, and does not at all concern perception. Furthermore, the images which are combined are ordinarily derived from sequences of quite definite perceptions. There is, in short, no evidence of causal con-

6. S. Freud, "The Interpretation of Dreams," *The Basic Writings of Sigmund Freud* (Random House, Inc., New York, 1938).

nection between the superposition of states corresponding to different values of an observable and combination of images in the preconscious. Consequently, this speculation seems to be only an *ad hoc* attempt to explain why we are unaware of an initial stage of vagueness when the physical "input" has a superposed character.

Suppose, in spite of the above evidence, that states of mind do satisfy a superposition principle. We must then investigate the further proposal of London and Bauer that the introspection of the observer effects a reduction of an initial superposition. As pointed out previously, this would be possible only if the state of the observer failed to evolve in accordance with the Schrödinger equation – furthermore, only if the evolution of the state of a composite system which includes a mind is a stochastic process. There have, in fact, been numerous speculations that mind is precisely that aspect of nature which is not governed by exact causal law. Gross evidence for such a characterization of mind can be found in human learning patterns, for consciousness and attentiveness accompany the exercise of a partially learned skill, whereas a fully mastered skill becomes mechanical and unconscious. It has been argued that this pattern in the life of a single individual has been followed on a large scale in the history of the species, so that processes like circulation, which once may have been accompanied by consciousness, are now unconscious, and processes like breathing are in transition from conscious to unconscious.[7] If this characterization of mind is correct, then it is reasonable to attribute the stochastic process of reducing a superposition to the mind's activity. The reduction of a superposition does not thereby become comprehensible from a common-sensical standpoint, but at least it is subsumed under a wider class of processes, which is a step towards scientific explanation. This line of analysis is weakened, though not decisively, by the fact that the immediate feelings associated with a fully determined observation are not different from those associated with an observation governed by probability; no more spontaneity or creativity on the part of the observer is *felt* in the second case than in the first. More decisive is the vast evidence of the evolutionary link of higher animals with the simplest organisms and even with inorganic matter. It is difficult to see how irreducibly stochastic behavior could be a structural characteristic, which could occur in a complex organism even though it is absent from all the components of the organism. Consequently, if there is an irreducibly stochastic element in the behavior or experience of higher animals, then one should expect a stochastic element in the primitive entities at the base of evolution. And if this is correct, then the Schrödinger equation, which is usually supposed

7. E. Schrödinger, *Mind and Matter* (Cambridge University Press, Cambridge, England, 1958); H. Bergson, *Creative Evolution* (Random House, Inc., New York, 1944).

to govern deterministically the state of a physical system except when it is being observed, can only be approximately valid.[8] This possibility should, in my opinion, be taken very seriously. However, it lies beyond the limits of this paper, in which quantum mechanics in its present form is assumed to hold exactly.

There is perhaps a feeling of fantasy about the foregoing discussion, in which states of mind are treated from the point of view of quantum mechanics. It is interesting to consider three suggestions which are free from this dubious extrapolation, but which, as is seen, are beset by other difficulties.

The first suggestion is to interpret the state vector ψ as a compendium of the observer's knowledge about the system, which is incomplete since he can only make statistical predictions regarding some of the observables. According to such an interpretation, the reduction of a superposition need not be a change in the physical system itself but only in the observer's knowledge about the system. The problem of the superposability of states of mind does not arise, because the interaction of a physical system with an observer does not result in a total state vector of the physical system plus observer, as proposed by London and Bauer, but merely in new knowledge on the part of the observer, which must be incorporated into an appropriate state vector expressing his revised total knowledge of $x + y$. This interpretation, which is *prima facie* very attractive, since it does little violence to our common view of the world, has been discussed frequently, and the essential reasons for rejecting it were given in Sec. II. Consequently, I only elaborate what was said there with a few familiar comments. If the state vector ψ is not a maximum specification of the system, then an explanation is required for the fact that further specification is either redundant (could be predicted from ψ) or inconsistent (leads to predictions definitely incompatible with some of the predictions based on ψ). The only plausible explanation is the common one that the measurement which yields information supplementary to ψ disturbs the system physically and changes its state. In other words, the uncertainty principle must be interpreted as an expression of a fundamental limitation upon simultaneous measurement rather than as a limitation upon the simultaneous actuality of complementary properties of a system. But there are various physical phenomena which can be understood quite well in terms of the latter interpretations of the uncertainty principle and not at all in terms of the former. For example, good statistical mechanical calculations

8. The foregoing argument, with appropriate modifications, has sometimes been used to show that the elementary entities in nature have rudimentary mental characteristics, as proposed by Leibniz and Whitehead.

of entropy, as in the Sakur–Tetrode equation for a Boltzmann gas, proceed by assigning to each physical state of a system a volume h^N in phase space (where N is the number of classical degrees of freedom). This assignation is comprehensible if, for every generalized coordinate q and its conjugate momentum p, the relation

$$\Delta q \Delta p \sim h$$

expresses an ontological limitation on the specificity of q and p; but since measurements of the entropy do not presuppose knowledge of q and p, the interpretation of this uncertainty relation as a limitation upon measurement provides no rationale for the treatment of phase space in this calculation of the entropy. In general, the uncertainty relations pervade all quantum mechanical explanations of phenomena, even when there is no question of measurement of complementary quantities, and it is difficult to see how this fact can be reconciled with the claim that the state vector can in principle be supplemented by additional specifications of a system.

The second suggestion is to suppose that the physical system itself, and not merely the state vector ψ, is in some sense derivative from the mind or experience of the observer. In effect, this suggestion replaces proposition (i) above – that the mind is to be considered a natural entity ontologically on a level with physical systems – by some variety of idealism or phenomenalism. Since adequate critiques of the various forms of these philosophical theories are evidently beyond the scope of this paper, I make only three brief comments stating general reasons for skepticism about them, the first being directed primarily against Kantian idealism and the second and third being directed primarily against phenomenalism. (a) The *Weltbild* of common sense and that of natural science characterize the individual human being as an entity of limited duration and limited spatial extent, whose existence is not necessary to the existence of the universe as a whole. Consequently, an ontology which considers the universe to be derivative from one human mind (or one field of experience) is faced with the problem of relating this transcendent mind to the limited and contingent creature which persistently appears in the *Weltbild*.[9] (b) The claims of idealism and phenomenalism regarding the status of physical entities is extremely programmatic. There are only a few instances of philosophers who have tried to show in detail how the common properties of physical systems can be regarded as combinations of ideas of the mind

9. This problem does not arise in Berkeley's form of idealism, in which there are many minds; but as is seen in the following section, a pluralism of minds generates a new set of problems if the reduction of a superposition is due to consciousness.

or groupings of experiences, notably Russell[10] and Carnap.[11] However, both Russell and Carnap abandoned their proposals because of the evident discrepancies between their constructs and the physical things as ordinarily characterized, and particularly because of the difficulty of attributing potentialities or "disposition properties" to their constructs. One cannot help feeling that the difficulty of exhaustively characterizing a physical entity in terms of ideas and experiences is indicative of an independent existence of the entity. (c) If the program of exhibiting an electron as a construct were somehow fulfilled, the description of the construct would be fantastically complex, presumably consisting of an infinite set of conditional statements such as "If a cloud chamber is prepared in such a manner, then the resulting photograph will (with a certain probability) have the following appearance." The complexity is multiplied by the fact that macroscopic objects such as cloud chambers, photographic plates, and Geiger counters would themselves be constructs referring to an infinite set of possible experiences. It is a remarkable fact, however, that extremely exact laws have been discovered for describing the motion of an electron, whereas none are known for describing the sequence of simple experiences. This fact is completely anomalous from the standpoint of a theory which takes human experience as ontologically primitive and regards electrons as complex constructs from actual and possible experiences; on the other hand, it is reasonable in a realistic ontology, in which electrons have independent existence and in which perceptions and other elements of experience are the result of complex interactions among physical objects and organisms.

The third suggestion is to agree with London and Bauer that the mind plays an essential role in the reduction of superpositions, but to deny that the superposition principle applies to states of mind. According to this suggestion, the reduction of a superposition occurs at the moment of interaction of the conscious observer with the physical system, rather than at a later stage.[12] The mind of the observer acts as a kind of filter system,[13] which forces the "input" to select one channel out of all those compatible with the superposition $\sum \psi_k u_k(x) v_k(y)$, and this selection is irreducibly stochastic. This suggestion, unlike the proposal of London and Bauer, is compatible with the psychological evidence that states of mind are not

10. B. Russell, *Our Knowledge of the External World* (W. W. Norton and Company, Inc., New York, 1929).
11. R. Carnap, *Der logische Aufbau der Welt* (Meiner, Berlin and Leipzig, 1928).
12. Since von Neumann said practically nothing about the work of consciousness, it is possible that this suggestion is consistent with his unarticulated ideas.
13. Cf. A. Landé, *From Dualism to Unity* (Cambridge University Press, Cambridge, England, 1960), pp. 8–12.

superposable. Its most obvious weakness is the difficulty of understanding why there can be no mental states reflecting the states of physical systems in which macroscopic observables have indefinite values; one cannot help suspecting that such peculiar states of physical systems do not exist, and that the present suggestion is a stratagem for disguising this fact. Some psychological objections can also be raised: for example, there is probably no sharp moment at which the observer becomes aware of the macroscopic variable. Supposing the threshold between physical input and mental registration to be the locus of the reduction of superpositions is thus *ad hoc* and implausible, though not decisively refuted by any psychological evidence regarding a *single* observer. However, when one turns to a consideration of several observers one finds crucial weaknesses in this suggestion, as in all proposals which attribute the reduction of superpositions to the mind.

IV. INTERSUBJECTIVE AGREEMENT

Suppose that the problems of Sec. III are set aside by accepting that the mind of the observer is a "black box" for reducing superpositions. A new set of problems arises when more than one observer interacts with the same physical systems, for it is important to explain how there can be agreement among them regarding the states of the physical systems and the value of observables.

According to London and Bauer (Ref. 5, p. 49) intersubjective agreement is possible because the measuring apparatus y is a macroscopic object. Consequently, the act whereby an observer becomes aware of the value g_i of the macroscopic observable G has only a negligible effect upon the apparatus. Another observer can then examine the apparatus and find that G has the same value g_i which was discovered by the first observer; and if the states of the apparatus y are correlated with those of the microscopic system x, then the two observers will come to the same conclusion regarding the value f_i of an observable F in x and regarding the state of x.

Unfortunately, this explanation is not adequate, because the quantum characterization of the measuring apparatus is at odds with the common-sensical opinion that mere looking has a negligible effect upon the apparatus. There is a great difference between the state $\sum \psi_k u_k(x) v_k(y)$ and the state $u_n(x) v_n(y)$, and this difference is precipitated by "mere looking." This objection is not discussed by London and Bauer, but an answer to it may clearly be sought in either of two directions: (A) A careful analysis of the superposition of states of a macroscopic system may show that the transition $\sum \psi_k u_k(x) v_k(y) \rightarrow u_n(x) v_n(y)$ is *for all practical purposes* an insignificant change, so that the measuring apparatus can appear the same to the initial and subsequent observers. (B) It may be claimed that the first observer affects the system $x+y$ nonnegligibly by changing it from an initial unstable state, the superposition $\sum \psi_k u_k(x) v_k(y)$, to a final stable state $u_n(x)v_n(y)$; and the stability of the final state then guarantees that subsequent observers will agree in their reading of the apparatus.

An answer of type (A) has been proposed by Bohm,[14] Ludwig,[15] Feyerabend,[16] and Daneri *et al.*[17] They claim that if the $v_k(y)$ are such that a macroscopic variable G has different values g_k in the states represented by them, then there is no way of distinguishing the superposition $\sum \psi_k u_k(x) v_k(y)$ from a mixture having the proportions $|\psi_n|^2$ of the $u_n(x)v_n(y)$. The argument in Sec. II against interpreting Ψ as a body of partial information (i.e., a mixture) remains valid in principle, but it is unimportant in practice if there exists no quantity G' susceptible of actual measurement such that the statistical predictions regarding G' depend upon the phases of the coefficients ψ_k. Since the quantities which experimenters actually measure in macroscopic systems are gross, in the sense that considerable variation of the microscopic constitution of the system makes no discernible difference in their values, it is difficult to see how the phases of the ψ_k can affect the outcome of an actual measurement. Suppose now that an initial observer z notes the value of the observable G, thereby reducing the superposition $\sum \psi_k u_k(x) v_k(y)$, with probability $|\psi_n|^2$ that the final state is represented by $u_n(x)v_n(y)$. A second observer z', who knows that z has performed the observation but who is ignorant of the outcome, will optimally describe the system $x+y$ by a mixture having proportions $|\psi_n|^2$ of the $u_n(x)v_n(y)$. But this is, precisely, the mixture which is indistinguishable, for the purpose of practical predictions, from the state of $x+y$ prior to the observation by z. In this sense, the measuring apparatus appears the same both to the initial and to the subsequent observers. Two objections can be brought against this line of reasoning, one sophisticated and one quite simple, the latter being the more decisive. The sophisticated objection is that two remarkable effects have been discovered recently, the spin-echo effect[18] and the Mössbauer

14. D. Bohm, *Quantum Theory* (Prentice-Hall, Inc., Englewood Cliffs, New Jersey, 1951).
15. G. Ludwig, *Die Grundlagen der Quantenmechanik* (Springer-Verlag, Berlin, 1954). In his recent work, an article in *Werner Heisenberg und die Physik unserer Zeit* (Friedreich Vieweg und Sohn, Braunschweig, Germany, 1961), Ludwig makes the more radical proposal that quantum mechanics is not exactly correct when applied to macroscopic systems.
16. P. Feyerabend, "On the Quantum-Theory of Measurement," *Proceedings of the Ninth Symposium of the Colston Research Society* (Butterworths Scientific Publications Ltd., London, 1957).
17. A. Daneri, A. Loinger, and G. M. Prosperi, Nucl. Phys. **33**, 297 (1962).
18. E. L. Hahn, Phys. Rev. **80**, 580 (1950); J. Blatt, Progr. Theoret. Phys. **22**, 745 (1959).

effect,[19] in which unexpected coherent contributions were obtained from the parts of macroscopic systems. No one claims that by means of these effects the phases in a superposition $\sum \psi_k u_k(x) v_k(y)$ can be determined, where again the $v_k(y)$ represent eigenstates in which a macroscopic observable has different values. Nevertheless, the discovery of unexpected coherence is a caution against underestimating the ingenuity of experimenters regarding phase relations in macroscopic systems. The simple objection is that agreement between two observers in reading their apparatus requires that at the conclusion of their observations they assign to $x + y$ the same state. From the legitimacy of their use of the same mixture to describe $x + y$ prior to their observations one can infer only that they make the same statistical predictions correctly characterizing an ensemble of similar situations. The agreement of the two observers in a *specific* reading of the apparatus would be a coincidence unless the examination of the first observer leaves the system $x + y$ in the state represented by a specific $u_n(x) v_n(y)$. Therefore, the first observer must have effected a change in $x + y$ which is not negligible from the standpoint of subsequent observers.

An answer of type (B) appears very reasonable if one accepts the proposal that consciousness is responsible for the reduction of a superposition. The state $\sum \psi_k u_k(x) v_k(y)$ is "unstable"[20] in the sense that the composite system $x + y + z$ undergoes the stochastic transition

$$(1) \qquad \sum \psi_k u_k(x) v_k(y) w_0(z) \to u_n(x) v_n(y) w_n(z),$$

where $w_0(z)$ represents the state of the observer immediately before the observation and $w_n(z)$ represents the state of the observer correlated with the nth state of the apparatus. According to the analysis of London and Bauer there is an intermediate stage in the transition (1), in which the composite system is in the superposed state $\sum \psi_k u_k(x) v_k(y) w_k(z)$, while the existence of such an intermediate stage is denied by the last suggestion made in Sec. III; according to either account, however, the subsystem $x + y$ is in different states at the initial and final stages of (1). By contrast, the state $u_n(x) v_n(y)$ is "stable" in the sense that the system $x + y + z'$, where z' is another observer, passes only through

$$(2) \qquad u_n(x) v_n(y) w_0'(z') \to u_n(x) v_n(y) w_n'(z'),$$

and throughout the process (2) the subsystem $x + y$ remains in the same state. However, a consideration of the causal relations between observers

19. *The Mössbauer Effect*, edited by H. Frauenfelder (W. A. Benjamin, Inc., New York, 1961).
20. Instability in this sense is completely different from the "meta-stability" of the apparatus y (e.g., the supersaturation of the vapor in a cloud chamber) which underlies the physical transition $u_n(x)\phi_0(y) \to u_n(x)v_n(y)$ described in Sec. II, Eqs. (6) and (7).

z and z' suffices to exhibit a difficulty in an answer of type (B). Suppose that both z and z' observe $x+y$ by photographing the apparatus y, z' being the first to take a photograph, but z being the first to develop and examine his film. From the assumption that reduction of a superposition occurs only when there is registration upon consciousness it follows that $x+y+z$ undergoes transition (1), and afterwards $x+y+z'$ undergoes transition (2). Thus the observation by z selects a specific image on the film of z' from among the range of images compatible with the original superposition; and this selection is a kind of causation, even though there need be no physical interaction between z and z'. An even worse discrepancy with ordinary physical causality can be adduced if the observations of their respective films by z and z' are events O and O' lying outside each other's light cones – which is certainly a possible experimental arrangement. Then, in some frames of reference, O precedes O', so that the observation by z is the cause of what z' observes, and in other frames of reference O' precedes O, thus reversing the causal relation. This is in conflict with the special theory of relativity, in which a causal relation is invariant under Lorentz transformations.

A possible reply to these objections is that the ordinary characteristics of causal relations need not apply to the relation between z and z', because the effect of an observation by one of them upon the experience of the other is undetectable. Whether z performs an observation or not, the ultimate result of the observation by z' is to find the system $x+y$ in a state represented by some $u_n(x)f_n(y)$, since this is what happens in the last stage of both processes (1) and (2). Thus z' can only tell that an initial superposition has been reduced, but he cannot know whether precipitating the reduction was his own work or the work of z. Hence z cannot use his power to reduce superpositions as a communication device for transmitting information to z'.[21] Furthermore, this conclusion permits various consistent, though desperate, resolutions of the conflict with relativity theory which arise when O and O' lie outside each other's light cones: for

21. The inability of z to communicate with z' by reducing a superposition is in no way due to the fact that only a single observation is involved. We may imagine that z and z' both photograph an indefinitely large number of pieces of apparatus, each measuring a different microscopic system in the state $\sum \psi_k u_k$. They may agree that in order to answer "yes" to a predesignated question z will examine all his films within some specified time interval, and in order to answer "no" he will refrain from examining them. In the former case, z causes each of the films of z' to pass into a definite state, though z' will not know which prior to examining them; and, in the latter case, each film will be in an indeterminate state until z' performs his observations. In both cases, however, the statistics of the films observed by z' can be described by exactly the same mixture. An essential reason for this statistical identity is that prior to any observation by either z or z' each of the films is only part of a composite system in a definite state and is not itself in a definite state.

instance, the temporal order of O and O' may be regarded as indeterminate (i.e., relative to the frame of reference) with respect to the transmission of detectable signals between these two events, but there may nevertheless be an absolute temporal order between them which determines whether the reduction of the superposition was due to z or to z'.

The foregoing arguments are familiar, particularly in connection with the paradox of Einstein, Podolsky, and Rosen,[22] and they suffice to dispel the suspicion that by considering observation upon systems which are spatially separated but in correlated states a contradiction can be exhibited in quantum mechanics. Nevertheless, there seems to be no one who seriously maintains the position now under consideration: that agreement among observers is due to a causal relation between their acts of observation, but a causal relation which is not circumscribed by the usual limitations of relativity theory. The consistency of this position does not seem to be adequate compensation for its counter-intuitive and *ad hoc* character.

It is difficult, at this point in the analysis, to resist the promptings of common sense, that the agreement between observers z and z' is due simply to a correspondence between the physical configurations of the emulsions on their films. This attractive but reactionary proposal can be resisted only by re-iterating what it implies: that the superposition in the microscopic system x is reduced when x interacts with the apparatus y, in contradiction to the linear dynamics of quantum theory.

Von Neumann has proposed an explanation of intersubjective agreement which is quite different from any considered above, but which in no way relaxes the rigorous application of quantum mechanical principles to physical systems. According to him, quantum theory is compatible with

the so-called principle of the psycho-physical parallelism – that it must be possible so to describe the extra-physical process of the subjective perception as if it were in reality in the physical world – i.e., to assign to its parts equivalent physical processes in the objective environment, in ordinary space. (Ref. 4, pp. 418–9.)

Whenever there is a measurement, the world is divided into two parts, one part comprising the system being observed and the other the observer. The boundary between these parts is largely arbitrary, for the observer may be understood to include the experimenter's body along with some of his laboratory equipment; or the observer may consist only of a consciousness together with some part of the nervous system. "That this boundary can be pushed arbitrarily far into the interior of the actual observer is the content of the principle of the psycho-physical parallelism." (Ref. 4, p. 420.) Von Neumann then proceeds to prove that if the

22. A. Einstein, B. Podolsky, and N. Rosen, Phys. Rev. **47**, 777 (1935); D. Bohm and Y. Aharonov, Phys. Rev. **108**, 1070 (1957).

formalism of quantum mechanics is applied to the observed part of the world, the predictions obtained are independent of the location of the boundary. The essence of his argument is that the expansion coefficients in Eqs. (1), (10), and (12) of Sec. II are all the same; so that whether the observed part of the world consists of x, or of $x+y$, or of $x+y+y'$, etc., the probability of finding the value f_i of an observable F of the microscopic x is always $|\psi_i|^2$. The agreement of different observers is asserted to be a corollary of this result, but the proof of this corollary is left by von Neumann as an exercise for the reader. (Ref. 4, p. 445.)

Von Neumann seems to be asserting that any observer can describe the mental processes of any other observer as if they were physical processes, in other words that one observer can treat all others behavioristically. It follows that from the point of view of observer z, the concurrence of z' regarding an observable G of the measuring apparatus is equivalent to a control reading of the apparatus by means of some auxiliary physical device. From z's standpoint, states of z' are correlated with those of the apparatus y in the same physical way that states of the apparatus are correlated with those of the microscopic system x. Hence z can effect a reduction of a superposition $\sum \psi_k u_k(x)v_k(y)w'_k(z')$ by observing G directly in y or indirectly via z' (e.g., by asking questions and noting the answers). Furthermore, once the superposition is reduced, a remeasurement of G in either way is certain to yield the same result. I believe that this is in outline the proof of the corollary which von Neumann left for the reader.

The obvious difficulty with von Neumann's argument is its irrelevance to the question of agreement between z and z' when both are considered as ultimate subjects. If both observe a physical system and independently effect the same reduction of a superposition, their agreement seems to be "pre-established harmony." If the first to make the observation reduced the superposition, leaving no option to the other, one is again confronted with the difficulties discussed above regarding the causal relation between the two observers. Consequently, unless "pre-established harmony" is accepted, the only *Weltbild* in which quantum theory is rigorously maintained appears to be one in which there is a single ultimate subject. This is a different conclusion from the common skeptical position regarding the existence of other minds, and in one respect it is stronger. It does not say that I am unsure other people have the same kind of subjective characteristics as myself, but rather that if they have minds then certainly their minds lack some of my powers, particularly the ability to reduce a superposition. This conclusion agrees with Wigner's analysis[23] of a thought-

23. E. P. Wigner, in *The Scientist Speculates,* edited by I. J. Good (William Heinemann, London, 1962).

experiment in which a friend is used as an instrument for detecting a photon:

> It is not necessary to see a contradiction here from the point of view of orthodox quantum mechanics, and there is none if we believe that the alternative is meaningless, whether my friend's consciousness contains either the impression of having seen a flash or of not having seen a flash. However, to deny the existence of the consciousness of a friend to this extent is surely an unnatural attitude, approaching solipsism, and few people will go in their hearts along with it.[24]

V. BOHR: THE COMPLEMENTARITY PRINCIPLE IN EPISTEMOLOGY

Throughout the first four sections of this paper it was implicitly assumed that intrinsic characteristics of things in nature are clearly distinguishable from characteristic which are relative to an observer. I do not think that the presentation of the position of von Neumann or of London and Bauer was distorted because of this assumption, and indeed the lucidity of their work is probably due partly to the fact that they accepted it. However, it may be suspected that the counter-intuitive conclusions to which they were led – particularly that a change in the state of a physical system is effected by the registration of the result of a measurement in a consciousness – resulted from a rigid distinction between objectivity and subjectivity. This suggestion is implied in the writing of Niels Bohr, for an essential feature of his philosophy and of all the cognate varieties of the "Copenhagen interpretation" is a flexibility in contrasting objectivity and subjectivity. In this section, I try to formulate Bohr's position briefly and to assess its explanation of the role of the observer in quantum mechanics. All that I state, both in exegesis and in criticism, is tentative, since his writing is both obscure and profound.

Bohr's philosophy consists essentially of the systematic formulation and application of the principle of complementarity. In its narrowest formulation, this principle is a renunciation of a single picture of the microphysical object in favor of a set of mutually exclusive descriptions appropriate in different circumstances. Bohr interprets quantum mechanics as a theory of quantum phenomena, and he uses the word 'phenomenon' "to refer exclusively to the observations obtained under specified circumstances, including an account of the whole experimental arrangement." (Ref. 3c, p. 64.) He insists that obscurity results from speaking of physical attributes of objects without reference to the measuring instruments

---

24. Dr. Howard Stein (personal communication) commented that if quantum mechanics implies solipsism then it must be self-contradictory, since manifestly no person could have discovered it alone!

with which they interact. Furthermore, even though the phenomena themselves cannot be accounted for by classical physical *explanations,* they
must inevitably be described in terms of classical physical *concepts,* because these concepts are indispensable in characterizing the measuring instruments. The indeterministic transition which occurs when, for example,
a position measurement is followed by a momentum measurement can be
understood straightforwardly in terms of the scheme of complementarity.
Because of wave–particle dualism and the relation between momentum
and wave-number, the experimental arrangements for measuring the position and momentum of a particle are mutually exclusive (Ref. 3c, pp.
41-7); but since any microphysical phenomenon inextricably involves a
type of experimental arrangement, the transition from a position to a
momentum determination consists of the rupture of one well-defined state
of a composite system and the initiation of another well-defined state of a
different composite system in a manner which cannot be controlled (Ref.
3c, p. 40). It seems possible for Bohr to avoid attributing the reduction of
a superposition to the consciousness of the observer, for the experimental
arrangement and the experimental result can be described completely in
terms of "everyday concepts, perhaps refined by the terminology of classical physics." (Ref. 3c, p. 26.) But whenever everyday and classical concepts can be unambiguously applied, it is possible to distinguish sharply
between the objects observed and the observer (Ref. 3c, p. 25). Intersubjective agreement is thus assured in the reading of the measuring apparatus; and hence the outcome of a quantum mechanical measurement is objectively determined when a spot is made on a photographic plate or a
Geiger counter discharges, etc., without dependence upon the registration
of the result in the observer's consciousness (Ref. 3c, p. 64 and p. 88).

The insistence upon a classical description of the measuring apparatus,
not as a convenient approximation but as a matter of principle, clearly
differentiates Bohr's interpretation of quantum mechanics from that of
von Neumann and of London and Bauer. It is tempting, therefore, to
accept Feyerabend's characterization[25] of Bohr's point of view as "positivism of higher order," that is to say, a positivism in which the "given"
for scientific construction consists not of subjective perceptions but rather
of descriptions in terms of classical physics. According to this interpretation, the senses in which microscopic and macroscopic objects are real are
quite different, for a microscopic object is "now characterized as a set of
(classical) *appearances* only, without any indication being given as to its
nature" (Ref. 25, p. 94). Occasional passages in Bohr's writings support
Feyerabend's interpretation that microphysical symbolism is only a convenient shorthand for describing intricate relations among macrophysical

25. P. Feyerabend, The Aristotelian Society Suppl. **32**, 75 (1958).

objects, which are taken to be the fundamental building blocks in his ontology. For example,

there can be no question of any unambiguous interpretation of the symbolism of quantum mechanics other than that embodied in the well-known rules which allow to predict the results to be obtained by a given experimental arrangement described in a totally classical way. (Ref. 3b, p. 701.)

Other passages, cited below, indicate that Feyerabend's formulation of Bohr's position is incomplete. However, the strength of this formulation should be acknowledged: that for the working physicist, particularly the experimentalist, "positivism of higher order" provides a practical compromise between microphysical realism and phenomenalism. In contrast to phenomenalism, which encounters difficulties in attempting to characterize ordinary objects, this position operates *ab initio* with everyday concepts and classical physical concepts, which suffice to characterize laboratory equipment and to permit unambiguous communication among observers.[26] And in contrast to microphysical realism, this position does not run the danger of postulating attributes and processes which are inaccessible to observation. Nevertheless, in spite of the advantages of this "positivism of higher order" as a working philosophy, it is doubtful that it can be maintained under critical scrutiny. Perhaps the worst flaw is that the interpretation of the microphysical formalism as a shorthand for macrophysical observations is a program, of which only a small part has been carried out. It is commonly said that every hermitian operator upon the space of state vectors of a given system represents an observable of the system.[27] Yet, only for a small number of the nondenumerable infinity of such operators has experimental apparatus capable of measuring the values of the corresponding observables been built or even designed. When such apparatus exists, as in the case of the position, linear momentum, angular momentum and energy operators, then it is reasonable to say that the corresponding eigenstates of the microscopic object are merely convenient ways of speaking about the various experimental results obtainable with the apparatus. But such an interpretation is not reasonable when no one has the least idea for designing the apparatus appropriate for a given operator. In the formalism of quantum mechanics the operators for which no measuring apparatus has been designed enter upon exactly the same footing as those for which apparatus exists, and there

26. This is one of the reasons why most of the logical positivists abandoned phenomenalistic languages in favor of physicalistic languages.
27. In careful treatments, this statement is made only of those hermitian operators for which a solution to the eigenvalue problem exists; cf. J. von Neumann, Ref. 4, pp. 167–9; and P. A. M. Dirac, *The Principles of Quantum Mechanics* (Clarendon Press, Oxford, 1947), 3rd ed., p. 38.

is no way of eliminating the former without curtailing the superposition principle, which would cripple the present formulation of quantum mechanics. This situation is understandable from the point of view of microphysical realism, which does not equate reality with accessibility for measurement, but it is incompatible with the interpretation of quantum mechanical symbolism as a shorthand for classical appearances.[28] Another flaw in "positivism of higher order" is its fundamental obscurity regarding the ontological status of macrophysical objects. If these objects have an objective existence, then they must have intrinsic properties independent of human observers. Their properties cannot be specified in terms of classical physics without limitation, because of the uncertainty principle; but if the intrinsic character of a macrophysical object includes a specification of the extent to which each classical quantity is applicable, presumably in terms of definite probability distributions, then the object is effectively in a quantum mechanical state. However, admitting that a macroscopic object is intrinsically characterized by a quantum mechanical state leads to the difficulties which were found in the interpretation of quantum mechanics of von Neumann and of London and Bauer: the state of a macroscopic object can be a superposition of states in which a macroscopic observable has distinct values, and the reduction of this superposition can be effected only by the consciousness of the observer. There are passages in which Bohr appears to avoid these difficulties by retreating from "positivism of higher order" to a novel combination of idealism and phenomenalism, in which even macroscpic objects have the status of organizations of experience and in which the concepts of classical physics become forms of perception (e.g., Ref. 3a, p. 1). Many questions of interpretation can be raised regarding these passages: for example, the expression "forms of perception" is reminiscent of Kant, but since Bohr never suggests that classical physical concepts have an *a priori* origin, it is not clear why "all experience must ultimately be expressed in terms of classical concepts" (Ref. 3a, p. 94). These questions of interpretation are not pursued here, however, since some general reasons for rejecting idealism and phenomenalism were stated at the end of Sec. III. The important point here is that an examination of the status of macrophysical objects shows "positivism of higher order" to be an unstable half-way position between

28. The above criticism is also stated by P. Feyerabend, "Problems of Microphysics," *Frontiers of Science and Philosophy* (University of Pittsburgh Press, Pittsburgh, 1962). He cites similar criticisms by E. Schrödinger, Nature **173**, 442 (1954), and P. Bridgman, *The Nature of Physical Theory* (Princeton University Press, Princeton, New Jersey, 1936). A possible answer is to interpret quantum mechanical states in terms of the scattering matrix, all elements of which are in principle measurable; see R. E. Cutkosky, Phys. Rev. **125**, 745 (1962), and the references given there. The extent to which such an interpretation is possible is controversial, and I have not studied the literature sufficiently to comment on the question.

a realistic interpretation of microphysical objects and an idealistic or phenomenalistic interpretation of all physical entities, and therefore, whether or not one ought to attribute this position to Bohr, it is not tenable.

I suspect that Bohr was aware of the difficulties inherent in a macrophysical ontology, and in his most careful writing he states subtle qualifications concerning states of macroscopic objects. For example,

The main point here is the distinction between the *objects* under investigation and the *measuring instruments* which serve to define, in classical terms, the conditions under which the phenomena appear. Incidentally, we remark that, for the illustration of the preceding considerations, it is not relevant that experiments involving an accurate control of the momentum or energy transfer from atomic particles to heavy bodies like diaphragms and shutters would be very difficult to perform, if practicable at all. It is only decisive that, in contrast to the proper measuring instruments, these bodies together with the particles would constitute the system to which the quantum-mechanical formalism has to be applied. (Ref. 3c, p. 50.)

Bohr is saying that from one point of view the apparatus is described classically and from another, mutually exclusive point of view, it is described quantum mechanically. In other words, he is applying the principle of complementarity, which was originally formulated for microphysical phenomena, to a macroscopic piece of apparatus. This application of complementarity on a new level provides an answer to the difficulty regarding macrophysical objects which confronted "positivism of higher order": the macrophysical object has objective existence and intrinsic properties in one set of circumstances (e.g., when used for the purpose of measuring) and has properties relative to the observer in another set of circumstances,[29] thereby evading the dilemma of choosing between realism and idealism. Two important conclusions follow from this discussion. The first is that Bohr's point of view is only partially represented by the macrophysical ontology which Feyerabend attributes to him; and accordingly, one should not dismiss as mere *façons de parler* his occasional statements that atoms are real and have individuality (e.g., Ref. 3a, p. 93). The second conclusion is that Bohr's extension of the principle of complementarity beyond its original function of reconciling apparently contradictory microphysical phenomena is not gratuitous, as critics have often claimed.[30] The internal logic of his position requires the application of complementarity to other domains, such as biology and psychology, to be more than merely a hopeful extrapolation of a conceptual device which proved successful in microphysics.

Perhaps the most explicit formulation of the generalized principle of complementarity is the following:

29. D. Bohm, Ref. 14, Chap. 23, states this generalization of the principle of complementarity more explicitly than Bohr.
30. For example, A. Grünbaum, J. Philosophy **54**, 713 (1957).

For describing our mental activity, we require, on one hand, an objectively given content to be placed in opposition to a perceiving subject, while, on the other hand, as is already implied in such an assertion, no sharp separation between object and subject can be maintained, since the perceiving subject also belongs to our mental content. From these circumstances follows not only the relative meaning of every concept, or rather of every word, the meaning depending upon our arbitrary choice of view point, but also that we must, in general, be prepared to accept the fact that a complete elucidation of one and the same object may require diverse points of view which defy a unique description. Indeed, strictly speaking, the conscious analysis of any concept stands in a relation of exclusion to its immediate application. (Ref. 3a, p. 96.)

As an example of the generalization of complementarity, Bohr states that the display of life and the conformity to physical laws are both proper descriptions of an organism, but are observed in mutually exclusive experimental arrangements. Similarly, he argues that determinism and free will are complementary aspects of the behavior of an organism. Behavioristic and introspective psychology are evidently complementary, and even within introspective psychology a scheme of explanation using complementarity is necessary:

... it must be emphasized that the distinction between subject and object, necessary for unambiguous description, is retained in the way that in every communication containing a reference to ourselves we, so-to-speak, introduce a new subject which does not appear as part of the content of the communication. (Ref. 3c, p. 101.)

Critics have pointed out, with considerable justification, the looseness of Bohr's arguments in these contexts: for example, that he presents no decisive evidence of conflict between organic behavior and conformity to physical law analogous to the conflict between the wave and particle aspects of light, and that the operational interferences of two types of observation is not always sufficient grounds, even in atomic physics, for denying the possibility of simultaneous sharp values of the respective quantities.[31] However, I am not here concerned with specific weaknesses of Bohr's remarks on biology and psychology, but rather with the general epistemological view which emerges from his remarks: that *given any domain of phenomena there is a standpoint from which this domain can be observed in mutually exclusive ways.* Several further quotations suffice to show the radical character of his position:

the notion of complementarity serves to symbolize the fundamental limitation, met with in atomic physics, of the objective existence of phenomena independent of the means of their observation. (Ref. 3c, p. 7.)

31. See, for example, P. Feyerabend, Ref. 28, and A. Grünbaum, Ref. 30.

Without entering into metaphysical speculations, I may perhaps add that an analysis of the very concept of explanation would, naturally, begin and end with a renunciation as to explaining our own conscious activity. (Ref. 3c, p. 11.)

Such considerations point to the epistemological implications of the lesson regarding our observational position, which the development of physical science has impressed upon us. In return for the renunciation of accustomed demands on explanation, it offers a logical means of comprehending wider fields of experience, necessitating proper attention to the placing of the object–subject separation. Since, in philosophical literature, reference is sometimes made to different levels of objectivity or subjectivity or even of reality, it may be stressed that the notion of an ultimate subject as well as conceptions like realism and idealism find no place in objective description as we have defined it. (Ref. 3c, pp. 78–9.)

It is clear that Bohr considers the distinction between subject and object to be a necessary condition for knowledge and communication; but he believes the separation of subject and object can be performed in arbitrarily many different ways, and never in such a way that an absolute subject or an absolute object exists. Evidently, there is a profound difference between Bohr's position and that of von Neumann and of London and Bauer, who acknowledge an ultimate subject in any act of observation, and this difference accounts for their different treatments of measurement in quantum mechanics.

As practical maxims for scientific activity some of Bohr's proposals are sound. It is reasonable, in any investigation, to delimit the objects of investigation and to be aware of the way in which the description of the objects depends upon the conditions of observation. It is also possible that the alternative descriptions of organisms, minds, cultures, etc. must be taken to be complementary, though the evidence regarding this is fragmentary and inconclusive. The only very weak point in the program of complementarity, from the standpoint of scientific investigation, is the absence of any analogue in the domains of biology and psychology to the systematic statistical relations between complementary descriptions in microphysics, and, of course, it is these statistical relations which establish the predictive power of quantum theory.

If the program of complementarity is taken to be a philosophical system, rather than a set of practical suggestions in science, then there are strong objections against it. The indefiniteness and arbitrariness of the distinction between subject and object seems to imply the renunciation of an ontological framework for locating the activity of knowing.[32] Knowledge

32. The affinity between Bohr's philosophy of complementarity and Hegel's dialectic is evident, even though Bohr proposes nothing comparable to the ultimate synthesis in the Absolute Ideal and though he explicitly disavows the conception of levels of reality (Ref. 3c, p. 79). There is, also, a strong affinity between Bohr and Kant, which to my knowledge has not been sufficiently stressed by commentators. It was noted above that

is certainly a phenomenon of immense complexity, and yet one cannot seriously doubt that it is a natural phenomenon, involving the interaction of an organism with various other natural entities. Since this interaction can occur in many different ways, the choice of which is partly under the control of the organism itself, it is understandable that the "placing of the object–subject separation" is variable. But if no ontological framework is presupposed, the object–subject separation could no longer be understood as a natural event, but only as a mode of organizing the content of experience. Bohr evidently believes that the renunciation of an ontological framework is imposed by "the old truth that we are both onlookers and actors in the great drama of existence" (Ref. 3a, p. 119). But a quite different philosophical conclusion could equally well be drawn from this old truth: that we must try to formulate a view of nature which accommodates all our experience, including experience of ourselves as onlookers in the world; and we must formulate a theory of knowledge which suffices to provide a rationale for this view of nature. In such a complete and coherent view entities capable of having subjective experience could be identified as having objective status in nature, and subjective experience would be shown to be capable of envisaging an objectively real world. Bohr has given no decisive evidence that such a view is attainable, and the alternative which he outlines is too indefinite to serve as a surrogate.

The foregoing objection against Bohr's epistemological position implicitly contains a criticism of his account of observation in quantum mechanics. If the arbitrariness of the subject–object relation is understood, as I would contend, merely as freedom to arrange different interactions between microscopic objects, measuring apparatus, and the human observer, then the question of the intrinsic properties of these entities must be considered. These intrinsic properties may be far from familiar and common-sensical: there may be limitations on spatio-temporal localizability, or it may be that composite systems have an intrinsic wholeness and unanalyzability into parts which was not found in classical physics, or it may be that the superposition principle applies to the states of macroscopic objects and even to states of mind. The point is that whatever these intrinsic properties may be, they must suffice to account for the transitions

there is a Kantian flavor in the recurrent characterization of the concepts of classical physics as "forms of perception." More significant, perhaps, is the similarity between Bohr's renunciation of an absolute ontological framework and Kant's renunciation of a treatment of the thing-in-itself by theoretical reason. It is reasonable to attribute to Bohr the belief that when one attempts to characterize objects intrinsically, the principle of complementarity engenders contradictions analogous to Kant's antinomies of pure reason.

which occur in quantum mechanical measurement as natural processes. The shifting of the separation between subject and object postpones for an arbitrary but finite number of steps the necessity for explaining these transitions in terms of the intrinsic characteristics of natural entities, but in principle it must finally be possible to provide such an explanation.

VI. CONCLUSIONS

The foregoing study has examined, in detail, two of the many accounts of observation in quantum mechanics – that of von Neumann, London, and Bauer, and that of Bohr. Both of these accounts maintain that the current formulation of quantum mechanics provides a rigorous description of physical reality, but they differ in their epistemological proposals and, therefore, the difficulties which they encounter are quite different. Von Neumann, London, and Bauer propose that objective changes of the state of physical systems occur when certain measurements are performed, and they explain these changes by referring to the subjective experience of an observer. This mutual involvement of physical and mental phenomena is counter-intuitive in the extreme, though without apparent inconsistency. However, there is no empirical evidence that the mind is endowed with the power, which they attribute to it, of reducing superpositions; and furthermore, there is no obvious way of explaining the agreement among different observers who independently observe physical systems. Thus, their interpretation of quantum mechanics rests upon psychological presuppositions which are almost certainly false. In Bohr's account, the change of state of a microphysical object under measurement is due to the change of the experimental arrangement for observing the object. Since the experimental arrangement can be described in physical terms, Bohr does not need to postulate an intrusion of consciousness into the course of physical phenomena. However, his explanation implies the renunciation of an intrinsic characterization of objects in favor of complementary descriptions from alternative points of view, and the flexibility which this conceptual device provides is excessive, for ultimately it requires the abandonment of any definite ontology. Some variant interpretations of quantum mechanics were also noted and criticized in the course of studying these two major accounts. Even though this survey of interpretations was not exhaustive, it did cast strong doubt upon the possibility of giving a coherent discussion of observation in quantum mechanics without modifying the theory itself.[33] It should be emphasized

33. Among the interpretations which have not been explicitly discussed above, that of Hugh Everett, III, Rev. Mod. Phys. **29**, 454 (1957) (accom anied by J. A. Wheeler's

again that the psychological and philosophical problems concerning observation which must be clarified do not arise from considerations external to physics – for instance, from an attempt to relate physics to other disciplines – but from an attempt to understand the conceptual apparatus of the current formulation of quantum theory.

It is possible that the conceptual problems of quantum mechanics will be resolved by discovering corrections to the physical theory itself, for example, by finding that the time-dependent Schrödinger equation is only an approximation to an exact nonlinear equation governing the evolution of the state of a system. If this proves to be true, then the reduction of a superposition could perhaps occur when the microscopic system interacts with the macroscopic apparatus, and no appeal to the consciousness of an observer for this purpose would be required.[34] The ontological problem of the relation between mind and matter would remain, but it would be posed in essentially the same way as in the period of classical physics and with no more urgency. Such a solution would be welcome to many physicists, for it would perpetuate the convenience of the "bifurcation of nature." Nevertheless, there may be reasons for regret if such a solution proves to be successful. "Small clouds" in an otherwise highly successful theory have often been precursors of great illumination – e.g., the difficulties in explaining blackbody radiation led to Planck's discovery of the quantum of action. The conceptual problem of the reduction of a super-

assessment) should be specially mentioned. According to Everett, quantum theory is complete without the postulate that changes of state of type 1 occur. He eliminates the problem of accounting for the reduction of superpositions by denying that a reduction occurs at *any* stage in a measurement. Furthermore, he accepts the radical ontological consequences of his premisses: ". . . *all* elements of a superposition (all 'branches') are 'actual', none any more 'real' than the rest. It is unnecessary to suppose that all but one are somehow destroyed, since all the separate elements of a superposition individually obey the wave equation with complete indifference to the presence or absence ('actuality' or not) of any other elements. This total lack of effect of one branch on another also implies that no observer will ever be aware of any 'splitting' process." Everett's interpretation, and particularly his treatment of the relation between appearance and reality, deserves a more detailed analysis than is given here, but its essential weakness, I believe, is the following. From the standpoint of any observer (or more accurately, from the standpoint of any "branch" of an observer) the branch of the world which he sees evolves stochastically. Since all other branches are observationally inaccessible to the observer, the empirical content (in the broadest sense possible) of Everett's interpretation is precisely the same as the empirical content of a modified quantum theory in which isolated systems of suitable kinds occasionally undergo "quantum jumps" in violation of the Schrödinger equation. Thus the continuous evolution of the total quantum state is obtained by Everett at the price of an extreme violation of Ockham's principle, the entities multiplied being entire universes.

34. E. P. Wigner, Am. J. Phys. **31**, 6 (1963); also the second of the two works by G. Ludwig cited in Ref. 15; and A. Komar, Phys. Rev. **126**, 365 (1962).

position is a "small cloud" in contemporary physical theory, in which the laws of physics are otherwise completely independent of the existence of minds. If this difficulty should eventually provide some insight into the mysterious coexistence and interaction of mind and matter, our present intellectual discomfort would be overwhelmingly compensated.

Acknowledgments. I am grateful to Professor E. P. Wigner for encouraging my philosophical investigations of problems in quantum theory and also for making several valuable specific suggestions. Dr. Howard Stein read an earlier draft of this paper with great care and made a number of comments which were incorporated into the final version. I also thank Professor Hilary Putnam for an important correction to my earlier account of observation in classical physics.

COMMENT

Section III discusses some possible connections between putative superpositions of psychological states of the observer and various psychological phenomena, but makes no proposal of controlled experiments. This omission is partly repaired in the article "Wave-Packet Reduction as a Medium of Communication," reprinted as Chapter 21 of this volume.

Although I cite Landé on filters with approbation in footnote 13, I do not accept his later claim (*New Foundations of Quantum Mechanics,* London: Cambridge University Press, 1965) that standard quantum mechanics can be derived from a few assumptions about symmetry and transitivity in successive filterings; see my "Basic Axioms of Microphysics" [*Physics Today* (September 1966), pp. 85–90]. In his reply, "Quantum Fact and Fiction, III" [*American Journal of Physics* 37 (1969), pp. 541–43], Landé augmented his set of assumptions, but his new assumption was correctly criticized by D. K. Nartonis, "Quantum Fact and Fiction" [*British Journal for the Philosophy of Science* 25 (1974), pp. 329–33].

2
Approximate measurement in quantum mechanics

Part I *

Mary H. Fehrs and Abner Shimony

This is the first of two papers showing that the quantum problem of measurement remains unsolved even when the initial state of the apparatus is described by a statistical operator and when the results of measurement have a small probability of being erroneous. A realistic treatment of the measurement of observables of microscopic objects (e.g., the position or the spin of an electron) by means of observables of macroscopic apparatus (e.g., the position of a spot on a photographic plate) requires the consideration of errors. The first paper considers measurement procedures of the following type: An initial eigenstate of the object observable leads to a final statistical operator of the object plus apparatus which describes a mixture of "approximate" eigenstates of the apparatus observable. It is proved that each of a large class of initial states leads to a final statistical operator which does not describe any mixture containing even one "approximate" eigenstate of the apparatus observable.

I. INTRODUCTION

Several writers [1] have tried to solve the quantum-mechanical problem of measurement through one or both of the following proposals: (a) describing the initial state of the measuring apparatus by a projection onto a subspace of the associated Hilbert space or by a statistical operator, thus taking into account the practical impossibility of knowing the exact quantum state of a macroscopic object; (b) recognizing that there may be some physical inaccuracy, such as a small error in the position of a pointer

These two papers originally appeared in *Physical Review D* 9 (1974), pp. 2317–20 and 2321–23. Reprinted by permission of The American Physical Society.

*This paper is based upon material contained in a thesis submitted by one of the authors (M.H.F.) to the Graduate School of Boston University in 1973 in partial fulfillment of the requirements for the degree of Doctor of Philosophy. It was written while one of the authors (A.S.) was a John Simon Guggenheim Memorial Fellow.

1. W. Heisenberg, *Physics and Philosophy* (Harper and Row, New York, 1962), pp. 53, 54; L. Landau and E. Lifshitz, *Quantum Mechanics* (Pergamon, London, 1958), pp. 21–24.

needle, in the final registration of the outcome of the measurement by the apparatus. These writers hope that in a measuring process which satisfies (a) and (b), any initial state of the object will result in a final statistical state of the object plus apparatus which is a mixture of exact or approximate eigenstates of the apparatus observable. If this hope were justified, then the quantum-mechanical problem of measurement might be resolved. The apparatus observable could be considered to have, at least with high probability, a definite though unknown value at the end of the physical process of interaction between the object and the apparatus. The consciousness of the observer would become aware of this definite value, and therefore would not have to be assigned the role of reducing a superposition.

This and the following paper on approximate measurement are a continuation of earlier work,[2-6] initiated by Wigner, strongly indicating that no satisfactory solution to the measurement problem can be obtained in the manner that has just been sketched.[7] The papers differ in their formulations of the approximate measuring procedure proposed in (b). In the present paper the following formulation is adopted: If the initial state of the object is an eigenstate of the object observable, then the final statistical state of the object plus apparatus can be described as a mixture of pure quantum states, all of which are "almost" eigenstates of the apparatus observable associated with the same eigenvalue. This is a less stringent conception of measurement than those treated in Refs. 2–6, and one might therefore conjecture that it permits any initial state of the object to eventuate in a final mixture of "almost" eigenstates of the apparatus observable. The falsity of this conjecture follows from a mathematical theorem which is proved in Sec. II and discussed in Sec. III. Even if the conjecture had been true, however, it is not clear that progress would have been made towards solving the problem of measurement; for unless each

2. E. P. Wigner, Am. J. Phys. **31**, 6 (1963). The relevant section of this essay is entitled "Critiques of the Orthodox Theory."
3. B. d'Espagnat, Nuovo Cimento Suppl. **4**, 828 (1966).
4. J. Earman and A. Shimony, Nuovo Cimento **54B**, 332 (1968).
5. H. Stein and A. Shimony, *Foundations of Quantum Mechanics,* edited by B. d'Espagnat (Academic, New York, 1971), p. 56 (especially pp. 64–66).
6. B. d'Espagnat, *Conceptual Foundations of Quantum Mechanics* (Benjamin, Menlo Park, 1971), pp. 331–339.
7. A quite different proposal for solving the measurement problem is based on the assumption that only some of the self-adjoint operators on the Hilbert space associated with the apparatus represent physical observables. See, for example, A. Daneri, A. Loinger, and G. M. Prosperi [Nucl. Phys. **33**, 297 (1962)] and K. Hepp [Helv. Phys. Acta **45**, 237 (1972)]. That proposal is not discussed in this or the following paper, but is discussed in Ref. 6, pp. 278–284.

pure state of the final mixture were an exact eigenstate of the apparatus observable, a reduction of a superposition would seem to be required in order to produce an objectively definite value of this observable. For this reason, the formulation of measurement analyzed in the sequel to the present paper probably has greater philosophical interest than the one analyzed here. A further conjecture, primarily of mathematical interest, is discussed in Sec. IV.

II. PROOF OF A THEOREM

In the following discussions, $\mathcal{K}_1$ and $\mathcal{K}_2$ denote the Hilbert spaces associated with the object and the apparatus, respectively, and $\mathcal{K}_1 \otimes \mathcal{K}_2$ denotes the space associated with the composite system, the object plus apparatus. Subscripted letters E and F are used for projection operators on $\mathcal{K}_1$ and $\mathcal{K}_2$, respectively, and correspondingly subscripted underlined letters, $\underline{E}$ and $\underline{F}$, are used for the subspaces onto which they project. However, the projection operators onto $\mathcal{K}_1$ and $\mathcal{K}_2$ themselves are in each case denoted by 1; and the one-dimensional subspace (ray) spanned by a nonzero vector u is denoted by $\langle u \rangle$, and the associated projection operator by P_u. $\underline{M}^{\perp}$ is the orthogonal complement of $\underline{M}$, and $M^{\perp}$ is the projection operator onto $\underline{M}^{\perp}$. An eigenvector of a statistical operator with a nonzero eigenvalue is called a "constitutive vector." A nonzero vector u is said to be "within ϵ of being contained in the subspace $\underline{M}$" if $\|u\|^{-1}\|M^{\perp}u\|$ is equal to or less than ϵ; and the same designation is applied to a subspace $\underline{N}$ if every nonzero vector in $\underline{N}$ is within ϵ of being contained in $\underline{M}$.

The theorem to be proved is the following.

Hypotheses:

(i) $\{\underline{E}_m\}$ *is a finite or denumerably infinite family of mutually orthogonal subspaces spanning* $\mathcal{K}_1$, $\{\underline{F}_m\}$ *is a family of mutually orthogonal subspaces of* $\mathcal{K}_2$, *and U is a unitary operator on* $\mathcal{K}_1 \otimes \mathcal{K}_2$;

(ii) *T is a statistical operator on* $\mathcal{K}_2$ *such that for every m and every* $v \in \underline{E}_m$, *the range of* $U(P_v \otimes T)U^{-1}$ *is within* ϵ_m *of being contained in* $\mathcal{K}_1 \otimes \underline{F}_m$, *where all* ϵ_m *are equal to or less than some fixed* ϵ, *all but N of the* ϵ_m *are 0, and* $N\epsilon^2$ *is less than* 1;

(iii) $U(P_u \otimes T)U^{-1}$ *has a constitutive vector within* ϵ_0 *of being contained in one of the subspaces* $\mathcal{K}_1 \otimes \underline{F}_m$, *say, m = k.*

Conclusion: u is within κ of being contained in $\underline{E}_k$, *where*

$$\kappa = (2N^{1/2}\epsilon + \epsilon_0)(1 - N\epsilon^2)^{-1};$$

the range of $U(P_u \otimes T)U^{-1}$ *is within* $\kappa + \epsilon$ *of being contained in* $\mathcal{K}_1 \otimes \underline{F}_k$.

The proof will make use of three lemmas.

Lemma 1. *Hypotheses:*
(a) *Same as* (i) *of the theorem;*
(b) *for each m, $U(E_m \otimes \langle \eta \rangle)$ is within ϵ_m of being contained in $\mathfrak{IC}_1 \otimes \underline{F}_m$, where all ϵ_m are equal to or less than some fixed ϵ, all but N of the ϵ_m are 0, and $N\epsilon^2$ is less than 1;*
(c) *u is a vector of $\mathfrak{IC}_1$ such that $U(u \otimes \eta)$ is within ϵ_0 of being contained in $\mathfrak{IC}_1 \otimes \underline{F}_k$.*
Conclusion: Same as the first part of the conclusion of the theorem.

Proof. Let $u' = u/\|u\|$ and $\eta' = \eta/\|\eta\|$. Then u' can be expressed as $\sum_i c_i u_i$, where u_i is a normalized vector belonging to $\underline{E}_i$ and the sum of the $|c_i|^2$ is unity. Relabel the subspaces so that $\epsilon_1, \ldots, \epsilon_N$ are the nonzero members of $\{\epsilon_m\}$. By (c)

$$1 - \epsilon_0^2 \leq \|(1 \otimes F_k)U(u' \otimes \eta')\|^2 = \left\| \sum_{i \neq k}^N c_i \chi_i + c_k \chi_k \right\|^2,$$

where χ_m is defined as $(1 \otimes F_k)U(u_m \otimes \eta')$. By (b), $\|\chi_m\|^2$ is equal to or greater than $(1 - \epsilon^2)$ for $m = k$ and equal to or less than ϵ^2 for $m \neq k$. Therefore, a lower bound h on $|c_k|$ is obtained by rewriting the foregoing inequality as

$$1 - \epsilon_0^2 - \left\| \sum_{i \neq k}^N c_i \chi_i \right\|^2 - 2 \operatorname{Re} \left(\sum_{i \neq k}^N c_i \chi_i, c_k \chi_k \right) \leq |c_k|^2 \|\chi_k\|^2,$$

and then replacing the third and fourth terms on the left-hand side by their respective lower bounds, $-(1 - h^2)N\epsilon^2$ and $-2(1 - h^2)^{1/2}N^{1/2}\epsilon$, and $\|\chi_k\|^2$ by its upper bound 1. The condition on h (with much information thrown away in this manner) is then

$$1 - \epsilon_0^2 - (1 - h^2)N\epsilon^2 - 2(1 - h^2)^{1/2}N^{1/2}\epsilon \leq h^2.$$

This may be rewritten as a condition on $(1 - h^2)^{1/2}$, which is an upper bound on $(1 - |c_k|^2)^{1/2}$. Using the resulting inequality together with the condition that $N\epsilon^2$ is less than 1, one finds that this upper bound is less than κ (the quantity defined in the conclusion of the theorem).

Lemma 2. *Hypotheses:*
(a) *Same as* (i) *of the theorem;*
(b) *T is a statistical operator on $\mathfrak{IC}_2$ having range $\underline{R}$ and such that for every m and every v in $\underline{E}_m$ the range of $U(P_v \otimes T)U^{-1}$ is within ϵ_m of being contained in $\mathfrak{IC}_1 \otimes \underline{F}_m$.*
Conclusion: $U(\underline{E}_m \otimes \underline{R})$ is within ϵ_m of being contained in $\mathfrak{IC}_1 \otimes \underline{F}_m$ for each m.

Lemma 3. *If u is a nonzero vector of* $\mathcal{K}_1$ *and A an operator on* $\mathcal{K}_2$, *then every eigenvector of* $P_u \otimes A$ *with nonzero eigenvalue a is of the form* $u \otimes \eta$, *where* η *is an eigenvector of A with the same eigenvalue a.*

Lemma 2 is a slight variation of the second lemma on p. 65 of Ref. 5, while lemma 3 is exactly the third lemma on that page.

The proof of the theorem now proceeds as follows. Let ξ be the constitutive vector of $U(P_u \otimes T)U^{-1}$ referred to in hypothesis (iii) of the theorem. Then $U^{-1}\xi$ is a constitutive vector of $P_u \otimes T$. Hence, by lemma 3, $U^{-1}\xi$ has the form $u \otimes \eta$, where η is a constitutive vector of T and therefore also a member of the range $\underline{R}$ of T. Therefore, (iii) implies that $U(u \otimes \eta)$ is within ϵ_0 of being contained in $\mathcal{K}_1 \otimes \underline{F}_k$, satisfying (c) of lemma 1. We have both hypotheses of lemma 2 and hence its conclusion, which implies (b) of lemma 1. Since (a) of lemma 1 is given as hypothesis (i), we now have all the hypotheses of lemma 1 and therefore its conclusion. But this is also the first part of the conclusion of the theorem. To prove the second part of the conclusion, let σ be any nonzero vector in the range of $U(P_u \otimes T)U^{-1}$. Then for some $\rho \in \mathcal{K}_1 \otimes \mathcal{K}_2$, $\sigma = U(P_u \otimes T)U^{-1}\rho$; and hence for some $\tau \in \mathcal{K}_2$, $T\tau \neq 0$, $\sigma = U(u \otimes T\tau)$. We can then write σ as the sum $U(E_k u \otimes T\tau) + U(E_k^{\perp} u \otimes T\tau)$, the first term being a nonzero member of the range of $U(P_{E_k u} \otimes T)U^{-1}$. Therefore

$$\frac{\|(1 \otimes F_k)^{\perp} U(u \otimes T\tau)\|}{\|U(u \otimes T\tau)\|} \leq \frac{\|(1 \otimes F_k)^{\perp} U(E_k u \otimes T\tau)\|}{\|U(u \otimes T\tau)\|}$$

$$+ \frac{\|(1 \otimes F_k)^{\perp} U(E_k^{\perp} u \otimes T\tau)\|}{\|U(u \otimes T\tau)\|}$$

$$\leq \frac{\|(1 \otimes F_k)^{\perp} U(E_k u \otimes T\tau)\|}{\|U(E_k u \otimes T\tau)\|} + \frac{\|U(E_k^{\perp} u \otimes T\tau)\|}{\|u \otimes T\tau\|}$$

$$\leq \epsilon_k + \kappa,$$

where the last step uses hypothesis (ii) and the first part of the conclusion of the theorem.

III. DISCUSSION OF THE THEOREM

A possible formulation of a procedure of measurement in quantum theory which satisfies proposals (a) and (b) of Sec. I is given by hypothesis (i) together with a somewhat modified version of hypothesis (ii), in which the condition that all but N of the $\{\epsilon_m\}$ are 0 is replaced by the condition that all the ϵ_m are much less than 1. In this formulation the subspaces $\{E_m\}$ are eigenspaces associated with distinct eigenvalues of some object observable $\mathcal{O}$. The subspaces $\{\underline{F}_m\}$ are eigenspaces associated with distinct eigenvalues of some apparatus observable $\mathcal{Q}$. The initial state of the apparatus is described by a statistical operator T, in recognition of the fact stated in proposal (a) that the exact quantum state of the apparatus is unknown. If $u \in \mathcal{3C}_1$, then $P_u \otimes T$ is a statistical operator describing the object plus apparatus at the beginning of the measurement process. $U(P_u \otimes T)U^{-1}$ is the statistical operator which evolves from $P_u \otimes T$ in a certain time interval, U being the unitary operator governing the evolution of any pure quantum state of the object plus apparatus during that interval. The modified hypothesis (ii) (with $\epsilon_m \ll 1$, but with no condition on the number of nonzero ϵ_m) asserts that the final statistical operator describes a mixture of quantum states which are all "almost" eigenstates of $\mathcal{Q}$ with the same eigenvalue. Hence, if one begins with the object observable having a definite value λ_m, then the final statistical state of the object plus apparatus is such that a subsequent measurement of the apparatus observable will yield a value from which the correct original value of $\mathcal{O}$ can be inferred with a high probability. Since $\mathcal{Q}$ is in practice a macroscopic observable, the measurement of $\mathcal{Q}$ could presumably be performed merely by looking, or in any case without the elaborate amplifying equipment needed to measure a microscopic observable.

The theorem of Sec. II shows that a procedure of measurement which satisfies this formulation, and also the condition of hypothesis (ii) that $N\epsilon^2$ is less than 1, will not fulfill the hopes of the writers in Ref. 1. For, according to the theorem, a final statistical state of the object plus apparatus which describes a mixture of exact or approximate eigenstates of the apparatus observable will come about only if the initial state of the object is almost an eigenstate of the object observable.

The mathematical results of Refs. 2, 4, and 5, and part of Ref. 3, are contained in the theorem by letting ϵ and ϵ_0 both be 0. The result of the other part of Ref. 3 is obtained by letting ϵ be 0, but taking ϵ_0 to be nonzero but much less than 1. That of Ref. 6 is obtained by imposing the following conditions: $0 \le \epsilon_0 \ll 1$, $0 \le \epsilon \ll 1$, and $N = 2$.

The theorem can surely be strengthened somewhat, since much information was thrown away in the course of proving it. However, the counterexample given in Sec. IV shows that the natural generalization of the theorem is false. It seems unlikely that any of the valid strengthened versions of the theorem would have any physical or philosophical implications of interest which are not already contained in the theorem as stated, unless a different formulation of the procedure of measurement is given.

Indeed, it is dubious that the results presented above have much philosophical significance, because a scheme of measurement in which the pure states of the final mixture are only "almost" eigenstates of the apparatus observable does not seem to ensure the objective existence of a definite value of the apparatus observable. A superposition of two non-null eigen-vectors v_1 and v_2 of the operator A, with distinct eigenvalues, does not represent a state in which the corresponding observable $\mathfrak{a}$ is definite, al-though possibly unknown. This is a matter of principle in the ordinary interpretation of quantum mechanics, and it is not altered when the norm of v_1 is much greater than the norm of v_2. Any attempt to dismiss as neg-ligible the contribution from a vector of small norm must be regarded as an alteration of principle, which would require justification. Without such an alteration of principle, the scheme of measurement of this paper would not be satisfactory – even apart from the theorem of Sec. II – be-cause it would apparently have to be supplemented by some means of reducing a superposition.

IV. FURTHER CONJECTURE

A natural generalization of the theorem of Sec. II is the following.

Conjectured theorem. *Hypotheses:*
(i′) *Same as* (i) *of the theorem*;
(ii′) *T is a statistical operator on $\mathfrak{K}_2$ such that for every m and every $v \in \underline{E}_m$, the range of $U(P_v \otimes T)U^{-1}$ is within ϵ_m of being contained in $\mathfrak{K}_1 \otimes \underline{F}_m$, $\epsilon_m \ll 1$;*
(iii′) *$U(P_u \otimes T)U^{-1}$ has at least one constitutive vector within ϵ_0 of being contained in the subspace $\mathfrak{K}_1 \otimes \underline{F}_m$, $\epsilon_0 \ll 1$.*
Conclusion: u is within κ of being contained in E_k, $\kappa \ll 1$.

A counterexample shows that this conjecture is false, even when the rela-tion $\ll$ is construed as stringently as one could reasonably demand in the hypotheses and as liberally as possible in the conclusion. Let $\{\underline{E}_m\}$, $m = 1, \ldots, N$, be a family of one-dimensional subspaces spanning $\mathfrak{K}_1$, and let the unit vector u_m span $\underline{E}_m$. Let $T = P_\eta$, where η is a unit vector of $\mathfrak{K}_2$. Let $\{\chi_m, \xi\}$ be a set of $N+1$ orthonormal vectors of $\mathfrak{K}_2$, and let $\underline{F}_m$ be spanned by χ_m for $m < N$, while $\underline{F}_N$ is spanned by χ_N and ξ. Finally, let v be an arbitrary unit vector of $\mathfrak{K}_1$ and (partially) define the operator U by

$$U(u_i \otimes \eta) = (N+1)^{-1} v \otimes \left[N\chi_i - \sum_{j \neq i} \chi_j + (N+2)^{1/2} \xi \right].$$

U as so far defined preserves inner products, and it may be extended to a unitarity. Clearly, hypotheses (i′) and (ii′) are satisfied if N is sufficiently large, with $\epsilon_m \leq (N+1)^{-1/2}$. Now let w be defined as $N^{-1/2} \sum_i^N u_i$. Then $U(w \otimes \eta)$ is in the range of $U(P_w \otimes T)U^{-1}$, and

$$U(P_w \otimes T)U^{-1} = (N+1)^{-1} v \otimes [N^{-1/2} \sum \chi_i + (N^2 + 2N)^{1/2} \xi],$$

which is within $(N-1)^{1/2}(N^2+N)^{-1/2}$ of being contained in $\mathcal{K}_1 \otimes \underline{F}_N$. Thus, hypothesis (iii') is satisfied, by taking N sufficiently large. But the conclusion is not satisfied, since

$$\|E^\perp w\|/\|w\| = (1-N^{-1})^{1/2},$$

which is as close to 1 as one desires.

The invalidity of the conjecture opens no new avenue for a solution to the problem of measurement, for if the hypotheses (i') and (ii') are satisfied, one can construct initial states of the object such that the final statistical state of the object plus apparatus is far from describing a mixture of approximate eigenstates of the apparatus observable. For this purpose it suffices to choose N greater than 2 but small enough that $N\epsilon_m^2$ is much less than 1 for each $m \le N$, and to let u be $N^{-1/2} \sum_i^N u_i$, where u_i is a unit vector belonging to $\underline{E}_i$. If one takes the $\mathcal{K}_1$ of the theorem of Sec. II to be the direct sum of the subspaces $\underline{E}_i$, with $i = 1, \ldots, N$, and ϵ_0 to be $\frac{1}{2}$, then that theorem implies that the final statistical state of the object plus apparatus does not contain a single constitutive vector within $\frac{1}{2}$ of being an eigenstate of the apparatus observable.

Acknowledgment. We wish to thank Professor B. d'Espagnat for his corrections of an earlier version of this paper and for hospitality to one of us (A.S.) at the Laboratoire de Physique Théorique et Hautes Energies at Orsay. We also wish to thank Professor A. Fine for illuminating correspondence on the conception of measurement in quantum mechanics.

COMMENT

The paper of Stein and Shimony cited in Ref. 5 is only in small part devoted to the quantum-mechanical measurement problem in the sense of the present paper. Its main purpose was to extend certain results of E. P. Wigner and of H. Araki and M. M. Yanase concerning limitations on quantum-mechanical measurement due to the existence of additive conserved quantities.

Part II

Abner Shimony

An approximate measurement procedure of the following type is considered: (i) An initial eigenstate of the object observable leads to a final statistical operator of the object plus apparatus describing a mixture of exact eigenstates of the apparatus observable; (ii) almost all the statistical weight of the mixture is assigned to eigenstates associated with one eigenvalue of the apparatus observable, which is uniquely determined by the initial value of the object observable. It is proved that each of a large class of initial states of the object leads to a final statistical operator which does not describe any mixture of exact eigenstates of the apparatus observable. The analysis also yields a proof of a theorem on measurement stated by Fine.

I. INTRODUCTION

In this paper the following formulation of an approximate measuring procedure is considered: If the initial state of the object is an eigenstate of the object observable with eigenvalue λ_m, then the final statistical state of the object plus apparatus can be described as a mixture of pure quantum states, all of which are exact eigenstates of the apparatus observable, and the total statistical weight in the mixture of those eigenstates associated with the eigenvalue μ_m is close to 1. (It is understood that $m \neq n$ implies both $\lambda_m \neq \lambda_n$ and $\mu_m \neq \mu_n$.) Hence, the value of the apparatus observable at the end of the interaction between the object and the apparatus is strongly correlated with the initial value of the object observable. This formulation of the procedure of measurement is more strictly in accordance with common sense than the formulation in a previous paper,[1] since a system in an exact eigenstate of the apparatus observable unequivocally *has* a sharp value of the apparatus observable, whereas it is not rigorously correct to speak of "having a sharp value" when the state is almost an eigenstate.[2]

Using the notation of Ref. 1, one can give the present formulation of approximate measurement in two conditions:

(a) $\{\underline{E}_m\}$ is a finite or denumerably infinite family of mutually orthogonal subspaces spanning $\mathfrak{K}_1$, $\{\underline{F}_m\}$ is a family of mutually orthogonal subspaces of $\mathfrak{K}_2$, and U is a unitary operator on $\mathfrak{K}_1 \otimes \mathfrak{K}_2$;

(b) T is a statistical operator on $\mathfrak{K}_2$ such that for every m and every $v \in \underline{E}_m$, $U(P_v \otimes T)U^{-1}$ can be expressed in the form $\sum_{n,r} a_{nr} P_{\chi_{nr}}$, where $\chi_{nr} \in \mathfrak{K}_1 \otimes \underline{F}_n$, and the a_{nr} are non-negative real numbers summing to 1 such that

$$\sum_{\substack{r \\ n \neq m}} a_{nr} = \epsilon_m \ll 1.$$

The theorem of Sec. II implies that if these two conditions are satisfied and if the number of subspaces $\underline{E}_m$ is greater than one, then there exist initial states of the object for which the final statistical state of the object

1. M. H. Fehrs and A. Shimony, preceding paper, Phys. Rev. D **9**, 2317 (1974).
2. Professor A. Fine emphasized this point in private correspondence.

plus apparatus is not expressible as a mixture of eigenstates of the apparatus observable.

II. A THEOREM ON MEASUREMENT

It will be convenient for proving the theorem of this section to use the Dirac bra and ket notation, in which $\langle\phi|\phi\rangle = 1$ implies that $|\phi\rangle\langle\phi|$ is the projection operator P_ϕ.

The theorem is the following.

Hypotheses:

(i) *u_1, u_2 are normalized orthogonal vectors of $\mathcal{K}_1$, $\{F_m\}$ is a family of mutually orthogonal subspaces of $\mathcal{K}_2$, U is a unitary operator on $\mathcal{K}_1 \otimes \mathcal{K}_2$, and T is a statistical operator on $\mathcal{K}_2$;*

(ii) *there exist orthonormal sets $\{\xi_{nr}^1\}$, $\{\xi_{nr}^2\}$ such that*

$$\xi_{nr}^j \in \mathcal{K}_1 \otimes F_n \quad for \ j = 1, 2,$$

and

$$U(P_{u_j} \otimes T)U^{-1} = \sum_{n,r} b_{nr}^j |\xi_{nr}^j\rangle\langle\xi_{nr}^j|;$$

and for some value of n,

$$\sum_r b_{nr}^1 \neq \sum_r b_{nr}^2.$$

Conclusion: If u is defined as $g_1 u_1 + g_2 u_2$, with both g_1 and g_2 nonzero, then there exists no orthonormal set $\{\psi_{nr}\}$ with $\psi_{nr} \in \mathcal{K}_1 \otimes F_n$ and no coefficients $\{b_{nr}\}$ such that $\sum_{n,r} b_{nr} = 1$ and

$$U(P_u \otimes T)U^{-1} = \sum_{n,r} b_{nr} |\psi_{nr}\rangle\langle\psi_{nr}|.$$

Proof. T can be written in the form

$$\sum_{i,s} a_i |\eta_{is}\rangle\langle\eta_{is}|,$$

where the η_{is} are orthonormal vectors of $\mathcal{K}_2$, and where a_i is a positive N_i-fold degenerate eigenvalue of T for each i, so that $\sum_i N_i a_i = 1$. Hence, if $U(u_j \otimes \eta_{is})$ is abbreviated by χ_{is}^j for $j = 1, 2$, then

$$U(P_{u_j} \otimes T)U^{-1} = \sum_{i,s} a_i |\chi_{is}^j\rangle\langle\chi_{is}^j|.$$

Since U is unitary, $\{\chi_{is}^j\}$ is a set of orthonormal vectors. The same statistical operator is thus expressed with respect to the two orthonormal sets $\{\chi_{is}^1\}$ and $\{\xi_{nr}^1\}$. Since the eigenvalues of a linear operator are invariant with respect to the choice of a basis, the coefficients $\{b_{nr}^1\}$ must be a permutation of the coefficients $\{a_i\}$ with proper multiplicities. The same is evidently also true for the coefficients $\{b_{nr}^2\}$ and $\{b_{nr}\}$. Consequently, the vectors ξ_{nr}^j can be relabeled $\bar{\xi}_{is}^j$ by appropriate permutation, and the vectors ψ_{nr} can be relabeled $\bar{\psi}_{is}$ in such a way that

$$U(P_{u_j}\otimes T)U^{-1} = \sum_{i,s} a_i |\bar{\xi}_{is}^j\rangle\langle\bar{\xi}_{is}^j|,$$

and

$$U(P_u\otimes T)U^{-1} = \sum_{i,s} a_i |\bar{\psi}_{is}\rangle\langle\bar{\psi}_{is}|.$$

In order that the condition $\sum_r b_{nr}^1 \neq \sum_r b_{nr}^2$ be satisfied for some value of n, there must be some value of i, say k, such that the number n_1 of $\{\bar{\xi}_{ks}^1\}$ belonging to $\mathfrak{IC}_1\otimes F_n$ is unequal to the number n_2 of $\{\bar{\xi}_{ks}^2\}$ belonging to $\mathfrak{IC}_1\otimes F_n$, and without loss of generality it may be assumed that $n_1 > n_2$. Then, by relabeling, we may write

$$\bar{\xi}_{ks}^j \in \mathfrak{IC}_1\otimes F_n, \quad s = 1, \ldots, n_j$$

$$\in \mathfrak{IC}_1\otimes \bigoplus_{m\neq n} F_m, \quad s = n_j+1, \ldots, N_k$$

for $j = 1, 2$.

Since each eigenvalue of the statistical operator $U(P_{u_j}\otimes T)U^{-1}$ is associated with an invariant subspace of the range of this operator,

(1) $$\chi_{kr}^j = \sum_{s=1}^{N_k} c_{rs}^j \bar{\xi}_{ks}^j, \quad j = 1, 2.$$

The coefficients $\{c_{rs}^j\}$, with fixed $j = 1, 2$ but with r and s varying from 1 to N_k, constitute a unitary matrix, so that

(2) $$\sum_{s=1}^{N_k} c_{rs}^j \bar{c}_{r's}^j = \delta_{rr'}, \quad j = 1, 2.$$

The statistical operator $U(P_u\otimes T)U^{-1}$ can be expressed in terms of the set $\{\bar{\psi}_{is}\}$ and also in terms of the sets $\{\chi_{is}^j\}$, so that

$$\sum_{i,s} a_i |\bar{\psi}_{is}\rangle\langle\bar{\psi}_{is}| = \sum_{i,s} a_i |g_1\chi_{is}^1 + g_2\chi_{is}^2\rangle\langle\bar{g}_1\chi_{is}^1 + \bar{g}_2\chi_{is}^2|.$$

If one considers only the terms associated with the eigenvalue a_k and makes use of Eqs. (1) and (2), one then obtains

$$\sum_{r=1}^{N_k} |\bar{\psi}_{kr}\rangle\langle\bar{\psi}_{kr}| = \sum_{r=1}^{N_k}\sum_{j=1}^{2}\sum_{j'=1}^{2} g_j\bar{g}_{j'} \sum_{s=1}^{N_k}\sum_{s'=1}^{N_k} c_{rs}^j \bar{c}_{rs'}^{j'} |\bar{\xi}_{ks}^j\rangle\langle\bar{\xi}_{ks'}^{j'}|.$$

Those members of $\{\bar{\psi}_{ks}\}$ which belong to $\mathfrak{IC}_1\otimes F_n$ are linear combinations of members of $\{\bar{\xi}_{ks}^1\}$ and $\{\bar{\xi}_{ks'}^2\}$ with $s \leq n_1$ and $s' \leq n_2$, respectively; and similarly for those members of $\{\bar{\psi}_{ks}\}$ which belong to

$$\mathfrak{IC}_1\otimes \bigoplus_{m\neq n} F_m.$$

Consequently, if g_1 and g_2 are nonzero, then a necessary condition for the foregoing equation to hold is that

$$(3) \qquad \sum_{r=1}^{N_k} c_{rs}^1 \bar{c}_{rs'}^2 = 0, \quad \text{for } s \leq n_1, \ s' > n_2.$$

But the N_k-tuples $\{c_{rs}^1\}$ (fixed s) and $\{c_{rs'}^2\}$ (fixed s'), $r = 1, \ldots, N_k$, can be considered as vectors in an N_k-dimensional complex vector space with an appropriate inner product. By Eqs. (2) and (3), the N_k-tuples such that $s \leq n_1$ and $s' > n_2$ constitute orthonormal vectors in this space. The number of them is

$$n_1 + (N_k - n_2) = N_k + (n_1 - n_2) > N_k.$$

But this is impossible, since the space is N_k-dimensional, and therefore the conclusion of the theorem follows.

III. DISCUSSION

In a procedure of measurement satisfying conditions (a) and (b) of Sec. I the hypotheses of the theorem are clearly satisfied whenver u_1 is chosen from one of the subspaces $\underline{E}_m$ and u_2 is chosen from another. Hence, if the number of the $\{\underline{E}_m\}$ is greater than 1, there exist initial states of the object leading to final statistical states of the object plus apparatus which cannot be expressed as mixtures of exact eigenstates of the apparatus observable. Consequently, the problem of measurement in quantum mechanics cannot be solved by imposing (a) and (b) as conditions of measurement.

Fine[3] has proposed a formulation of measurement more general than that discussed in Sec. I. Taking the operators O and A on the Hilbert spaces $\mathcal{H}_1$ and $\mathcal{H}_2$, respectively, to represent the object and apparatus observables, he gives the following two definitions (here slightly rewritten).

Definition 1. If Q is a self-adjoint operator on a Hilbert space $\mathcal{H}$ then the statistical operators W and W' are Q-distinguishable if and only if $\text{Tr}(WP_q) \neq \text{Tr}(W'P_q)$ for some projection operator P_q in the spectrum of Q.

Definition 2. If W_a is a statistical operator on $\mathcal{H}_2$, then a unitary operator U on $\mathcal{H}_1 \otimes \mathcal{H}_2$ is a W_a-measurement of O by means of A if and only if the O-distinguishability of W_o, W_o' implies the A-distinguishability of $U(W_o \otimes W_a)U^{-1}$ and $U(W_o' \otimes W_a)U^{-1}$.

3. A. Fine, Phys. Rev. D **2**, 2783 (1970).

The procedure of measurement envisaged by Fine's second definition may give extremely little information regarding the initial eigenvalue of an object observable by means of a single interaction of the object with a measuring apparatus. For this reason, his conception of measurement is very different from those considered in Ref. 1 and in Sec. I of this paper. However, his conception of measurement is legitimate as a procedure for determining some statistical information (in general less than the statistical state) about an ensemble of objects by means of an arbitrarily large number of measurements using a certain type of apparatus.

Concerning his conception of measurement, Fine asserts, but does not give a complete proof, of this theorem: There are no W_a-measurements U such that $U(W_o \otimes W_a)U^{-1}$ is a mixture of eigenstates of $1 \otimes A$ for all initial states W_o. (He assumes that the object observable O has at least two distinct eigenvalues.) His theorem is a consequence of the theorem of Sec. II. Suppose u_1 and u_2 are two eigenvectors of O associated with different eigenvalues, and hence O-distinguishable, and suppose that $U(P_u \otimes W_a)U^{-1}$ and $U(P_{u_2} \otimes W_a)U^{-1}$ are $1 \otimes A$-distinguishable and are both expressible as mixtures of eigenstates of $1 \otimes A$. Then the hypotheses of the theorem of Sec. II are satisfied. Therefore $U(P_u \otimes W_a)U^{-1}$ is not a mixture of eigenstates of $1 \otimes A$, if u is the superposition $g_1 u_1 + g_2 u_2$ with nonzero coefficients g_1 and g_2. On the other hand, if vectors u_1 and u_2 with the assumed properties do not exist, then evidently Fine's theorem would also hold.

Acknowledgments. Discussions with Professor B. d'Espagnat and Dr. A. Frenkel, and correspondence with Professor A. Fine have been very stimulating in this work. The author also gratefully acknowledges financial support by the John Simon Guggenheim Memorial Foundation.

COMMENT

Harvey Brown (1986, p. 862) gave a very simple proof of a theorem which reaches essentially the same conclusion as the theorem in Section II, but with an additional premiss, which he calls "real unitary evolution" (RUE). RUE states that the initial statistical operator of I + II (object + apparatus) corresponds to a real mixture of pure states ϕ_n of I + II with weights p_n, which then evolves into a final real mixture of pure states $U\phi_n$ (U unitary) with the same weights. (It is not clear to me that anything need be assumed other than the initial real mixture if I + II is isolated, since the unitary evolution of each ϕ_n would then guarantee the final real mixture.) Because Brown's theorem uses an additional premiss I shall refer to it as "the weak insolubility theorem" and to mine as "the strong insolubility theorem."

Brown states that the weak insolubility theorem was effectively proved by Fine (1970), mentioned in Ref. 3. I agree, and Fine (1987, p. 504) agrees also. Fine, however, fails to mention that the theorem which he announces at the beginning of his Section IV, "There are no W_a measurements U such that (4) holds for all initial object states W_0" (1970, p. 2785), is equivalent to the strong insolubility theorem, which he does not prove; only in the course of his argument does the tacit assumption of RUE emerge, showing that he has actually proved the weak theorem.

Apart from these textual and mathematical matters, there is a question of philosophical significance. If, indeed, RUE is acceptable, then the weak insolubility theorem would contain all that is philosophically significant in the strong theorem, and the interest of the latter would be merely mathematical. But if the apparatus, system II, is macroscopic, as it surely must be to function as desired in an actual experiment, then RUE is surely false. The reason is that the energy levels of macroscopic systems are so densely spaced that no shielding can prevent the entanglement of system II with the environment. The argument for this thesis has been presented with some quantitative detail by Zeh (1970). But if system II + the environment is in an entangled state, then the statistical operator for II by itself is not describable by a real mixture, as Brown himself notes (p. 859 and his Ref. 7). It follows that the statistical operator for I + II is also not describable by a real mixture, and hence RUE fails.

There is one possible fallback for Brown and Fine: to take system II to be the entire universe except for the object. Then there is nothing for I + II to be entangled with, and hence objectively I + II is in a pure though unknown state. But if so, then the suppositions made in describing the measurement situation guarantee that both I and II separately are objectively in pure quantum states, so that the insolubility of the measurement problem becomes an entirely trivial consequence of quantum dynamics (as André Mirabelli has pointed out in discussions). Hence the fallback position would deprive both Fine's and my insolubility theorems of philosophical interest.

REFERENCES

Brown, H. (1986), *Foundations of Physics* 16: 857–70.
Fine, A. (1970), *Physical Review D* 2: 2783–87.
Fine, A. (1987), "With Complacency or Concern: Solving the Quantum Measurement Problem." In R. Kargon and P. Achinstein (eds.), *Kelvin's Baltimore Lectures and Modern Theoretical Physics,* pp. 491–506. Cambridge, MA: MIT Press.
Zeh, D. (1970), *Foundations of Physics* 1: 69–76.

3
Proposed neutron interferometer test of some nonlinear variants of wave mechanics

A family of nonlinear variants of the Schrödinger equation was defined by Bialy-nicki-Birula and Mycielski by adding terms of the form $F(|\psi|^2)$ to the Hamiltonian. It is proposed that this family be tested by observing whether a phase shift occurs when an absorber is moved from one point to another along the path of one of the coherent split beams in a neutron interferometer. If F is b times a logarithmic function, which is the most important case, a null result with apparatus now available would impose an upper bound on b of 1.5×10^{-12} eV, more than two orders of magnitude smaller than the bound estimated by the above authors on the basis of the Lamb-shift measurement.

Various authors have suggested that the Schrödinger equation is only an approximation of the true nonlinear wave equation.[1-4] These suggestions are largely motivated by the fact that nonlinear equations can have solutions which are qualitatively different from those of the standard linear equation – e.g., nonspreading wave packets in the absence of a potential (Ref. 4), and superpositions in which all terms but one die away asymptotically in time (Ref. 3).

The important family of nonlinear wave equations investigated by Bialynicki-Birula and Mycielski (Ref. 4) consists of those of the form

(1) $$i\hbar\frac{\partial\psi(\vec{r}, t)}{\partial t} = \left(-\frac{\hbar^2}{2m}\Delta + U(\vec{r}, t)\right)\psi + F(|\psi|^2)\psi,$$

where $\vec{r}$ is the position vector in an n-dimensional configuration space and F is real valued. Any square-integrable solution $\psi(\vec{r}, t)$ of an equation of the form (1) shares a number of properties with solutions of the linear Schrödinger equation, e.g., (i) the norm $(\langle\psi|\psi\rangle)^{1/2}$, defined in the standard way, is preserved in time; (ii) in the absence of the potential,

This work originally appeared in *Physical Review A* 20 (1979), pp. 394–96. Reprinted by permission of The American Physical Society.
1. L. de Broglie, *Non-linear Wave Mechanics* (Elsevier, Amsterdam, 1960).
2. D. Leiter, Ann. Phys. (N.Y.) **51**, 561 (1969).
3. P. Pearle, Phys. Rev. D **13**, 857 (1969).
4. I. Bialynicki-Birula and J. Mycielski, Ann. Phys. (N.Y.) **100**, 62 (1976).

invariance under the full Galilei group holds, with ψ transforming as in the linear theory; (iii) an equation of continuity, $\partial\rho/\partial t = -\nabla\cdot\vec{j}$, holds, where the density ρ and the current density $\vec{j}$ are defined as in the linear theory. On the other hand, most of the nonlinear equations have the undesirable feature of generating correlations between two particles even when there is no interaction potential between them. In order to eliminate this feature, the authors postulate that if a system consists of noninteracting subsystems, then the solution of an admissible equation can be constituted for the system by taking the product of arbitrary solutions of this equation for separate subsystems. Using the elementary properties of the logarithm of a product, they prove that the only equation of form (1) which satisfies this "separability condition" is

$$(2) \qquad i\hbar\frac{\partial\psi(\vec{r}, t)}{\partial t} = \left(-\frac{\hbar^2}{2m}\Delta + U(\vec{r}, t)\right)\psi - b\ln(a^n|\psi|^2)\psi.$$

Of the two constants a and b in Eq. (2), the latter is by far the more interesting physically. In order to preserve the norm of solutions, b must be real; and it must be universal if the "separability condition" is to hold for any pair of systems with their interaction switched off. The dimension of a must be length, but its value is not physically significant, since changing it can be compensated by multiplying the wave function by a phase harmonically dependent on time, or, equivalently, by adding a constant to U. Furthermore, there is no need for the conventionally chosen value of a to be universal any more than a constant added to the potential in the linear theory would have to be, since the value of a does not enter into the nonlinearity. The physically interesting values of b are positive, since only these permit the construction of free-particle wave packets which do not spread. In view of the great success of standard quantum mechanics, the value of b must be small, and Bialynicki-Birula and Mycielski estimate that the Lamb-shift measurement sets an upper limit on b:

$$(3) \qquad b < 4 \times 10^{-10}\,\text{eV}.$$

The main purpose of this paper is to propose an experiment which would either exhibit the presence of a nonlinear correction to the Schrödinger equation or (the more likely outcome) set an upper bound on b several orders of magnitude lower than that of (3).

Although this paper is primarily concerned with the logarithmic nonlinear equation (2), there are two reasons for giving some consideration to more general nonlinearities. In the first place, one could postulate equations of the form (1) for one-particle systems only and leave open the question of the form of a nonlinear Schrödinger equation for more complex systems. With this restriction the question of separability does not even arise. In the second place, the test which will be proposed for Eq. (2) can be adapted to the general equation (1) (restricted to a single particle), provided that some additional experimental precautions are taken.

The experiment is to be performed using a neutron interferometer,[5-12] in which the wave packet of a single neutron is split into two coherent beams I and II of negligible overlap which propagate several centimeters before recombination. The most suitable apparatus for the experiment is the two-crystal interferometer of Ref. 7, because the space it provides between splitting and recombination is uninterrupted by a crystal.

Essentially the experiment consists of observing the phase shift Δ due to moving a partial absorber from one position P to another position P' in the path of beam I (see Fig. 1). Since the beams are well collimated and nearly monochromatic, quantum mechanics predicts that Δ is negligible. The nonlinear theories, however, imply that the translation of the absorber causes non-negligible phase shifts in each ray of beam I. In order to test Eq. (1), where the function F is general, beam I must be prepared (by initial stops) so that its amplitude is nearly uniform over a transverse cross section, thus ensuring approximately the same phase shift in each ray. It will be seen that the condition of uniform amplitude is unnecessary when F is logarithmic, as in Eq. (2). For the present, uniformity is assumed, so that $F(|\psi|^2)$ serves as an effective potential depending only upon the distance s along the path of beam I. In fact, since the beam is well collimated, $|\psi|$ has a constant value between P and P': it is the initial amplitude $|\psi_I|$ if the absorber is inserted at P', and $|\alpha\psi_I|$ if the absorber (with amplitude attentuation factor α) is inserted at P. Elsewhere the contribution of the nonlinear term is the same whether the absorber is inserted at P or P'. Since the nonlinear term is presumably small compared to the total energy E of the neutron, and since U is zero in the region of interest, the WKB approximation immediately yields a very good expression for the phase shift:

5. H. Rauch, W. Treimer, and U. Bonse, Phys. Lett. **47A**, 369 (1974).
6. W. Bauspiess, U. Bonse, and W. Graeff, J. Appl. Crystallogr. **9**, 68 (1976).
7. A. Zeilinger, C. G. Shull, M. A. Horne, and G. L. Squires, in *Proceedings of the International Neutron Interferometer Conference, Grenoble, 1978,* edited by U. Bonse and H. Rauch (Oxford University, New York, 1979).
8. D. M. Greenberger and A. W. Overhauser, Rev. Mod. Phys. **51**, 43 (1979).
9. A. W. Overhauser and R. Colella, Phys. Rev. Lett. **33**, 1237 (1974).
10. S. A. Werner, R. Colella, A. W. Overhauser, and C. F. Eagan, Phys. Rev. Lett. **35**, 1053 (1975).
11. S. A. Werner, R. Colella, and A. W. Overhauser, Phys. Rev. Lett. **34**, 1475 (1975).
12. H. Rauch. A. Zeilinger, G. Badurek, A. Wilfing, W. Bauspiess, and U. Bonse, Phys. Lett. **54A**, 425 (1975).

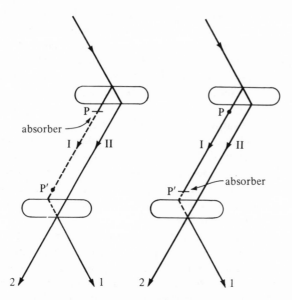

Figure 1. Dashed line indicates the attenuation of the amplitude of a beam by a factor α. Output beams I and II are directed towards the neutron counters.

$$(4) \qquad \Delta \simeq \int_{P}^{P'} ds \left(\frac{2m}{\hbar^2}[E-F(|\alpha\psi_I|^2)]^{1/2} - \frac{2m}{\hbar^2}[E-F(|\psi_I|^2)]^{1/2} \right)$$

$$\simeq (d/\hbar)(m/2E)^{1/2}[F(|\psi_I|^2) - F(|\alpha\psi_I|^2)]$$

$$\simeq (\tau/\hbar)[F(|\psi_I|^2) - F(|\alpha\psi_I|^2)],$$

where d is the distance from P to P' and τ is the time for the center of the wave packet to travel this distance.

If $F(x)$ is taken to be $b\ln(a^3x)$, as in Eq. (2), then

$$(5) \qquad \qquad \Delta \simeq (2/\hbar)\tau b \ln|\alpha|.$$

Because Eq. (5) is obtained by taking the logarithm of the ratio of $a^3|\psi_I|^2$ to $a^3|\alpha|^2|\psi_I|^2$, the constant a cancels out, and, furthermore, the assumption made earlier that the amplitude $|\psi_I|$ is approximately uniform throughout most of the transverse cross section of beam I is not needed. Equation (5) is also immediately derived by using the fact noted in Ref. 4 that if $\psi(\vec{r}, t)$ is a solution to Eq. (2), so also is

$$\alpha\psi(\vec{r}, t) \exp[(i/\hbar)bt \ln|\alpha|^2].$$

Although Eqs. (4) and (5) express the phase shift Δ implied by the non-linear equations (1) and (2), no investigation has been made of the implications of the latter equations for the diffraction of neutrons by crystals and for other matters relevant to the performance of the neutron interferometer. The quantum-mechanical theory of this performance is intricate (Ref. 7 and further references given there), and the introduction of a nonlinear term would cause additional complications. It is reasonable to suppose, however, that when the nonlinear term is very small, as has been assumed, then nonlinear wave mechanics predicts the same general features of interferometer performance as the standard linear theory, despite their differences with respect to the magnitudes of phase shifts. If this supposition is not true, then the large body of data concerning coherent recombination of split neutron beams would *ipso facto* constitute evidence against the nonlinear theories.

We anticipate that within experimental error the phase shift Δ will be zero, and we shall now calculate the upper bound thereby imposed upon b, using the specifications of the MIT two-crystal neutron interferometer. The path length between the two crystals is 4.5 cm for 1.5-Å wavelength neutrons. To obtain an attenuation of the amplitude by $|\alpha| = e^{-1/2}$, which is optimum for the experiment, there exist convenient absorbers of thickness 2 mm or less. Hence the distance between the two loci P and P' of insertion of the absorber can be 4 cm, so that $\tau = 1.5 \times 10^{-5}$ sec. The fringe contrast and count rate exhibited in Ref. 7 indicate that an uncertainty in Δ of 2° can be achieved in an experiment of reasonable duration. Therefore, a null result would imply

(6) $$b = \hbar\Delta/\tau < 1.5 \times 10^{-12} \text{ eV}.$$

This upper limit could be lowered by increasing the neutron count, which requires the expenditure of more time or the achievement of a higher incident flux,[13] or by building a longer interferometer.

Equation (1) is more difficult to test than Eq. (2), not only because of the need to assume the uniformity of $|\psi_I|$, but also because of the generality of the function F, which need not even be monotonic. However, it would be shown that F varies little over a wide range of its argument – and hence differs little from a constant that could be incorporated into U – if null phase shifts were found with each of a fairly large number of

13. It should be noted that in present practice the flux of neutrons through the interferometer is only a few hundred per minute. Since the mean transit time is of the order of 10^{-5} sec, one can reasonably assume that the wave packets of neutrons passing through the interferometer are nonoverlapping, and thus one has no complications from interactions among them. Clearly, the flux could be considerably increased without losing this benefit.

absorbers having different attenuation factors α. Previous instances of coherent recombination of split neutron beams already constitute strong evidence against Eq. (1) with any F which is neither extremely small nor logarithmic, since the profile of intensity in the transverse cross section of a diffracted neutron beam is known to be, in general, highly nonuniform.[14]

If the phase shift Δ is (unexpectedly) found to be nonzero, then the path distance from P to P' should be varied, since both Eqs. (4) and (5) imply that Δ is proportional to this length.

Acknowledgments. The author is grateful to Professor M. A. Horne and Professor E. P. Wigner for valuable discussions of theoretical questions, and to Professor Horne and Professor C. G. Shull for essential information about neutron interferometers. This work was supported in part by the NSF.

14. C. G. Shull, J. Appl. Crystallog. **6**, 257 (1973).

COMMENT

The minus sign before the last term on the right-hand side of Eq. (2) is a correction of the originally printed plus sign, but this error did not affect the relation between the fundamental constant b and the phase shift Δ in Inequality (6).

It is less easy to test Eq. (1) than I indicated at the end of the paper, because only the logarithmic nonlinearity of Eq. (2) has the experimentally useful consequence that the phase shift depends only on ratios of values of the wave function. The tests that have been performed concern Eq. (2).

The proposed experiment was performed at the Neutron Diffraction Laboratory at M.I.T. by C. G. Shull, D. K. Atwood, J. Arthur, and M. A. Horne (1980, *Physical Review Letters* 44: 765). They obtained an upper bound on the fundamental constant b of 3.4×10^{-13} eV, which was the best limit on a nonlinear modification of the Schrödinger equation until that time. A different test, observing the Fresnel diffraction pattern of cold neutrons, was performed at the Institut Laue-Langevin by R. Gähler, A. G. Klein, and A. Zeilinger (1981, *Physical Review A* 23: 1611). They obtained an even lower value of 3.3×10^{-15} eV for the upper bound on b, which is more than five orders of magnitude lower than the bound inferred by Bialynicki-Birula and Mycielski from Lamb-shift data. An entirely different family of nonlinear modifications of the Schrödinger equation was studied by S. Weinberg, with proposals for precise measurements of the effects of nonlinearities (1989, *Annals of Physics* 194: 336; 1989, *Physical Review Letters* 62: 485). Weinberg's experimental proposals

were carried out by several groups, including J. S. Bollinger, D. J. Hein-
zen, M. W. Itano, S. L. Gilbert, and D. J. Vineland (1989, *Physical Review Letters* 63: 1031) and T. E. Chupp and R. J. Hoare (1990, *Physical
Review Letters* 64: 2261). No nonlinear effects were observed, and very
small upper bounds were set upon the magnitude of nonlinear terms.
N. Gisin demonstrated that nonlinear equations of the type studied by
Weinberg imply the possibility of superluminal communications when
they are applied to two-particle systems (1989, *Helvetica Physica Acta*
62: 363; 1990, *Physics Letters A* 143: 1).

4

Desiderata for a modified quantum dynamics[1]

I. THE MOTIVATION FOR MODIFYING QUANTUM DYNAMICS

A cluster of problems – the "quantum mechanical measurement problem," the "problem of the reduction of the wave packet," the "problem of the actualization of potentialities," and the "Schrödinger cat problem" – is raised by standard quantum dynamics when certain assumptions are made about the interpretation of the quantum mechanical formalism. Investigators who are unwilling to abandon these assumptions will be motivated to propose modifications of the quantum formalism. Among these, many (including Professor Ghirardi and Professor Pearle) have felt that the most promising locus of modification is quantum dynamics, and they have suggested stochastic modifications of the standard deterministic and linear evolution of the quantum state (*PSA 1990,* A. Fine, M. Forbes, and L. Wessels (eds.), East Lansing, Michigan: Philosophy of Science Association, 1991). Others who have followed this avenue of investigation are F. Károlyházy, A. Frenkel, and B. Lukács (Károlyházy et al. 1982), N. Gisin (1984, 1989), A. Rimini and T. Weber (in Ghirardi et al. 1986), L. Diósi (1988, 1989), and J. S. Bell (1987, pp. 201–12). At a workshop at Amherst College in June 1990 Bell remarked that the stochastic modification of quantum dynamics is the most important new idea in the field of foundations of quantum mechanics during his professional lifetime. My own attitude is somewhat cautious and exploratory. The stochastic modification of quantum dynamics ought to be examined intensively, but the possibility should be kept in mind that it may fail, in which case the aforementioned assumptions about the interpretation of quantum mechanics will have to be re-assessed. Many, and perhaps all, of the investigators listed above share this exploratory and empiricist attitude.

It will be useful for later discussion to review briefly the standard dynamics of quantum mechanics and to state its implications for a schema-

This work originally appeared in *PSA 1990*, vol. 2, pp. 49–59. Copyright © 1991 by the Philosophy of Science Association. Reprinted by permission of the publisher.
1. This work was partially supported by the National Science Foundation, Grant No. 8908264.

tized formulation of the measurement process. According to quantum mechanics, the state of a physical system is represented by a normalized vector of an appropriate Hilbert space (the representation being many-one, for two normalized vectors which are complex scalar multiples of each other represent the same state). If the system is closed, then there is a family of time-dependent unitary (and hence linear) operators $U(t)$, with the property $U(t_1)U(t_2) = U(t_1 + t_2)$, such that if $s(0)$ is a vector representing the state of the system at time 0, then the state at an arbitrary time t is represented by $s(t)$, where

(1) $$s(t) = U(t)s(0).$$

Eq. (1), which is slightly more general than the familiar time-dependent Schrödinger equation, is the fundamental dynamical principle of non-relativistic quantum mechanics. Its relativistic counterpart, the Tomonaga–Schwinger equation, will not be needed for our purposes. What is most important for the problem of measurement is the linearity of $U(t)$:

(2) $$U(t)(c_1 s_1 + c_2 s_2 + \cdots + c_n s_n) = c_1 U(t)s_1 + c_2 U(t)s_2 + \cdots + c_n U(t)s_n,$$

for any vectors $s_1, s_2, \ldots, s_n$ and any scalars $c_1, c_2, \ldots, c_n$.

Suppose now that any object of interest and an apparatus employed to measure some property of the object together constitute a closed physical system (perturbations from the rest of the universe being negligible). Then by the dynamical principle of quantum mechanics there is a family of unitary operators $U(t)$ governing the temporal evolution of the states of object-plus-apparatus, as in Eqs. (1) and (2). The apparatus is to serve the purpose of revealing the value of a property of the object which is represented by the hermitian operator A, where

(3) $$As_i = a_i s_i \quad (a_i \neq a_j \text{ if } i \neq j)$$

for some basis s_i in the object's Hilbert space. Then there must be some vector v_o in the Hilbert space of the apparatus (representing a "neutral" apparatus state) such that for some time t the vector $U(t)(s_i \otimes v_o)$ is an eigenvector of an operator representing a property of the apparatus, with an eigenvalue b_i from which one can infer a_i. A highly idealized version of this schema of measurement is one in which for each i there is a normalized vector v_i in the Hilbert space of the apparatus such that

(4) $$U(t)(s_i \otimes v_o) = s_i \otimes v_i$$

and

(5) $$Bv_i = b_i v_i \quad (b_i \neq b_j \text{ if } i \neq j).$$

In general, however, the initial state of the object will not be represented by a single one of the eigenvectors of A, but by a superposition of the form

(6) $$s(0) = c_1 s_1 + \cdots + c_n s_n,$$

with the sum of the absolute squares of the scalar coefficients c_i being unity and with more than one of them non-zero. Then the state of object-plus-apparatus at time t is represented by

(7) $$U(t)((c_1 s_1 + \cdots + c_n s_n) \otimes v_o) = c_1 U(t)(s_1 \otimes v_o)$$
$$+ \cdots + c_n U(t)(s_n \otimes v_o),$$

which is a superposition of n vectors, each representing a state in which the property of the apparatus has a different value b_i. It is at this point that the problems mentioned in the first paragraph are revealed. The purpose of a measurement is to obtain information about a property of an object (typically a microscopic object which cannot be directly scrutinized) by means of a correlation established between that property and a property of the apparatus. But in the state represented in Eq. (7) the apparatus property does not have a definite value, and hence the purpose of the measurement has not been achieved. Thus the "measurement problem" is posed. Furthermore, Eq. (7) shows that the peculiar indefiniteness of a physical property is not confined to microscopic objects (as in Eq. (6)), but is manifested by a property of a macroscopic apparatus on account of the linearity of the dynamical evolution of object-plus-apparatus. In particular, when the notorious experimental arrangement of Schrödinger (1935) is analyzed quantum mechanically, then in the final state of the experiment it is indefinite whether the cat is alive or dead – the "Schrödinger cat problem." These problems arising in the context of physical measurement may be considered to be special cases of a more general problem of "the actualization of potentialities," for it is obscure how actual events – such as the emission or absorption of photons, or the replication of a macro-molecule, or the firing of a neuron – can occur if quantum dynamics typically gives rise to states in which these events are merely potential because of the indefiniteness of relevant properties.

The following assumptions concerning the interpretation of the quantum mechanical formalism have the consequence of making the foregoing problems so serious that it is difficult to envisage their solution without some modification of the formalism itself. The assumptions themselves are strongly supported by physical and philosophical considerations, and therefore a high price would be paid by sacrificing one of them in order to hedge standard quantum mechanics against modifications.

(i) The quantum state of a physical system is an objective characterization of it, and not merely a compendium of the observer's knowledge of it, nor merely an intellectual instrument for making predictions concerning observational outcomes.

(ii) The objective characterization of a physical system by its quantum state is complete, so that an ensemble of systems described by the same quantum state is homogeneous, without any differentiations stemming from "hidden variables."

(iii) Quantum mechanics is the correct framework theory for all physical systems, macroscopic as well as microscopic, and hence it specifically applies to measuring apparatuses.

(iv) At the conclusion of the physical stages of a measurement (and hence, specifically, before the mind of an observer is affected), a definite result occurs from among all those possible outcomes (potentialities) compatible with the initial state of the object.

I shall very briefly point out how these assumptions preclude some of the proposals that have been made for solving the problem of measurement and related problems. Assumption (i) stands in the way of an instrumentalist interpretation of the quantum mechanical formalism. Such an interpretation could accommodate an expression of the form of Eq. (7), with many terms corresponding to different observational outcomes, just as well as a characterization of the final state of object-plus-apparatus by a single term; either expression would merely be an instrument for anticipating an observational outcome or the probabilities of various outcomes. Some arguments against such an instrumentalist interpretation are given in Shimony (1989), along with references to other discussions. Assumption (ii) rejects a hidden variables interpretation of quantum mechanics, according to which the indefiniteness of the value a_i in Eq. (6) and of b_i in Eq. (7) applies only to ensembles and not to individual members of the ensembles. The main consideration in favor of Assumption (ii) is the incompatibility proved by Bell (1987, pp. 14–21 and 29–39) between quantum mechanics and local hidden variables theories, but Bell himself emphasizes that there is still an option of non-local hidden variables theories, which he does not regard as completely repugnant (1987, pp. 173–80). Assumption (iii) rules out all variants of the Copenhagen interpretation, which rejects the impasse of Eq. (7) by rejecting the application of quantum mechanics to the apparatus of measurement. In favor of this Assumption is the immense success of the general physical program of understanding macroscopic systems in terms of microscopic parts, conjoined with the immense success of quantum mechanics in the microscopic domain. Wigner (1971, pp. 14–15) emphasizes particularly that we have no theory at present for dealing with the interaction of a quantum system and a classical system. Finally, Assumption (iv) precludes the "many-worlds" interpretation, in which all terms on the right hand side of Eq. (7) are considered to be equally real. The conceptual difficulties of this point of view, which effectively denies the distinction between actuality and potentiality, have been analyzed by many writers, for example Bell (1987, pp. 93–100) and Stein (1984).

Henceforth in this paper I shall not question Assumptions (i)–(iv), even though, as stated in the first paragraph, an eventual re-assessment is not ruled out. Given these Assumptions, however, one finds Eq. (7) intolerable as a description of the final physical state of the measurement process. There must be a further stage in which a selection is made from the superposition, and that further stage must be physical. A modification of quantum dynamics is thereby required.

2. PROPOSED DESIDERATA

a. *The proposed modification of quantum dynamics should not be restricted to situations of measurement, for such a restriction would inject an anthropocentric element into fundamental physical theory.* This desideratum would preclude von Neumann's (1955, p. 351) postulate of a special process of reduction occurring when a physical variable is measured, unless that postulate could be shown to be a special case of a more general dynamical law. All the authors mentioned in the first paragraph are committed to satisfying this desideratum.

b. *The modified dynamics must agree very well with quantum dynamics in the domain of successful application of the latter.* This desideratum is primarily the demand of experimental adequacy of a proposed new theory, for if standard quantum dynamics makes very accurate predictions of such phenomena as resonances and beats, then the new theory would have to agree closely with quantum dynamics in order to fit these phenomena. One may anticipate, however, some additional content in this desideratum: that the modified dynamics be related to standard quantum dynamics in a systematic way, by some limiting principle, which would be analogous to the "correspondence principle" relating quantum to classical mechanics.

c. *If the proposed modified dynamics is applied to a measurement situation, it should predict definite outcomes in a "short" time, where the vague word "short" is made quantitative by the known reaction time of the experimental apparatus.* This desideratum strongly favors a stochastic modification of quantum dynamics over a deterministic non-linear modification. If the composite system object-plus-apparatus is governed by a non-linear dynamical equation, then one would not preserve Eq. (7), in which a final superposition mirrors the initial superposition of eigenvectors of the object variable A; and one could easily imagine a continuous dwindling away of all coefficients except one, which asymptotically would approach absolute value unity. But it is very difficult to construct a plausible dynamical equation for which this asymptotic behavior occurs in a finite time interval. Difficult is not impossible, however, and Pearle (1985) actually succeeded in constructing such an equation (but it is tailored to the measurement of a particular object variable, contrary to desideratum a). The non-deterministic "jumps" of a stochastic dynamical theory – whether they are sporadic and finite, as in the Spontaneous Localization theory of Ghirardi, Rimini, and Weber (1986), or infinitesimal, as in the Continuous Spontaneous Localization theory of Pearle (1989) – are a promising means for achieving definite measurement outcomes rapidly.

    d. *If a stochastic dynamical theory is used to account for the outcome of a measurement, it should not permit excessive indefiniteness of the outcome, where "excessive" is defined by considerations of sensory discrimination.* This desideratum tolerates outcomes in which the apparatus variable does not have a sharp point value, but it does not tolerate "tails" which are so broad that different parts of the range of the variable can be discriminated by the senses, even if very low probability amplitude is assigned to the tail. The reason for this intolerance is implicit in Assumption (iv) of Section 1. If registration on the consciousness of the observer of the measurement outcome is more precise than the "tail" indicates, then the physical part of the measurement process would not yield a satisfactory reduction of the initial superposition, and a part of the task of reducing the superposition would thereby be assigned to the mind. For this reason, I do not share the acquiescence to broad "tails" that Pearle advocates (1990, pp. 203–4), with the concurrence of Bell and Penrose (*ibid.*, p. 213, footnote 30).

    e. *The modified dynamics should be Lorentz invariant.* This desideratum has not been achieved by any of the proposed stochastic theories, and it evidently will be very difficult to satisfy. A discussion both of the difficulties and of some progress towards solving them is given by Pearle (1990, pp. 204–12) and Fleming (1989).

    f. *The modified dynamics should not lose the "peaceful coexistence" with special relativity that standard quantum mechanics possesses – that is, the impossibility of capitalizing upon the entanglement of the state of spatially separated systems to send a superluminal message.* Gisin (1989) has shown that a large class of non-linear deterministic modifications of quantum dynamics violate this desideratum. His argument provides a consideration supplementary to desideratum c for preferring stochastic theories.

    g. *The modified dynamics should preclude the gestation of Schrödinger's cat, and in general the occurrence – even for a brief time – of states of a system in which a macroscopic variable is indefinite.* This desideratum is less strongly entrenched than the others discussed so far, because one could presumably achieve agreement with our failure to observe such states by supposing that they are highly unstable and decay very rapidly into states where macroscopic variables have sharp variables (to the extent required by desideratum d). Incidentally, it is fascinating, and perhaps fruitful, to consider the experimental search for very short-lived superpositions of radically differing states in mesoscopic systems.

h. *The modified dynamics should be capable of accounting for the occurrence of definite outcomes of measurements performed with actual apparatus, not just with idealized models of apparatus.* The Spontaneous Localization theory of Ghirardi, Rimini, and Weber (1986) has been criticized for not satisfying this desideratum. In the measuring apparatus that they consider, the macroscopic variable which is correlated with a variable of a microscopic object is the center of mass of a macroscopic system, and spontaneous localization ensures that within about ten nanoseconds this variable will be quite sharp. (At the 1990 PSA meeting I incorrectly stated that the macroscopic system had to be rigid in order to obtain such rapid localization, and my error was pointed out by Professor Ghirardi.) Albert and Vaidman (Albert 1990, pp. 156–8) note that the typical reaction of a measuring apparatus in practice is a burst of fluorescent radiation, or a pulse of voltage or current, and these are hard to subsume under the scheme of measurement of the Spontaneous Localization theory.

3. THE "QUANTUM TELEGRAPH": A PROMISING LOCUS OF INVESTIGATION

A great weakness in the investigations carried out so far in search of modifications of quantum dynamics is the absence of empirical heuristics. To be sure there is one grand body of empirical fact which motivates all the advocates of stochastic modifications of quantum dynamics and most of the advocates of non-linear modifications: that is, the occurrence of definite events, and in particular, the achievement of definite outcomes of measurement. But this body of fact is singularly unsuggestive of the details of a reasonable modification of quantum mechanics. What is needed are phenomena which are suggestive and even revelatory. No more promising phenomena for this purpose have been found than the intermittency of resonant fluorescence of a three-level atom.

H. Dehmelt (1975) proposed to study fluourescent radiation from a trapped atom (confined to a small region by techniques of which he was a pioneer) exposed to two laser beams, one labeled "strong" and one "weak." The first is tuned to the frequency of a transition from the ground state 0 to an excited state 1, and the second to the frequency of a transition from 0 to an excited state 2. The 1–0 transition is dipole-allowed, so that the state 1 has a lifetime of about 10^{-8} s, whereas the 2–0 transition is dipole-forbidden and the lifetime of 2 is about 1 s. Dehmelt anticipated that there would be fairly long periods (of the order of a second) in which the atom undergoes cycles of excitation and spontaneous emission about 10^{-8} s in duration. During such a period the radiation from the single trapped atom would be visible to the naked eye; Cook (1990, p. 367) says, "With a 10× magnifying lens a point source of this strength would appear as bright as one of the stars in the Big Dipper"! Every few seconds, however, Dehmelt conjectured, the atom would absorb a photon from the weak laser beam and would be excited to state 2, where it would remain for a fairly long period, "shelved", in his descriptive term. Consequently, the fluorescent radiation from this three-level atom would be intermittent, with a pattern of alternating light and dark periods that has been described as the "quantum telegraph." (Of course, unlike the dots, dashes, and spaces of Morse telegraphy, the periods of light and dark would be of random durations.)

Dehmelt's reasoning seemed implicitly to accept the idea of quantum jumps from one state to another. It is reminiscent of the old Bohr theory of atomic transitions (1913), though to an advocate of stochastic modification of quantum dynamics it could be construed as an intimation of the theory of the future. In any case, it was criticized for neglecting the superposition principle and the linearity of quantum dynamics, which seem to be inconsistent with "shelving" (Pegg, Loudon, and Knight 1986). But if the atom is always in a superposition of states 0, 1, and 2 except when a photon is detected (at which point emission has occurred with certainty and the state is "reset" to 0), then it is straightforward to show that there is negligible probability of a dark period longer by an order of magnitude than the natural lifetime of state 1. It follows that the phenomenon of the quantum telegraph should not appear.

Dehmelt's intuition was confirmed by experiment (Bergquist et al. 1986, Nagourney et al. 1986, Sauter et al. 1986, Itano et al. 1987). These results are among the most dramatic in the history of optics. And they have given rise to a number of sophisticated analyses, attempting to show the consistency of the quantum telegraph with standard quantum mechanics (e.g., Cohen-Tannoudji and Dalibard 1986, Porrati and Putterman 1989, Erber et al. 1989, Cook 1990). For the most part these analyses agree with each other, but there are some differences in emphasis and detail. I shall summarize the main ideas without examining the differences, since this procedure will suffice for the heuristic purposes of the present paper.

First, the system of interest is taken to be the atom together with the scattered part of the radiation field (which is discriminated sufficiently from the incident laser beams because of the precisely defined directions of these beams). The states of interest will be represented by superpositions of the form

(8) $|\Psi(t)\rangle = c_1(t)|1\rangle \otimes |0\rangle_F + c_2(t)|2\rangle \otimes |0\rangle_F + \Sigma\, c_{kp}(t)|0\rangle \otimes |kp\rangle_F,$

where $|0\rangle$, $|1\rangle$, $|2\rangle$ respectively represent the ground state and the two relevant excited states of the atom, $|0\rangle_F$ represents the state of the scattered radiation field with no photons, and $|kp\rangle_F$ represents a state with a single photon of wave vector k (of variable direction but with magnitudes restricted by the energies of the 1–0 and 2–0 transitions) and polarization p. These states evolve from an initial state consisting only of the first two terms of Eq. (8), with coefficients $c_1(0)$ and $c_2(0)$.

Second, for simplicity it is assumed that a perfect photo-detector is in place, to respond to any photon in one of the permitted modes kp. A detection of a photon would "reset" the state to a superposition of the first two terms of Eq. (8).

In addition to such normal "positive" measurements, there is a recognition of "null measurements": i.e., the non-detection of a photon by the perfect photo-detector after a time interval which is long compared to the lifetime of the short-lived state 1. Non-detection has the effect of projecting the vector of Eq. (8) onto the two-dimensional space spanned by $|1\rangle \otimes |0\rangle_F$ and $|2\rangle \otimes |0\rangle_F$, so that the last term of Eq. (8) is projected out and the first two terms are preserved but with renormalization.

The questions of exactly when the projection occurs, and what the state looks like before the projection is fully accomplished, are evaded by making use of the enormous difference in order of magnitude of the lifetimes of states 1 and 2. The statistics of light and dark periods are therefore insensitive to answers to these two questions.

Once this projection is accomplished, the usual unitary evolution of the state will automatically account for a rapid diminution of the coefficient $c_1(t)$ relative to $c_2(t)$, thereby greatly extending the period of darkness to a length comparable to the natural lifetime of state 2.

Finally, an epistemic concept of probability is invoked. For example, Porrati and Putterman (1988, p. 3014) write, "In our picture the measurement of a period of time during which no photons are recorded changes our information about the system and thus the wave function. This null measurement increases the probability of successive periods of darkness." Cook (1990, p. 407) uses the locution "Bayesian transitions" to describe the consequences of null measurements, and he contrasts his point of view with Dehmelt's original suggestion as follows: "It is interesting that the quantum formalism attributes electron shelving to the lack of fluorescence, whereas the intuitive picture of the process attributed the lack of fluorescence to electron shelving."

A strenuous objection must be brought against the foregoing scheme of ideas, in spite of the elegance of the theoretical analysis based upon them and the agreement of this analysis with experiment. The scheme takes for granted that a photo-detector definitely has or has not registered the arrival of a photon in a certain interval of time. This assertion does not make a commitment to a definite instant beginning the interval and a definite instant ending it; the time–energy uncertainty relation and the operational uncertainties of the detector can be fully respected. The point is rather that a reduction of the wave packet has been assumed at the level of a macroscopic measuring apparatus, and the analogue of Schrödinger's cat – that is, a superposition of photon detected and photon not detected – has tacitly been excluded. This assumption underlies the Bayesian locutions about probabilities conditional upon the occurrence or non-occurrence of a certain event. Of course, working physicists regularly assume that at the level of macroscopic apparatus the superposition principle does not preclude definite outcomes. The opportunistic employment of the superposition principle in the early stages of a physical process and its suspension at the final stage is, in fact, a part of the ordinary practice of quantum mechanics, and as Bell forcefully reminded us (1990, p. 18), "ORDINARY QUANTUM MECHANICS (as far as I know) IS JUST FINE FOR ALL PRACTICAL PURPOSES." But the purpose of Section 1 of this paper was to review the argument that the opportunistic employment of the superposition principle is not understood from the standpoint of first principles.

My proposal is to avoid a merely "practical" explanation of the quantum telegraph in terms of ordinary quantum dynamics, but instead to let this remarkable phenomenon guide us heuristically to a modified dynamics. Two propositions seem to me to suggest themselves quite strongly. The first is that a stochastic modification of quantum dynamics is a natural way to accommodate the jumps from a period of darkness to a period of fluorescence. The second is that the natural locus of the jumps is the interaction of a physical system with the electromagnetic vacuum. Whether stochasticity is exhibited when the system in question is simple and microscopic, like a single atom, or only when it is macroscopic and complex, like the phosphor of a photo-detector, is not suggested preferentially by the quantum telegraph, for the simple reason that the single trapped atom and the photo-detector are both essential ingredients in the phenomenon. But whichever choice is made points to a stochastic modification of quantum dynamics that has little to do with spontaneous localization. There is hope, therefore, for a stochastic theory that will escape the criticisms leveled by Albert and Vaidman against the localization theories of Ghirardi, Rimini, and Weber, and of Pearle. I must admit, however, that the envisaged theory which I prefer to those of Professors Ghirardi et al. and of Pearle has one serious disadvantage relative to theirs – it does not exist, whereas theirs do!

4. TWO CONCLUDING REMARKS

The search for a reasonable modification of quantum dynamics was motivated by a cluster of problems arising from the linearity of the standard time evolution operators. The implications of a modified dynamics, however, may reach far beyond the original motivation. In particular, a stochastic modification of quantum dynamics can hardly avoid introducing time-asymmetry. Consequently, it offers an explanation at the level of fundamental processes for the general phenomenon of irreversibility, instead of attempting to derive irreversibility from some aspect of complexity (which has the danger of confusing epistemological and ontological issues). Thus a stochastic modification of quantum dynamics is a promising way to satisfy the thesis of R. Penrose (1986) that the problem of the reduction of the wave packet is inseparable from the problem of irreversibility.

Finally, to the list of eight desiderata listed in Section 2 for a modification of quantum dynamics I want to add a ninth, highly personal one: that a satisfactory theory be found by someone during my lifetime.

REFERENCES

Albert, D. Z. (1990), "On the Collapse of the Wave Function", in *Sixty-Two Years of Uncertainty: Historical, Philosophical, and Physical Inquiries into the Foundations of Quantum Mechanics*, A. I. Miller (ed.). New York: Plenum Press, pp. 153–65.
Bell, J. S. (1987), *Speakable and Unspeakable in Quantum Mechanics*. Cambridge, U.K.: Cambridge University Press.
Bell J. S. (1990). "Against 'Measurement'", in *Sixty-Two Years of Uncertainty: Historical, Philosophical and Physical Inquiries into the Foundations of Quantum Mechanics*, A. I. Miller (ed.). New York: Plenum Press, pp. 17–31.
Bergquist, J. C., Hulet, R. G., Itano, W. M., and Wineland, J. J., (1986), "Observation of Quantum Jumps in a Single Atom", *Physical Review Letters* 57: 1699–1702.
Bohr, N. (1913), "On the Constitution of Atoms and Molecules", *Philosophical Magazine* 26: 1–25.

Cohen-Tannoudji, C. and Dalibard, J. (1986), "Single-Atom Laser Spectroscopy. Looking for Dark Periods in Fluorescence Light", *Europhysics Letters* 1: 441–8.

Cook, R. J. (1990). "Quantum Jumps", in *Progress in Optics XXVII*, E. Wolf (ed.). Amsterdam: Elsevier Science Publishers, pp. 361–416.

Dehmelt, H. (1975). "Proposed $10^{14}\Delta\nu > \nu$ Laser Fluorescence Spectroscopy on Tl^+Mono-Ion Oscillator II", *Bulletin of the American Physical Society* 20: 60.

Diósi, L. (1988), "Quantum Stochastic Processes as Models for State Vector Reduction", *Journal of Physics* A 21: 2885–98.

Diósi, L. (1989), "Models for Universal Reduction of Macroscopic Quantum Fluctuations", *Physical Review* A 40: 1165–74.

Erber, T., Hammerling, P., Hockney, G., Porrati, M., and Putterman, S. (1989), "Resonance Fluorescence and Quantum Jumps in Single Atoms: Testing the Randomness of Quantum Mechanics", *Annals of Physics* 190: 254–309.

Fleming, G. (1989), "Lorentz Invariant State Reduction, and Localization", in *PSA 1988*, A. Fine and J. Leplin (eds.). East Lansing, Michigan: Philosophy of Science Association, pp. 112–26.

Ghirardi, G. C., Rimini, A., and Weber, T. (1986), "Unified Dynamics of Microscopic and Macroscopic Systems", *Physical Review* D 34: 470–91.

Gisin, N. (1984), "Quantum Measurements and Stochastic Processes", *Physical Review Letters* 52: 1657–60.

Gisin, N. (1989). "Stochastic Quantum Dynamics and Relativity", *Helvetica Physica Acta* 62: 363–71.

Itano, W. M., Bergquist, J. C., Hulet, R. G., and Wineland, D. J. (1987), "Radiative Decay Rates in Hg^+ from Observations of Quantum Jumps in a Single Ion", *Physical Review Letters* 59: 2732–5.

Károlyházy, F., Frenkel, A., and Lukács, B. (1986), "On the Possible Role of Gravity in the Reduction of the Wave Function", in *Quantum Concepts in Space and Time*, R. Penrose and C. Isham (eds.). Oxford: Clarendon Press, pp. 109–28.

Nagourney, W., Sandberg, J., and Dehmelt, H. (1986), "Shelved Optical Electron Amplifier: Observation of Quantum Jumps", *Phyical Review Letters* 56: 2797–9.

Pearle, P. (1985), "On the Time It Takes a State Vector to Reduce", *Journal of Statistical Physics* 41: 719–27.

Pearle, P. (1989), "Combining Stochastic Dynamical State-Vector Reduction with Spontaneous Localization", *Physical Review* A 39: 227–39.

Pearle, P. (1990), "Toward a Relativistic Theory of Statevector Reductions", in *Sixty-Two Years of Uncertainty: Historical, Philosophical, and Physical Inquiries into the Foundations of Quantum Mechanics*, A. I. Miller (ed.). New York: Plenum Press, pp. 193–214.

Pegg, D. T., Loudon, R., and Knight, P. L. (1986), "Correlations in Light Emitted by Three-Level Atoms", *Physical Review* A 33: 4085–91.

Penrose, R. (1986), "Gravity and State Vector Reduction", in *Quantum Concepts in Space and Time,* R. Penrose and C. Isham (eds.). Oxford: Clarendon Press, pp. 129–46.

Porrati, M. and Putterman, S. (1989), "Coherent Intermittency in the Resonant Fluorescence of a Multilevel Atom", *Physical Review* A: 3010–30.

Sauter, T., Neuhauser, D., Blatt, R., and Toschek, P. E. (1986), "Observation of Quantum Jumps", *Physical Review Letters* 57: 1696–8.

Shimony, A. (1989), "Search for a Worldview Which Can Accommodate Our Knowledge of Microphysics", in *Philosophical Consequences of Quantum Theory: Reflections on Bell's Theorem,* J. T. Cushing and E. McMullin (eds.). Notre Dame: University of Notre Dame Press, pp. 25–37.

Stein, H. (1984), "The Everett Interpretation of Quantum Mechanics: Many Worlds or None?", *Noûs* 18: 635–52.

von Neumann, J. (1955), *Mathematical Foundations of Quantum Mechanics.* Princeton: Princeton University Press.

Wigner, E. P. (1971), "The Subject of Our Discussions", in *Foundations of Quantum Mechanics,* B. d'Espagnat (ed.). New York: Academic Press, pp. 1–19.

5
Filters with infinitely many components

With the use of a suitable assumption about the structure of the class of experimental filters, it is shown that the sequence of alternating replicas of two filters is their greatest lower bound, as Jauch suggests. A generalization of his suggestion yields the greatest lower bound of a denumerable set of filters. The criteria of admissibility of filters are briefly discussed.

When Jauch first introduces the term "proposition of a physical system" in his book (Ref. 1, p. 73), he says that it refers to a kind of experimental filter, passage through which constitutes the truth of the proposition while nonpassage constitutes its falsity. He does not attempt to state precise criteria for the admissibility of filters. It is clear, however, that strictly operationalistic criteria are not intended, since he classifies as filters certain arrangements which require an infinite sequence of operations. Specifically, if $\hat{a}$ and $\hat{b}$ are two filters corresponding to propositions a and b, respectively, then the infinite sequence of alternating replicas of $\hat{a}$ and $\hat{b}$, $\hat{a}\hat{b}\hat{a}\hat{b}\hat{a}...$, is considered to be a filter.

His motivation for admitting such nonoperationalistic filters is clear.[1] He wishes the set of propositions to have a sufficiently strong structure to permit the formulation of all or much of the standard mathematical apparatus of quantum mechanics. It is hard to see how Jauch's program could be launched without assuming that the set of propositions is a lattice under the implication relation $\subseteq$. He therefore assumes that every pair of propositions a, b has a greatest lower bound (g.l.b.) $a \cap b$. Since the filters $\hat{a}$ and $\hat{b}$ may be incompatible, in the sense that a system which passes $\hat{a}$ and then $\hat{b}$ may not then pass a replica of $\hat{a}$ (as in the multiple Stern–Gerlach experiment), one cannot consistently represent $a \cap b$ by compounding any finite number of replicas of $\hat{a}$ and $\hat{b}$. Jauch is therefore led to the ingenious suggestion that the nonoperational filter $\hat{a}\hat{b}\hat{a}\hat{b}\hat{a}...$ appropriately represents $a \cap b$.

This work originally appeared in *Foundations of Physics* 1 (1971), pp. 325–28. Reprinted by permission of Plenum Publishing Corp.
1. For more general remarks about the impossibility of operationalist definitions of terms in theoretical physics see Ref. 2.

The purpose of this note is to make three comments on Jauch's suggestion.

1. Without further information, there is no guarantee that the filter $\hat{a}\hat{b}\hat{a}\hat{b}\hat{a}\dots$ is the g.l.b. of $\hat{a}$ and $\hat{b}$. In other words, it is possible that there exists a filter $\hat{c}$ with the following properties: All systems passing $\hat{c}$ will pass $\hat{a}$, all systems passing $\hat{c}$ will pass $\hat{b}$, and some systems passing $\hat{c}$ will not pass $\hat{a}\hat{b}\hat{a}\hat{b}\hat{a}\dots$. A direct experimental demonstration that no such $\hat{c}$ exists is out of the question, because of the infinite process involved in $\hat{a}\hat{b}\hat{a}\hat{b}\hat{a}\dots$ and also because of the well-known difficulty of refuting an existential proposition. Hence, Jauch's suggestion that $\hat{a}\hat{b}\hat{a}\hat{b}\hat{a}\dots$ represents $a \cap b$ can be justified only on the basis of sufficiently strong information about the structure of the class of filters. For instance, one might assume that the class of filters is isomorphic to the set of projections on some Hilbert space, with passage through filter $\hat{f}$ implying passage through filter $\hat{g}$ if and only if the corresponding projections satisfy $F \subseteq G$, and with the composite filter $\hat{f}$ followed by $\hat{g}$ corresponding to the product GF. With this assumption, the theorem given in the appendix can be used to infer that $\hat{a}\hat{b}\hat{a}\hat{b}\hat{a}\dots$ is the g.l.b. of $\hat{a}$ and $\hat{b}$. Undoubtedly, the same conclusion can be reached from weaker assumptions about the structure of the class of filters. The point, however, is that some explicit assumption is needed.

2. Jauch wishes the set of propositions of a system to be not only a lattice but a σ-lattice, that is, every denumerable set of propositions $\{a_i\}$ has a g.l.b. $\cap a_i$. He does not try to construct a filter corresponding to $\cap a_i$, but argues instead as follows (Ref. 1, p. 75), "There is no empirical correlate in the infinite conjunction implied in the definition of $\cap_I a_i$, since one can perform only a finite number of experiments. If we extend this definition to infinite sets, we transcend the proximately observable facts and we introduce *ideal elements* into the description of physical systems." However, once infinite sequences of filters are considered to be acceptable filters, there is a plausible candidate for a representative of $\cap a_i$, namely

$$\hat{a}_1\hat{a}_2\hat{a}_1\hat{a}_3\hat{a}_1\hat{a}_2\hat{a}_1\hat{a}_4\hat{a}_1\hat{a}_2\hat{a}_1\hat{a}_3\hat{a}_1\hat{a}_2\hat{a}_1\dots.$$

This sequence can be characterized informally as being the limit of the finite sequences $\hat{b}_n$, where $\hat{b}_1 = \hat{a}_1$ and $\hat{b}_{n+1} = \hat{b}_n \hat{a}_{n+1} \hat{b}_n$. Again, whether this sequence correctly represents $\cap a_i$ can be ascertained only if the structure of the class of filters is sufficiently specified. For example, if the assumption considered in comment 1 is used (i.e., the filters are isomorphic to the projections on some Hilbert space), then the theorem of the appendix implies that $\lim \hat{b}_n$ exists and is the g.l.b. of the $\{\hat{a}_i\}$. Hence, Jauch's suggestion that infinite sequences of filters are acceptable filters may permit him to retain without qualification his initial statement that all propositions refer to filters, instead of treating some propositions as ideal elements to which no filters correspond.

3. The class of filters envisaged by Jauch (and also by most other writers who make statements about possible experiments) contains not only those which have been built or for which blueprints exist, but also others which are realizable "in principle." Whether an infinite sequence of filters should be accepted as a filter depends upon the meaning of "in principle." It certainly is unwise to expect a unique explication of this phrase. Rather, one can reasonably ask what experimental arrangements are possible *modulo* a certain set of physical and epistemological assumptions. When these assumptions are explicitly formulated, which is rarely if ever done thoroughly in the literature on foundations of quantum mechanics, it becomes possible to discuss critically whether they are too restrictive or too weak for specific purposes. The current controversy over the existence of superselection rules (cf. Ref. 3) indicates the need for explicitness on these matters.

APPENDIX

Theorem.[2] *If $\{E_i\}$ is a sequence of projections on a Hilbert space $\mathcal{H}$, and if $F_1 = E_1$ and $F_{n+1} = F_n E_{n+1} F_n$, then $\{F_i\}$ converges to a limit F which is the g.l.b. of $\{E_i\}$.*

Proof. Clearly, each F_n is bounded and the bound is uniform. Furthermore, by a trivial induction $F_n^* = F_n$. Using the definition of F_n and the properties of projections, we obtain for every $f \in \mathcal{H}$

$$(f, F_{n+1}f) \le (F_n f, F_n f) \le (E_n F_{n-1}f, E_n F_{n-1}f) = (f, F_n f).$$

Hence, by the theorem of Vigier (Ref. 4) that every bounded monotonic sequence of symmetric operators on a Hilbert space strongly converges to a symmetric operator, $\{F_i\}$ converges to a symmetric limit F, which is evidently bounded.

To show that F is a projection operator, first note that $\{F_i^2\}$ converges to F^2, for $\|F_n^2 - F^2\| \le \frac{1}{2}[\|F_n + F\|\|F_n - F\| + \|F_n - F\|\|F_n + F\|]$, which converges to 0. But it is easily checked that $(f, F_{n+1}f) \le (f, F_n^2 f) \le (f, F_n f)$ for all $f \in \mathcal{H}$, so that $\lim F_n^2 = \lim F_n$. Hence, $F^2 \equiv F$, and F is therefore a projection.

---

2. I am grateful to Prof. Stanley Gudder for pointing out that a special case of this theorem (when each E_i is one of two projections E, F) is proved in Ref. 5.

To show that F is the g.l.b. of $\{E_i\}$, it now suffices to show that $\Delta_F = \bigcap M_i$, where Δ_F is the range of F and M_i is the range of E_i. That $\bigcap M_i \subseteq \Delta_F$ is trivial. To show the inclusion in the other direction, suppose that $Ff = f$. For sufficiently large n, $\|F_n f - f\| \le \frac{1}{2}\epsilon^2 \|f\|$. Let $f = f' + f''$, where $f' \in M_1$ and $f'' \in M_1^\perp$, and suppose that $\|f''\| > \epsilon \|f\|$. Then $\|F_n f\| = \|F_n f'\| \le (1 - \epsilon^2)^{1/2} \|f\|$, and therefore

$$\|f\| \le \|F_n f\| + \|F_n f - f\| \le (1 - \epsilon^2)^{1/2} \|f\| + \frac{1}{2}\epsilon^2 \|f\| < \|f\|$$

if $\epsilon > 0$. Contradiction is avoided only if $f'' = 0$, so that $f \in M_1$. Now repeat the argument, letting $f = f' + f''$, where $f' \in M_2$ and $f'' \in M_2^\perp$; in order to avoid a contradiction, $f \in M_2$. Continuing in this way, one obtains $f \in M_i$ for all i. Hence, $f \in \bigcap M_i$ and $\Delta_F = \bigcap M_i$.

Acknowledgment. I wish to thank Prof. Jauch for interesting discussions of the questions raised in this note.

REFERENCES

[1] Josef M. Jauch, *Foundations of Quantum Mechanics* (Addison-Wesley, Reading, Massachusetts, 1968).
[2] Howard Stein, On the conceptual structure of quantum mechanics, in *Paradigms and Paradoxes,* ed. R. G. Colodny (University of Pittsburgh Press, Pittsburgh, 1972), pp. 367–438.
[3] R. Mirman, Analysis of the experimental meaning of coherent superposition and the nonexistence of superselection rules, *Phys. Rev.* **D1**, 3349 (1970), and references given there.
[4] F. Riesz and B. Sz.-Nagy, *Functional Analysis* (Ungar, New York, 1955), p. 263.
[5] John von Neumann, *Functional Operators,* Vol. II (Princeton University Press, Princeton, 1950), p. 55.

COMMENT

In considering the quantum-mechanical structure of the set of "propositions of a physical system," I am not seeking a novel logic of discourse. On this matter, see John Stachel, "Do Quanta Need a New Logic?", in *From Quarks to Quasars: Philosophical Problems of Modern Physics,* ed. R. Colodny (University of Pittsburgh Press, Pittsburgh, 1986), pp. 229–347.

6

Proposed neutron interferometer observation of the sign change of a spinor due to 2π precession

Michael A. Horne and Abner Shimony

If a spin-$\frac{1}{2}$ particle is placed in a constant uniform magnetic field for a time t during which, classically, the Larmor precession angle would be θ, then according to quantum mechanics the components of the spin part of the wave function undergo phase changes of $\pm\theta/2$. Hence, corresponding to a classical precession of 2π radians, there is a sign change of the spinor, whatever the initial polarization may be. Bernstein[1] pointed out that this sign change can be observed by coherently splitting a fermion beam, subjecting one part to the magnetic field, and coherently recombining the parts. Evidently, the same interference effect can be achieved by subjecting the two parts of the beam to fields which would classically cause precessions of π radians in opposite senses. He noted advantages in studying neutral rather than charged particles, and outlined a proposal for an interferometry experiment with neutrons.

The recent invention[2] and application[3] of a remarkable single-crystal neutron interferometer greatly increases the feasibility of observing the sign change.[4] In the experiment of Ref. 3 a neutron beam is split, and the two parts follow paths of different elevations. The interference of the recombined parts exhibits the effect of gravitational potential difference

1. H. J. Bernstein (1967, *Phys. Rev. Lett.* 18: 1102). A similar proposal was made by Y. Aharonov and L. Susskind (1967, *Phys. Rev.* 158: 1237). Criticisms of the latter paper, emphasizing the importance of distinguishing dynamical from kinematical rotations, were made by G. C. Hegerfeldt and K. Kraus (1968, *Phys. Rev.* 170: 1185) and by A. Frenkel (1971) in B. d'Espagnat (ed.), *Foundations of Quantum Mechanics* (New York: Academic Press).
2. H. Rauch, W. Treimer, and U. Bonse (1974), *Phys. Lett.* 47A: 369).
3. R. Colella, A. W. Overhauser, and S. A. Werner (1975, *Phys. Rev. Lett.* 34: 1472).
4. The possibility of using the single-crystal neutron interferometer for observing the effect of a precession of 2π radians was noted by A. G. Klein and G. I. Opat (1975, *Phys. Rev. D* 11: 523), but they commented that it would be "difficult to set up and align" (p. 524), and proposed instead an ingenious neutron diffraction experiment. However, the remarkable detection of the gravitational potential difference by means of the single-crystal interferometer provides grounds for reversing their pessimistic judgment.

between the two elevations. In the experiment which we propose, the paths could be kept at the same elevation, but phase differences would be produced by appropriate magnetic fields. It is possible to insert without touching the crystal two parallel solenoids of reasonable size, with electric current in opposite senses, through which the separated parts of the neutron beam could respectively pass. The axes of the solenoids would be along the paths AB and CD respectively (or, equally well, along AC and BD) in Fig. 1 of Ref. 3. With this arrangement, and neglecting the fringing field, the contribution which the two precessions together make to the relative phase of the two parts, assuming that the initial beam is polarized, is

(1) $$\beta_s = 4\pi\lambda\mu h^{-2}MHL,$$

where λ is the neutron wavelength, μ is the magnetic moment of the neutron, h is Planck's constant, M is the neutron mass, H is the magnetic field intensity, and L is the length of each solenoid. (This equation should be compared with Eq. 2 of Ref. 3.) For one polarization state, β_s is added to the relative phase β of Ref. 3, and for the orthogonal polarization state it is subtracted. Consequently, if the initial beam is unpolarized, the effect of the precessions on the counting rates I_2 and I_3 of neutron counters situated as in Fig. 1 of Ref. 3 can be expressed as

(2a) $$I_2 = \gamma - \alpha \cos \beta_s \cos \beta,$$

(2b) $$I_3 = \alpha(1 + \cos \beta_s \cos \beta),$$

which can be compared with Eqs. 4 of Ref. 3. It is noteworthy that for certain values of β these counting rates do not vary with β_s. Taking $\lambda = 1.44$ Å and $L = 2$ cm, the field required to produce $\beta_s = \pi$ is 24 gauss. Doubling this intensity corresponds to $\beta_s = 2\pi$ and hence, theoretically, to the restoration of the zero-field values of I_2 and I_3. We hope that the thermal and magnetic effects of the current upon the crystal are controlable, or at least that they can be taken into account by monitoring with X-ray interference, as described in Ref. 3. In monitoring these effects it may help to observe I_2 and I_3 with the current flowing in the same sense in both solenoids. Actual solenoids only approximately produce the uniform magnetic fields assumed in the derivations of Eqs. (1) and (2). When Eqs. (2) are corrected to take into account the finite ratio of radius to length of the solenoid, the strong contrast in counting rate with varying β_s will be diminished. The contrast is best preserved by reducing the radius of each solenoid, and therefore also the transverse dimensions of the neutron beam (which are 3 mm × 6 mm in Ref. 3), with resulting increase of the time required for each data point.

Acknowledgment. We are very grateful to Profs. G. Zimmerman and W. Franzen for their suggestions and encouragement. One of us (A.S.) wishes to thank the National Science Foundation for support of research.

COMMENT

This paper was submitted to *Physical Review Letters* in 1975 and rejected, because two papers had already been submitted reporting the successful execution of experiments essentially the same as the one proposed: S. A. Werner, R. Colella, A. W. Overhauser, and C. F. Eagen (1975, *Physical Review Letters* 35: 1053); H. Rauch, A. Zeilinger, G. Badurek, A. Wilfing, W. Bauspiess, and U. Bonse (1975, *Physics Letters* 54A: 425). Furthermore, in these experiments permanent magnets were used instead of current-carrying solenoids, thus avoiding the thermal effects mentioned in our paper. The only value in printing our paper at this late date is to call attention to the consequences of using an unpolarized neutron beam, which are taken into account by the factor $\cos \beta$ in our Eqs. (2a) and (2b) but not mentioned in the two cited papers.

PART B

Quantum entanglement and nonlocality

7
Experimental test of local
hidden-variable theories

I

The study of hidden-variable interpretations of quantum mechanics was transformed by two papers of Bell. In his 1966 paper in *Reviews of Modern Physics* [1] he analysed the argument of von Neumann [2] and the more powerful arguments of Jauch and Piron [3] and of Gleason [4], all leading to similar conclusions about the nonassignability of simultaneous values to all the observables of a nontrivial quantum system; and he showed that these arguments do not exclude what may be called "contextualistic" hidden-variable theories, in which the value of an observable O is allowed to depend not only upon the hidden state λ, but also upon the set C of compatible observables measured along with O. In Bell's 1964 paper in *Physics* [5] (written later than the other, in spite of the earlier publication date) he gave a new kind of argument against a hidden-variable interpretation of quantum mechanics, by showing that no hidden-variables theory which satisfies a certain condition of locality can agree with all the statistical predictions of quantum mechanics.

The main purpose of this lecture is to discuss an experiment, designed by Clauser, Horne, Holt and myself [6], which is based on Bell's proof of the incompatibility of quantum mechanics and local hidden-variable theories. Before doing this, however, I wish to make several comments.

First, there is a historical comment. The great paper of Einstein, Podolsky and Rosen [7] of 1935 concludes that quantum mechanics is an incomplete theory, without suggesting that changes of the theory are prerequisites to the job of completion. They make no reference to von Neumann's argument, which had been published three years earlier, although they could hardly have been unaware of it. A possible reason for their disregard of von Neumann's argument is that Einstein had no confidence that the statistical predictions of quantum mechanics were all correct [8], and therefore he felt no obligation to seek a hidden-variable theory in exact

This work originally appeared in *Foundations of Quantum Mechanics,* Proceedings of the International School of Physics 'Enrico Fermi' (1971), pp. 182–94. Copyright © Società Italiana di Fisica. Reprinted by permission.

statistical agreement with quantum mechanics. If this is the reason, however, it misses the point of von Neumann's theorem. For von Neumann proved, on the basis of a certain set of premises, that no simultaneous assignment of definite values to all the observables of a quantum-mechanical system is possible – which implies that one cannot speak consistently of the statistical predictions of a hidden-variable theory concerning all observables. It is conceivable that Einstein was critical of one or more of von Neumann's premises, as were Bell [1], Siegel [9], and Kochen and Specker [10] thirty years later. However, in the absence of specific evidence that this is so, I doubt it. Rather, I suspect that the formal reasoning of von Neumann was very alien to Einstein's mode of physical thinking, and that therefore he failed to see how corrosive von Neumann's conclusion was of the proposal to "complete" quantum mechanics.

A second comment concerns the logical relation between Bell's two papers. His paper in *Physics* studies a situation of the Einstein–Podolsky–Rosen type, consisting of a pair of spatially separated subsystems. Bell restricts his attention to those hidden-variable theories which satisfy a condition of locality, namely that the value of an observable O_1 of subsystem 1 should depend only upon the hidden state λ of the composite system and not upon the choice of an observable O_2 which is simultaneously measured in subsystem 2, and conversely. However, in his other paper he acknowledged the force of the arguments of Gleason and of Jauch and Piron against all noncontextualistic hidden-variable theories. It is reasonable, therefore, to suppose that the hidden-variable theories considered in the paper in *Physics* are contextualistic as well as local, and accordingly to strengthen the condition of locality as follows: O_1 is a function $O_1(\lambda, C_1)$, where C_1 is the set of compatible observables of subsystem 1 which are being observed simultaneously with O_1, and similarly O_2 is a function $O_2(\lambda, C_2)$.

An interesting question arises when one takes the subsystems 1 and 2, as Bell does, to be spin-$\frac{1}{2}$ systems with configuration variables neglected. Then every quantum-mechanical observable of system 1 is of the form $O_1 = cI^{(1)} + d\sigma^{(1)} \cdot \hat{n}_1$, where c and d are real numbers, $I^{(1)}$ is the identity operator associated with subsystem 1, the components of $\sigma^{(1)}$ are its Pauli spin matrices, and $\hat{n}_1$ is a unit vector. In a contextualistic local hidden-variable theory the value assigned to O_1 is $O_1(\lambda, C_1)$; but, except in the trivial case of $d = 0$, C_1 can only consist of observables which are functions of $\sigma^{(1)} \cdot \hat{n}_1$, and therefore the dependence of O_1 upon compatible observables measured simultaneously with it is vacuous. In other words, $O_1(\lambda, C_1)$ is simply $O_1(\lambda)$, and similarly $O_2(\lambda, C_2)$ is simply $O_2(\lambda)$. Thus, because of the exteme simplicity of the subsystems, the contextualistic local hidden-variable theories seem to degenerate in this case into noncontextualistic ones. But since the latter have already been excluded by the analyses of Jauch and Piron and of Gleason, one may wonder what after all has been accomplished by the ingenious analysis of Bell's paper in *Physics* [11]. The answer is this. The theorems of Gleason and of Jauch and Piron refer to the entire set of observables, which cannot all be consistently assigned simultaneous values if the set has the structure prescribed by quantum mechanics. But no experimental procedure exists for measuring most of these observables, as critics of the operational character of quantum mechanics have pointed out with pleasure. Bell's analysis in *Physics,* however, concerns observables of the form $O_1 \otimes O_2$, where $O_1 = c_1 I^{(1)} + d_1 \sigma^{(1)} \cdot \hat{n}_1$ and $O_2 = c_2 I^{(2)} + d_2 \sigma^{(2)} \cdot \hat{n}_2$, and observables of this form are truly measurable. In other words, Bell exhibits a discrepancy between quantum mechanics and any local hidden-variable theory within the domain of questions which are experimentally answerable.

The remaining comments concern the derivation of the inequality which Bell presented in his lecture [12]. His assumption that the subsystems are spin-$\frac{1}{2}$ particles is evidently irrelevant to the derivation of the inequality, though it is important for his comparison with quantum-mechanical predictions. To obtain the inequality it suffices to consider two subsystems 1 and 2 which are spatially well separated, but otherwise arbitrary. Let A_a be a family of observables of subsystem 1 parametrized by a, and B_b a family of observables of subsystem 2 parametrized by b, with no restrictions placed upon the index sets to which a and b belong. For Bell's demonstration in his lecture the values of A_a and B_b need only be required to lie in the interval $[-1, 1]$ (whereas the proofs in ref. [5, 6] required each A_a and B_b to have only -1 or 1 as possible values). The appropriate apparatus for measuring A_a may simultaneously yield values for a set of observables C_1, so that in a local contextualistic hidden-variable theory the outcome of the measurement of A_a will be $A_a(\lambda, C_1)$; and similarly the outcome of the measurement of B_b will be $B_b(\lambda, C_2)$. In spite of locality, correlations can exist between A_a and B_b because of their common dependence upon λ. When a probability distribution μ is given upon the space Γ of hidden states, there is a well-defined correlation function:

$$P_{\text{hv}}(a, b) = \int_{\Gamma} A_a(\lambda, C_1) B_b(\lambda, C_2) \, d\mu.$$

Bell's derivation yields the inequality

(1) $\qquad |P_{\text{hv}}(a, b) - P_{\text{hv}}(a, c)| + P_{\text{hv}}(d, c) + P_{\text{hv}}(d, b) - 2 \leq 0,$

a convenient direct consequence of which is

(2) $\qquad S_{\text{hv}} \equiv P_{\text{hv}}(a, b) - P_{\text{hv}}(a, c) + P_{\text{hv}}(d, c) + P_{\text{hv}}(d, b) - 2 \leq 0.$

It often happens in practical cases that addition and subtraction of the parameters are well-defined operations, and also that $P(a, b)$ depends only upon $a-b$ and hence can be written $P(a-b)$. Then if $d-b=\beta$, $d-c=\gamma$, $a-d=\alpha$, we have the following useful form of Bell's inequality:

(3) $S_{hv} \equiv P_{hv}(\beta) + P_{hv}(\gamma) + P_{hv}(\alpha+\beta) - P_{hv}(\alpha+\gamma) - 2 \leq 0.$

2

Since inequalities (1) and (2) are implied by all local hidden-variable theories (and inequality (3) is implied by any for which $P(a, b)$ depends only on $a-b$), Bell's analysis makes it possible to design an experiment for testing the entire family of local hidden-variable theories. The system considered by Bell, consisting of a pair of spin-$\frac{1}{2}$ particles in the singlet state, is impractical for this purpose, since it appears to be very hard to produce [13] and would be difficult to make measurements upon if produced. A far more convenient procedure is to produce photon pairs and to examine correlations of their polarizations. To some extent this has already been done in 1950 by Wu and Shaknov [14], who examined the Compton scattering of pairs of photons produced in positronium annihilation. When Horne and I began our work, and Clauser independently began his, we thought that the Wu–Shaknov experiment could be used, perhaps with minor modifications, to provide a test of the family of local hidden-variable theories. We found, however, that this experiment provided insufficient information about the polarization correlation. The angular distribution of photons scattered by electrons at rest is predicted by quantum mechanics to be a function of photon polarization (Klein-Nishina formula). But from the direction of scattering of a single photon one can only infer the probabilities of its initial polarization along one of two perpendicular axes, and on the average the difference between these two probabilities is small. As a result, when one modifies the Wu-Shaknov experiment so as to give experimental meaning to $P(a, b)$ - e.g., by letting a represent one partition of the scattering sphere into two regions, with the values ± 1 assigned respectively to photons scattered into the two regions, and letting b represent another such partition – one finds that quantum mechanics predicts an upper bound of less than 0.2 for $P(a, b)$ [15]. Evidently, however, a discrepancy between quantum mechanics and the inequality (1) does not result unless the quantum-mechanical prediction for $P(a, b)$ has absolute value larger than $\frac{1}{2}$ for some values of the parameters a and b. Hence, if Compton scattering is used to provide information about polarization, there is no crucial test between quantum mechanics and local hidden-variable theories [16].

The experimental difficulty of measuring the polarization correlation of a photon pair can be avoided by using low-energy photons. Snider and Pritchard brought to our attention the experiment of Kocher and Commins [17], who studied the pairs of photons emitted in the $6\,^1S_0 \to 4\,^1P_1 \to 4\,^1S_0$ cascade in calcium, which are in the appropriate energy range. The photons were focused by lenses so as to impinge normally on polarizers of the Polaroid type. Coincidences were counted when the polarizer axes were parallel and again when they were perpendicular to each other. Although the Kocher–Commins experiment does not suffice to provide a test of local hidden-variable theories, various modifications of it will do so [6].

It is necessary to show first that if appropriate relative orientations of the polarizer axes are chosen, there is a discrepancy between the quantum-mechanical and the local hidden-variable predictions concerning the polarization correlations [18]. If A_a and B_b are operators representing appropriately oriented linear polarizers and Ψ represents the quantum-mechanical state of the photon pair which is emitted in the cascade and is incident upon the polarizers, then the quantum-mechanical correlation function is $P_{\text{qm}}(a, b) = (\Psi, A_a B_b \Psi)$. It will be instructive to calculate both the wave function and the operators under certain idealizing assumptions and then to remove the idealizations.

In a $J = 0 \to J = 1 \to J = 0$ atomic cascade, with no angular momentum exchanged with the nucleus [19], the angular wave function of the emitted photon pair is

(4) $\quad \Psi_0 = 1/\sqrt{3}\,[Y_{11}^{(1)}(\hat{n}_1) Y_{1-1}^{(1)}(\hat{n}_1) - Y_{10}^{(1)}(\hat{n}_1) Y_{10}^{(1)}(\hat{n}_2) + Y_{1-1}^{(1)}(\hat{n}_1) Y_{11}^{(1)}(\hat{n}_2)],$

where $\hat{n}_1$ and $\hat{n}_2$ are the directions of propagation of the first and second photons and $Y_{jm}^{(1)}$ is the vector spherical function of total angular momentum j, magnetic quantum number m, and parity -1. The vector spherical functions can be expanded into sums of terms [20], each of which is a product of orbital-angular-momentum eigenstates $Y_{lm'}$ and a spin angular momentum eigenstate $\chi_{m''}$:

$$Y_{11}^{(1)} = 30^{-1/2} Y_{20}\chi_1 - 10^{-1/2} Y_{21}\chi_0 + 5^{-1/2} Y_{22}\chi_{-1} + (2/3)^{-1/2} Y_{00}\chi_1, \text{ etc.}$$

The $\chi_{m''}$ can in turn be expressed in the standard way in terms of states of linear polarization:

$$\chi_1 = -(\chi_x + i\chi_y) = \begin{pmatrix} -1 \\ -i \\ 0 \end{pmatrix}, \qquad \chi_0 = \begin{pmatrix} 0 \\ 0 \\ 1 \end{pmatrix}$$

$$\text{and} \quad \chi_{-1} = (\chi_x - i\chi_y) = \begin{pmatrix} 1 \\ -i \\ 0 \end{pmatrix}.$$

Now make the first idealization, which is to assume that stops with infinitesimal apertures are inserted between the source and the polarizers, so that $\hat{n}_1$ is restricted to be $\hat{z}$ and $\hat{n}_2$ is restricted to be $-\hat{z}$. (Note that the identification of the first and second photons can be made unambiguous by the use of appropriate frequency filters to the right and left of the source along the z-axis.) Then a simple calculation shows that the wave function incident upon the polarizers is

$$
(5) \qquad \Psi_{\text{ideal}} = \text{normalization} \times \left[\begin{pmatrix} 1 \\ 0 \\ 0 \end{pmatrix} \begin{pmatrix} 1 \\ 0 \\ 0 \end{pmatrix} + \begin{pmatrix} 0 \\ 1 \\ 0 \end{pmatrix} \begin{pmatrix} 0 \\ 1 \\ 0 \end{pmatrix} \right].
$$

The second idealization is to suppose that perfect polarization analysers exist, which permit all normally incident photons polarized along the axis of polarization, and none polarized perpendicular to this axis, to pass. Let the observable A_a have the value 1 if the first photon of a pair passes through such an analyser when it is oriented with axis of polarization in the xy-plane making an angle a with the x-axis, and let A_a have the value -1 if the photon fails to pass. The representation for A_a in a linear polarization basis is

$$
A_a = \begin{pmatrix} \cos 2a & \sin 2a & 0 \\ \sin 2a & -\cos 2a & 0 \\ 0 & 0 & 1 \end{pmatrix},
$$

though the 33-element of this matrix is arbitrary because of the nonexistence of longitudinal photons. That this expression is correct is checked by noting that $\cos a \chi_x + \sin a \chi_y$ and $-\sin a \chi_x + \cos a \chi_y$ are eigenvectors with respective eigenvalues ± 1. B_b has a similar meaning and a similar representation. With these idealizations

$$
(6) \qquad P_{\text{qm}}^{\text{ideal}}(\alpha) = (\Psi_{\text{ideal}}, A_a B_b \Psi_{\text{ideal}}) = \cos 2\alpha,
$$

where $\alpha = a - b$.

Evidently the first idealization must be removed in order to have a non-vanishing count of photon pairs. Instead of requiring $\hat{n}_1$ to be $\hat{z}$ and $\hat{n}_2$ to be $-\hat{z}$, suppose that all photons are gathered with $\hat{n}_1 \in \Omega_1$ and $\hat{n}_2 \in \Omega_2$, where Ω_1 and Ω_2 are cones oriented along the z-axis with the source as their common vertex [21]. The resulting angular wave function Ψ is identical with Ψ_0 of eq. (4) for $\hat{n}_1 \in \Omega_1$ and $\hat{n}_2 \in \Omega_2$ but is 0 for other $\hat{n}_1, \hat{n}_2$.

It is also convenient to insert a pair of lenses to make the photons normally incident upon the polarizers. The action of the lenses will be represented by the operator D (see Appendix), and the resulting angular wave function is $\Psi' = D\Psi$.

The second idealization must also be removed, since all actual polarizers are imperfect. Let ϵ_M^i ($i = $ I, II) be the probability that a photon polarized along the axis of polarization of the ith polarizer will pass it (non-passage being due either to reflection or to absorption), and let ϵ_m^i be the probability that a photon polarized perpendicular to this axis will pass. (For calcite polarizers $\epsilon_m^i \approx 10^{-5}$, and values of ϵ_M^i as high as 0.94 are obtainable by antireflection coating.) Then for photons fairly close to normally incident upon the polarizer face the observable A_a, whose value is ± 1 according as the photon passes or fails to pass the appropriately oriented polarizer, is replaced by

$$
(7) \quad \bar{A}_a = \begin{pmatrix} 2\epsilon_M^I \cos^2 a + 2\epsilon_m^I \sin^2 a - 1 & 2(\epsilon_M^I - \epsilon_m^I)\cos a \cdot \sin a & 0 \\ 2(\epsilon_M^I - \epsilon_m^I)\cos a \cdot \sin a & 2\epsilon_M^I \sin^2 a + 2\epsilon_m^I \cos^2 a - 1 & 0 \\ 0 & 0 & 1 \end{pmatrix}.
$$

The vectors $\cos a\chi_x + \sin a\chi_y$ and $-\sin a\chi_x + \cos a\chi_y$ are eigenvectors of $\bar{A}_a$ with respective eigenvalues $2\epsilon_M^I - 1$ and $2\epsilon_m^I - 1$, which in each case equals the probability of passing minus the probability of not passing. There is a similar expression for $\bar{B}_b$. One then finds, after a rather lengthy calculation,

$$
\begin{aligned}
(8) \quad P_{qm}(\alpha) = P_{qm}(a, b) &= (\Psi', \bar{A}_a \bar{B}_b \Psi') \\
&= [1 - (\epsilon_M^I + \epsilon_m^I)][1 - (\epsilon_M^{II} + \epsilon_m^{II})] \\
&\quad + (\epsilon_M^I - \epsilon_m^I)(\epsilon_M^{II} - \epsilon_m^{II})F_1(\theta)\cos 2\alpha,
\end{aligned}
$$

where $\alpha = a - b$, θ is the half-angle of each of the cones Ω_1 and Ω_2 and

$$
F_1(\theta) = \frac{(7 - 3\cos\theta - 3\cos^2\theta - \cos^3\theta)^2}{12(8 - 16\cos\theta + 9\cos^2\theta - 2\cos^4\theta + \cos^6\theta)}.
$$

$F_1(\theta)$ is a monotonically decreasing function which equals 1 at $\theta = 0$ and falls off slowly to 0.99 at 30° and more rapidly to 0.91 at 50° and 0.51 at 90°. Note that even if ϵ_M^i were 1 and ϵ_m^i were 0 ($i = I, II$), the range of $P_{qm}(\alpha)$ when $\theta > 0$ would not be from -1 to 1 as for $P_{qm}^{ideal}(\alpha)$. The reason is that the total angular momentum 0 of the two-photon system is accomplished by the coupling of orbital and spin angular momentum, and increasing the orbital contribution (by increasing θ) has the effect of damping the correlations of polarization. Nevertheless, even for large enough θ to permit a good counting rate, the range of $P_{qm}(\alpha)$ is sufficient to yield a discrepancy with Bell's inequality. For, if we assume for simplicity that $\epsilon_m^I = \epsilon_m^{II} = 0$ and $\epsilon_M^I = \epsilon_M^{II} = \epsilon$, then:

$$
\begin{aligned}
(9) \quad S_{qm} &\equiv P_{qm}(\beta) + P_{qm}(\gamma) + P_{qm}(\alpha + \beta) - P_{qm}(\alpha + \gamma) - 2 \\
&= 2\epsilon\{(\epsilon - 2) + \epsilon F_1(\theta)[\cos 2\beta + \cos 2\gamma + \cos 2(\alpha + \beta) \\
&\qquad\qquad - \cos 2(\alpha + \gamma)]\}
\end{aligned}
$$

and this is a maximum when $\alpha = 45°$, $\beta = -22\frac{1}{2}°$, $\gamma = 22\frac{1}{2}°$, namely

$$
2\epsilon[(\epsilon - 2) + \sqrt{2}\epsilon F_1(\theta)].
$$

Letting, for example, $\epsilon = 0.9$ and $\theta = 30°$, we obtain $S_{qm} = 0.306$, in clear discordance with the prediction of any local hidden-variable theory.

All the foregoing arguments hold in the case of a $J = 0 \rightarrow J = 1 \rightarrow J = 1$ cascade (and also a $J = 1 \rightarrow J = 1 \rightarrow 0$ cascade if the $m = 1, 0, -1$ states of the initial ensemble are equally populated), except that in the resulting eqs. (8) and (9) $F_1(\theta)$ must be replaced by $F_2(\theta)$, which is a function monotonically decreasing from 1 at $\theta = 0$, but more rapidly than does $F_1(\theta)$ [22].

The test between local hidden-variable theories and quantum mechanics consists in measuring the experimental counterpart of S_{qm} and S_{hv}, namely,

(10) $\qquad S_{exp} \equiv P_{exp}(\beta) + P_{exp}(\gamma) + P_{exp}(\alpha + \beta) - P_{exp}(\alpha + \gamma) - 2,$

where $P_{exp}(\beta)$ is the experimental value of the polarization correlation of photon pairs when the axes of polarization of the two polarizers made an angle β with each other. If it were possible to detect photons incident upon a polarizer which fails to pass it, then $P_{exp}(\beta)$ could be measured quite directly:

(11) $\qquad P_{exp}(\beta) = \dfrac{R_{pp}(\beta) + R_{ff}(\beta) - R_{pf}(\beta) - R_{fp}(\beta)}{R_{pp}(\beta) + R_{ff}(\beta) + R_{pf}(\beta) + R_{fp}(\beta)},$

where $R_{pp}(\beta)$ is the coincidence counting rate of photon pairs in which both partners pass the appropriate polarizers, $R_{ff}(\beta)$ is the coincidence counting rate of photon pairs in which both partners fail to pass, and $R_{pf}(\beta)$ and $R_{fp}(\beta)$ are the two coincidence counting rates in which one partner passes and one does not. One could try to realize this plan by using Wollaston prisms, which are calcite crystals cut so that photons normally incident upon one face emerge in one ray (ordinary ray) if polarized along the axis of polarization, and in the other ray (extraordinary ray) if polarized along the perpendicular axis. If ideal Wollaston prisms existed, all incident photons would emerge, and furthermore would emerge in the "correct" ray, so that all incident photon pairs would be sorted without loss into four channels (ordinary-ordinary, extraordinary-ordinary, ordinary-extraordinary, and extraordinary-extraordinary). In actuality, however, many photons would be lost by reflection from the prism faces. Furthermore, there are technical difficulties in achieving complete separation of the ordinary and extraordinary rays, and also complications for the experimentalist, who would have to use four coincidence counters. Consequently, it is preferable to use ordinary polarizers, and to determine $P_{exp}(\beta)$ by an indirect procedure.

To do this, the following counting rates must be measured: $R(\beta)$, which is the coincidence counting rate when both polarizers are in place with relative orientation β; R_1, which is the rate when the first polarizer is in place and the second is removed; R_2, which is the rate when the second polarizer is in place and the first is removed; and R, which is the rate when both are removed. (By cyclindrical symmetry R_1 and R_2 are expected to be independent of the orientation of the polarizer left in place, but this expectation can be checked experimentally.) Now clearly

$$R_{pp}(\beta) = R(\beta), \quad R_{pf}(\beta) = R_1 - R(\beta), \quad R_{fp}(\beta) = R_2 - R(\beta),$$

$$R_{ff}(\beta) = R - R_{pp}(\beta) - R_{pf}(\beta) - R_{fp}(\beta) = R + R(\beta) - R_1 - R_2,$$

$$R = R_{pp}(\beta) + R_{ff}(\beta) + R_{fp}(\beta) + R_{pf}(\beta).$$

Hence,

(12) $$P_{exp}(\beta) = ([4R(\beta) - 2R_1 - 2R_2]/R) + 1$$

and

(13) $$S_{exp} = 4[R(\beta) + R(\gamma) + R(\alpha + \beta) - R(\alpha + \gamma) - R_1 - R_2]/R.$$

In this way the decisive quantity S_{exp} is measurable.

Holt at Harvard is studying the cascade $9\,^1P_1 \to 7\,^3S_1 \to 6\,^3P_0$ in mercury, and Clauser, Commins and Freedman are studying the $6\,^1S_0 \to 4\,^1P_1 \to 4\,^1S_0$ cascade in calcium. In both experiments coincidences of photons of the correct frequencies have already been seen, and, consequently, we should know within a few months whether we come to bury local hidden-variable theories or to praise them.

It would not be accurate, however, to say that the experiment is completely decisive. The quantities S_{qm} and S_{hv} have been defined in terms of probabilities of passage of photon pairs through appropriately oriented polarizers, whereas the quantity S_{exp} is defined in terms of detector coincidences. Since the efficiency of detectors in the visible spectrum is 20%, there is a possibility that the detected photon pairs constitute a biased selection of the pairs passing the polarizers. The efficiency of photodetectors is much higher, of course, for high-energy photons, but (as pointed out in discussing the Wu–Shaknov experiment) the inefficiency of polarizers for these photons precludes a decisive polarization correlation experiment at high energies. The problem of comparing passage rates with detector rates could be eliminated if, in the future, detectors of 80% or greater efficiency were developed for photons in the visible spectrum. For then one could redefine $A_a = \pm 1$ and $B_b = \pm 1$ to mean detection or nondetection of a photon (rather than passage or nonpassage, as above), and S_{qm} and S_{hv} would be accordingly redefined. One could find θ and ϵ such that the redefined S_{qm} would be positive, while the derivation of $S_{hv} \le 0$ would remain valid. Hence the measurement of S_{exp} would directly test the predictions of quantum mechanics and of local hidden-variable theories.

With the current photodetectors, however, there seems to be no recourse but to make a plausible assumption: that if a pair of photons reaches the photodetectors, the probability of a coincidence count is independent of the placement of the polarizers – that is, independent of whether none or one or both are in place, and also of the orientation of those which are in place. This assumption implies that the counting rate $R(\beta)$ is proportional to the rate of joint passage through polarizers with relative orientation β. It follows then that S_{hv} and S_{qm}, though defined in terms of the passage of photons through the polarizers, constitute rival predictions of the measurable quantity S_{exp}. If S_{exp} turns out to be positive, a determined advocate of local hidden-variable theories could attribute the result to the falsity of the assumption. This special pleading would not be entirely unreasonable. However, if S_{exp} is not only positive but in very close agreement with S_{qm}, then the advocate of local hidden-variable theories would appear to be obsessive, for he would be claiming that somehow the spatially separated detectors receive the output of the polarizers, which satisfies the hidden-variable inequality, and select from this output in such a way that quantum-mechanical counting rates result.

Another manoeuvre is available to the advocate of local hidden-variable theories. He might say that the dependence of A_a upon b and of B_b upon a need not violate locality in the sense of relativity theory. For after the orientations a and b are established by the experimenter, there is sufficient time for each polarizer to "learn" the orientation of the other without using signals faster than light. However, Clauser has suggested an ingenious and possibly achievable device for changing a and b by means of Kerr cells while the photons are in flight. Then A_a could depend on b and B_b on a only if a violation of relativity theory occurred.

3

In conclusion I shall comment on two other forms of hidden-variable theories than those considered so far.

i) There are stochastic hidden-variable theories, in which the value of an observable O is not completely determined by λ or by λ together with the set C of observables measured simultaneously with O. Instead, λ (or λ and C) suffices only to determine a probability distribution of possible values of O. It appears that the hidden-variable theories considered by de Broglie [23] are of this form. Bell's condition of locality can easily be extended so as to apply to a stochastic hidden-variable theory: if A_a and B_b are families of observables of two spatially separated components, then locality requires that the probability of a specified value of A_a depends on λ (and on C_1 in a contextualistic theory) but not on b and C_2, and similarly for B_b. Bell showed in his lecture that his inequality is implied by stochastic local hidden-variable theories as well as by deterministic local hidden-variable theories. Consequently, if the outcome of the polarization correlation experiment is $S_{\text{exp}} > 0$, the family of stochastic local hidden-variable theories is also disconfirmed.

ii) There are nonlocal hidden-variable theories [24]. Certainly no empirical evidence refutes this family of theories, and in particular the result $S_{\text{exp}} > 0$ in the polarization experiment would not be a disconfirmation. No *a priori* objection against nonlocality would be decisive, because the history of science shows how little obligation Nature has to conform to our *a priori* conceptions. There are, however, methodological objections against this family of theories at present. The family is too large, and some member of it seems to be compatible with any experimental data whatsoever. Before it can be taken seriously, some heuristic principles must be exhibited for restricting attention to a subfamily which is small enough to have definite empirical consequences. Furthermore, as long as there is strong evidence that locality in the sense of relativity theory holds in the domain of what now is observable, it is highly artificial to suppose that it fails for hidden variables, for relativity theory is a theory of space-time, not of special systems. It surely would be contrary to the intentions of Einstein, who was the most profound advocate of hidden variables, to abandon locality for their sake without compelling reasons.

APPENDIX

The operator D representing the pair of lenses is a product $D_{\hat{z}} D_{-\hat{z}}$, where $D_{\hat{z}}$ operates upon the wave function of the first photon and rotates the propagation vector $\hat{n}$ into $\hat{z}$ while leaving unchanged a vector orthogonal to both; $D_{-\hat{z}}$ has a similar action for the second photon but with $\hat{z}$ replaced by $-\hat{z}$:

$$D_{\pm\hat{z}} = \begin{pmatrix} \pm\cos^2\varphi\cos\theta + \sin^2\varphi & \pm\sin\varphi\cos\varphi\cos\theta - \cos\varphi\sin\varphi & \mp\cos\varphi\sin\theta \\ \pm\cos\varphi\sin\varphi\cos\theta - \sin\varphi\cos\varphi & \pm\sin^2\varphi\cos\theta + \cos^2\varphi & \mp\sin\varphi\sin\theta \\ \pm\cos\varphi\sin\theta & \pm\sin\varphi\sin\theta & \pm\cos\theta \end{pmatrix}.$$

The following formulae are useful in the derivation of eq. (8):

$$D_{\hat{z}} Y_{1\pm1}^{(1)}(\theta,\varphi) = \tfrac{1}{4}(3/\pi)^{1/2} \begin{pmatrix} \mp(\cos^2\varphi\cos\theta + \sin^2\varphi) + i\cos\varphi\sin\varphi(1-\cos\theta) \\ \pm\cos\varphi\sin\varphi(1-\cos\theta) - i(\cos^2\varphi + \sin^2\varphi\cos\theta) \\ 0 \end{pmatrix},$$

$$D_{\hat{z}} Y_{10}^{(1)}(\theta,\varphi) = -\tfrac{1}{2}(3/2\pi)^{1/2} \begin{pmatrix} \cos\varphi\sin\theta \\ \sin\varphi\sin\theta \\ 0 \end{pmatrix},$$

88 *Quantum entanglement and nonlocality*

$$D_{-\hat{z}}Y^{(1)}_{1\pm1}(\theta',\varphi') = \tfrac{1}{4}(3/\pi)^{1/2}\begin{pmatrix} \pm(\cos^2\varphi'\cos\theta'-\sin^2\varphi')+i\cos\varphi'\sin\varphi'(1+\cos\theta') \\ \pm\cos\varphi'\sin\varphi'(1+\cos\theta')+i(\sin^2\varphi'\cos\theta'-\cos^2\varphi') \\ 0 \end{pmatrix},$$

$$D_{-\hat{z}}Y^{(1)}_{10}(\theta',\varphi') = \tfrac{1}{2}(3/2\pi)^{1/2}\begin{pmatrix} \cos\varphi'\sin\theta' \\ \sin\varphi'\sin\theta' \\ 0 \end{pmatrix}.$$

Note added in proofs. Clauser, Commins and Freedman at Berkeley have now determined S_{exp} and found it to be in agreement with S_{qm} but in disagreement (by at least five standard deviations) with S_{hv} – a strong disconfirmation of local hidden-variable theories.

REFERENCES

[1] J. S. Bell: *Rev. Mod. Phys.*, **38**, 447 (1966).
[2] J. von Neumann: *Mathematical Foundations of Quantum Mechanics* (Princeton, 1955), translated from *Mathematische Grundlagen der Quantenmechanik* (Berlin, 1932).
[3] J. M. Jauch and C. Piron: *Helv. Phys. Acta,* **36**, 827 (1963).
[4] A. M. Gleason: *Journ. Math. Mech.,* **6**, 885 (1957).
[5] J. S. Bell: *Physics,* **1**, 195 (1964).
[6] J. F. Clauser, M. A. Horne, A. Shimony and R. A. Holt: *Phys. Rev. Lett.,* **23**, 880 (1969).
[7] A. Einstein, B. Podolsky and N. Rosen: *Phys. Rev.,* **47**, 777 (1935).
[8] Private communications from Prof. N. Rosen, and also from Dr. E. Guth, who reported conversations with B. Podolsky.
[9] A. Siegel: in *Differential Space, Quantum Systems and Predictions,* edited by B. Rankin (Cambridge, 1966).
[10] S. Kochen and E. P. Specker: *Journ. Math. Mech.,* **17**, 59 (1967).
[11] This question was raised by Prof. J. Bub in a lecture at the Boston Colloquium for the Philosophy of Science, April 14, 1970.
[12] J. S. Bell: "Introduction to the Hidden-Variable Question," in *Foundations of Quantum Mechanics,* ed. B. d'Espagnat (Academic Press, New York, 1971), pp. 171–81.
[13] Possible ways of producing such pairs are discussed by A. Peres and P. Singer: *Nuovo Cimento,* **15**, 907 (1960).
[14] C. S. Wu and I. Shaknov: *Phys. Rev.,* **77**, 136 (1950).
[15] This follows from a discussion in M. A. Horne's thesis, Boston University (1970) (unpublished).
[16] L. Kasday reports on a test of local hidden-variable theories by means of a variant of the Wu–Shaknov experiment: "Experimental Test of Quantum Predictions for Widely Separated Photons," in *Founations of Quantum Mechanics,* ed. B. d'Espagnat (Academic Press, New York, 1971), pp. 195–210. He explicitly points out, however, his need for an assumption that the results

obtained in a Compton-scattering experiment are related correctly by quantum mechanics to the results of an ideal linear polarization analyser. Although this assumption is reasonable, it could be challenged by an advocate of local hidden-variable theories who wishes to account for experimental results favoring quantum mechanics.

[17] C. A. Kocher and E. D. Commins: *Phys. Rev. Lett.*, **18**, 575 (1967).
[18] This discrepancy is exhibited in detail in Horne's thesis (ref. [15]) by a quite different method from the one used here.
[19] Calculations by R. A. Holt, to be presented in his thesis, Harvard University, 1972, show that there is no discrepancy between the quantum-mechanical predictions and Bell's inequality in the case of cascades from atoms having nuclei with nonzero spin.
[20] A. I. Akhiezer and V. B. Berestetsky: *Quantum Electrodynamics*, Sect. 4 (Washington, D.C.).
[21] The assumption of a point source is also an idealization. For a discussion of the more realistic case of a line source, see Holt's thesis, ref. [19].
[22] See ref. [6] and ref. [15].
[23] L. de Broglie: *Nonlinear Wave Mechanics* (New York, 1960); also *Ondes électromagnétiques et photons* (Paris, 1968), and other publications.
[24] For example, D. Bohm: *Phys. Rev.*, **85**, 166, 180 (1952).

COMMENT

The question raised in the third paragraph of Section 1 was cleared up by a conversation with Prof. Peter G. Bergmann around 1980. He recalled a discussion with Einstein and Valentine Bargmann around 1938 at the Institute for Advanced Study, during which Einstein took von Neumann's book from the shelf and pointed to premise B' of von Neumann's theorem (in Section 1 of Chapter IV): "If $\mathfrak{R}, \mathfrak{S}, \ldots$ are arbitrary quantities and $a, b, \ldots$ real numbers, then $\mathrm{Exp}(a\mathfrak{R} + b\mathfrak{S} + \cdots) = a\,\mathrm{Exp}(\mathfrak{R}) + b\,\mathrm{Exp}(\mathfrak{S}) + \cdots$." Einstein then said that there is no reason why this premise should hold in a state not acknowledged by quantum mechanics if $\mathfrak{R}, \mathfrak{S}$, etc. are not simultaneously measurable. Einstein's criticism is essentially the same as those of Siegel, Jauch and Piron, Bell, and Kochen and Specker nearly thirty years later.

8
An exposition of Bell's theorem

The purpose of this lecture is to give a self-contained demonstration of a version of Bell's theorem and a discussion of the significance of the theorem and the experiments which it inspired. The lecture should be comprehensible to people who have had no previous acquaintance with the literature on Bell's theorem, but I hope that explicitness about premisses and consequences will make it useful even to those who are familiar with the literature.

All versions of Bell's theorem are variations, and usually generalizations, of the pioneering paper of J. S. Bell of 1964, entitled "On the Einstein–Podolsky–Rosen Paradox." All of them consider an ensemble of pairs of particles prepared in a uniform manner, so that statistical correlations may be expected between outcomes of tests performed on the particles of each pair. If each pair in the ensemble is characterized by the same quantum state ϕ, then the quantum mechanical predictions for correlations of the outcomes can in principle be calculated when the tests are specified. On the other hand, if it is assumed that the statistical behavior of the pairs is governed by a theory which satisfies certain independence conditions (always similar to the Parameter and Outcome Independence conditions stated below, though the exact details vary from version to version of Bell's theorem), then it is possible to derive a restriction upon the statistical correlations of the outcomes of tests upon the two particles. The restriction is stated in the form of an inequality, known by the collective name of "Bell's Inequality." Each version of Bell's theorem exhibits a choice of ϕ and of the tests upon the two particles such that the quantum mechanical predictions of correlations violates one of the Bell's Inequalities. The theorem therefore asserts that *no physical theory satisfying the specified independence conditions can agree in all circumstances with the predictions of quantum mechanics.* The theorem becomes physically most striking when the experimental arrangement is such that relativistic locality *prima facie* requires that the independence conditions be satisfied.

This work originally appeared in A. Miller (ed.), *Sixty-two Years of Uncertainty,* New York: Plenum Publishing Corp., 1990. Reprinted by permission of the publisher.

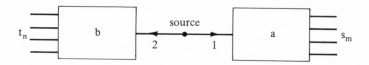

Figure 1. An ensemble of particle pairs $1+2$ is emitted in a uniform manner from the source. Particle 1 enters an analyzer with a controllable parameter a, and the possible outcomes are s_m ($m = 1, 2, ...$). Particle 2 enters an analyzer with controllable parameter b, and the possible outcomes are t_n ($n = 1, 2, ...$).

Because such arrangements are in principle possible (and, in fact, actually realizable, if certain reasonable assumptions are made), one can restate Bell's Theorem more dramatically as follows: *no local physical theory can agree in all circumstances with the predictions of quantum mechanics.* I shall now present a schematic arrangement which will allow the foregoing sketch to be filled out in detail.

Figure 1 shows a source from which particle pairs, labeled 1 and 2, are emitted in a uniform manner. The complete state of a pair $1+2$ is denoted by k, where k belongs to a space K of complete states. No assumption is made about the structure of K, except that probability measures can be defined on it. Because of the uniform experimental control of emission, it is reasonable to suppose that there is a definite probability measure w defined over K which governs the ensemble of pairs; but the uniformity need not be such that w is a delta-function, i.e., that every pair of the ensemble is in the same complete state k. Particle 1 enters an analyzer with a controllable parameter a, which the experimenter can specify, for instance, by turning a knob. Likewise, particle 2 enters an analyzer with a controllable parameter b. The possible outcomes of the analysis of 1 are s_m ($m = 1, 2, ...$), and for mathematical convenience all these values are assumed to lie in the interval $[-1, 1]$. The possible values of the analysis of 2 are t_n ($n = 1, 2, ...$), and these values are assumed to lie in the same interval. It will be assumed that when the parameters a and b and the complete state k are all specified, then the probabilities of the various single and joint outcomes of analysis are well-defined. Specifically,

$p^1(m/k, a, b)$ is the probability of the outcome s_m of the analysis of particle 1, given the complete state k and the parameters a and b;

$p^2(n/k, a, b)$ is the probability of the outcome t_n of the analysis of particle 2, given the complete state k and the parameters a and b;

$p(m, n/k, a, b)$ is the probability of joint outcomes s_m and t_n, given the complete state k and the parameters a and b;

$p^1(m/k, a, b, n)$ is the probability of the outcome s_m of the analysis of particle 1, given the complete state k, the parameters a and b, and the outcome t_n of the analysis of particle 2;

$p^2(n/k, a, b, m)$ is the probability of the outcome t_n of the analysis of particle 2, given the complete state k, the parameters a and b, and the outcome s_m of the analysis of particle 1.

The general principles of probability theory, with no further assumptions, impose the following product rule:

$$p(m, n/k, a, b) = p^1(m/k, a, b)p^2(n/k, a, b, m)$$
$$= p^2(n/k, a, b)p^1(m/k, a, b, n).$$

We now have sufficient notation to make explicit the independence conditions which were mentioned in the sketch above, and which were first made explicit by Jarrett (1984).

Parameter Independence.

$p^1(m/k, a, b)$ is independent of b, and hence may be written as
$\quad p^1(m/k, a)$,
$p^2(n/k, a, b)$ is independent of a, and hence may be written as
$\quad p^2(n/k, b)$.

Outcome Independence.

$p^1(m/k, a, b, n) = p^1(m/k, a, b)$,
$p^2(n/k, a, b, m) = p^2(n/k, a, b)$.

The conjunction of Parameter Independence and Outcome Independence implies the following factorization, which is crucial in the argument ahead:

(1) $$p(m, n/k, a, b) = p^1(m/k, a)p^2(n/k, b).$$

Eq. (1) is often called "Bell's locality condition," but even though I have used this nomenclature myself, I now think that it is misleading, and a more neutral name is preferable.

Expectation values can be defined explicitly in terms of the outcomes s_m and t_n and appropriate probabilities:

$E^1(k, a) = \sum_m p^1(m/k, a)s_m$ is the expectation value of the outcome of analysis of particle 1, given complete state k and parameter a;

$E^2(k, b) = \sum_n p^2(n/k, b)t_n$ is the expectation value of the outcome of analysis of particle 2, given complete state k and parameter b;

$E(k, a, b) = \sum_{m, n} p(m, n/k, a, b)s_m t_n$ is the expectation value of the product of the outcomes of analysis of the two particles, given k, a, and b.

These definitions, together with Eq. (1), immediately yield the following:

(2)
$$E(k, a, b) = E^1(k, a)E^2(k, b).$$

I shall now state and prove a simple mathematical lemma, which will bring us close to one of Bell's Inequalities.

Lemma. *If x', y', x'', and y'' all belong to the interval $[-1, 1]$, then S belongs to the interval $[-2, 2]$, where $S = x'y' + x'y'' + x''y' - x''y''$.*

The proof I shall now give will not be the inelegant one which I presented in Erice, but the elegant argument which N. David Mermin suggested after the lecture. The first step is to note that S is linear in each of its four variables and hence takes on its extreme values at corners of the domain, i.e. at $(x', y', x'', y'') = (\pm 1, \pm 1, \pm 1, \pm 1)$. Clearly, at a corner the value of S must be an integer between -4 and 4. But S can also be written as

$$S = (x' + x'')(y' + y'') - 2x''y''.$$

Since the two quantities in parentheses can only be 0 or ± 2, and the last term is ± 2, S cannot have values ± 3 or ± 4 at the corners.   Q.E.D.

The lemma is applied to our physical problem by identifying x' with $E^1(k, a')$, y' with $E^2(k, b')$, x'' with $E^1(k, a'')$, and y'' with $E^2(k, b'')$. Since each of the outcomes s_m and t_n lies in $[-1, 1]$, so also do these four expectation values, so that the conditions of the lemma are satisfied. The conclusion of the lemma is then also satisfied, and when Eq. (2) is combined with the conclusion, the result is

(3)
$$-2 \le E(k, a', b') + E(k, a', b'') + E(k, a'', b') - E(k, a'', b'') \le 2.$$

Now integrate Inequality (3) over the space K, using the probability distribution w throughout as a weighting, and we obtain

(4)
$$-2 \le E_w(a', b') + E_w(a', b'') + E_w(a'', b') - E_w(a'', b'') \le 2,$$

where we have used the normalization condition

(5)
$$\int_K dw = 1,$$

and we have defined the ensemble expectation value $E_w(a, b)$ as

(6)
$$E_w(a, b) = \int_K E(k, a, b) \, dw.$$

Inequality (4) is Bell's Inequality, or, more accurately, it is the version of Bell's Inequalities which emerges in the present exposition.[1]

1. This version of Bell's Inequality was first derived by Clauser, Horne, Shimony, and Holt (1969) in the special case where $p^1(m/k, a, b)$ and $p^2(n/k, a, b)$ are allowed to have only

It is noteworthy that except for the assumption that probability measures can be defined on K there are no assumptions about the structure of the space K of complete states and no characterization of the complete states k. Also, no assumptions have been made about the probability measure w over K, except that the same w is used in integrating each of the terms in Inequality (4). Physically this one assumption would not be justified if the choice of the parameters a and b affected the emission of particle pairs by the source. That the w governing the ensemble of particle pairs emitted by a source is independent of the parameters of the analyzers is an independence condition distinct from Parameter Independence and Outcome Independence, which were used above, but somewhat similar to Parameter Independence.

In order to complete the proof of Bell's Theorem it is essential to find a realization of the schema of Figure 1 in which the quantum mechanical predictions are in conflict with Inequality (4). One realization which is easy to analyze takes particles 1 and 2 to be photons propagating respectively in z and $-z$ directions and prepared by the source in the polarization state

(7) $$\phi = 2^{-1/2}[u_x(1)u_x(2) + u_y(1)u_y(2)],$$

and takes the analyzers to be linear polarization filters placed perpendicular to the z-axis in the paths of photons 1 and 2 respectively. The parameter a is the angle from the x-axis to the transmission axis of the first polarization filter, and b is similarly defined for the second filter. In Eq. (7) $u_x(1)$ is a normalized vector representing a quantum state of linear polarization along the x-axis for photon 1; and $u_y(1)$, $u_x(2)$, and $u_y(2)$ have analogous meanings. ϕ is a superposition of a state in which both 1 and 2 are polarized along the x-axis and another state in which both 1 and 2 are polarized along the y-axis. Obviously, ϕ is a quantum state in which neither photon 1 nor photon 2 has a definite polarization with respect to the x-y axes, and yet the results of polarization measurements with respect to these axes are strictly correlated, for if photon 1 passes through a filter with transmission along x, so also will photon 2; and if photon 1 fails to pass through such a filter, photon 2 will likewise fail.

The two outcomes of analysis of photon 1 are passage and non-passage through the polarization filter, and these outcomes will conventionally be

the values 1 and 0 (so called "deterministic" hidden variables theories). A derivation without this restriction was first given by Bell (1971) and in another way by Clauser and Horne (1974). The procedure in my lecture, making use of the simple mathematical lemma, was inspired by Clauser and Horne, although their lemma was different. Alain Aspect pointed out to me after the lecture that a proof exactly like mine, with the same lemma, is in the unpublished part of his doctoral thesis (1983).

assigned the numerical values 1 and −1 respectively (these are the s_m of Fig. 1). Likewise, passage and non-passage of photon 2 through its filter will be assigned 1 and −1 respectively (the t_n of Fig. 1). In order to calculate the quantum mechanical expectation value of the product of the outcomes, which will be the counterpart of the expectation value of Eq. (4), it is essential to find an appropriate self-adjoint operator S_a corresponding to analyzing photon 1 with a filter having a transmission axis at the angle a, and an analogous self-adjoint operator T_b corresponding to analyzing photon 2. S_a is determined by the requirements that it be linear on the two-dimensional space of polarization states of photon 1 and have eigenvalues 1 and −1 respectively for states of linear polarization along the directions specified by a and $a + \pi/2$ respectively:

(8) $$S_a u_a = u_a,$$

(9) $$S_a u_{a+\pi/2} = -u_{a+\pi/2}.$$

The states u_a and $u_{a+\pi/2}$ are obtained by rotating $u_x(1)$ and $u_y(1)$ by the angle a:

(10) $$u_a = \cos a u_x(1) + \sin a u_y(1),$$

(11) $$u_{a+\pi/2} = -\sin a u_x + \cos a u_y(1).$$

It is then straightforward to compute the effect of S_a on $u_x(1)$ and $u_y(1)$:

(12) $$S_a u_x(1) = \cos 2a u_x(1) + \sin 2a u_y(1),$$

(13) $$S_a u_y(1) = \sin 2a u_x(1) - \cos 2a u_y(1).$$

The operator T_b is constructed in the same way, and

(14) $$T_b u_x(2) = \cos 2b u_x(2) + \sin 2b u_y(2),$$

(15) $$T_b u_y(2) = \sin 2b u_x(2) - \sin 2b u_y(2).$$

The quantum mechanical counterpart of Eq. (6) is obtained by taking the expectation value of the operator product $S_a T_b$ in the quantum mechanical state ϕ of Eq. (7):

(16) $$\begin{aligned} E_\phi(a, b) &= \langle\phi|S_a T_b|\phi\rangle = \tfrac{1}{2}\langle u_x(1)u_x(2) + u_y(1)u_y(2)| \\ &\quad [\cos 2a u_x(1) + \sin 2a u_y(1)][\cos 2b u_x(2) + \sin 2b u_y(2)] \\ &\quad + [\sin 2a u_x(1) - \cos 2a u_y(1)][\sin 2b u_x(2) - \cos 2b u_y(2)]\rangle \\ &= \cos 2(b-a). \end{aligned}$$

If we now choose a', b', a'', b'' to be respectively $\pi/4$, $\pi/8$, 0, and $3\pi/8$, then

(17) $$E_\phi(a', b') = E_\phi(a', b'') = E_\phi(a'', b') = -E_\phi(a'', b'') = 0.707,$$

and therefore

$$E_\phi(a', b') + E_\phi(a', b'') + E_\phi(a'', b') - E_\phi(a'', b'') = 2.828,$$

in disaccord with Inequality (4) (Bell's Inequality).   Q.E.D.

More than ten experimental tests of Bell's Inequality have been performed by examining the correlation of linear polarizations of photon pairs, as outlined in the preceding paragraph, and several other tests have also been carried out.[2] In all these experiments the analyzers are separated by distances of the order of a meter or more, so that no obvious mechanism would exist whereby Parameter Independence or Outcome Independence would be violated. But it is also highly desirable to exclude the possibility of a mechanism which is not obvious, and this exclusion can be achieved only if the events of analysis have space-like separation and hence cannot be directly connected causally according to Relativity Theory. Only the experiment of Aspect, Dalibard, and Roger (1982) has realized this desideratum. In their experiment the choice between the values a' and a'' of the analyzer of photon 1, and between the values b' and b'' of the analyzer of photon 2, is effected by acousto-optical devices which switch from one value to the other in 10 nanoseconds; whereas the switch-analyzer assembly for photon 1 is separated from that for photon 2 by about 13 meters, which can be traversed by a relativistically permitted signal in no shorter time interval than 40 nanoseconds. One would therefore antecedently expect both Parameter and Outcome Independence to hold, and moreover the distribution w over the space K of complete states to be independent of the parameters. Aspect et al. found, however, that their measured expectation values $E(a, b)$ violated Inequality (4) by 5 standard deviations, but were in good agreement with the predictions of quantum mechanics. If one disregards certain loopholes (which will be discussed below), then this experiment constitutes a spectacular confirmation of quantum mechanics at a point where it seems to be endangered, as well as a spectacular demonstration that there is some nonlocality in the physical world.

Since Bell's Inequality is violated by the results of Aspect et al., and the Inequality follows from Parameter and Outcome Independence together with the independence of w from the parameter values, one of these three premises must be false, and it is important to locate the false one. The natural way to obtain this information is to examine the implications of quantum mechanics, which after all was brilliantly confirmed

---

2. Summaries of experiments up to 1978 are given by Clauser and Shimony (1978) and later ones by Redhead (1987), pp. 107ff. Two important recent tests of Bell's Inequality, using photon pairs produced by parametric down-conversion, are Ou and Mandel (1988) and Shih and Alley (1988).

by Aspect et al., as well as by most of the other experiments inspired by Bell's Theorem.

Outcome Independence is violated by the quantum mechanical predictions based upon ϕ of Eq. (7). Suppose that the angles a and b of the two polarization filters are both taken to be 0, i.e., their transmission axes are both along the x direction. The conditional probability of photon 2 passing through its filter if photon 1 passes through its filter is 1, but it is 0 if photon 1 fails to pass through its filter. Since these two conditional probabilities are different from each other, it is impossible for both of them to equal the unconditioned probability that photon 2 will pass through its filter (which, in fact, is obviously $\frac{1}{2}$). Thus Outcome Independence, as defined above, is violated. It should be noted that a violation of Outcome Independence is predicted on the basis of any quantum state which is "entangled" (in Schrödinger's locution), that is, not expressible as a product of a quantum state of particle 1 and a quantum state of particle 2. For any entangled state of a two-particle system can be written in the form [3]

$$\psi = \sum_i c_i u_i(1) v_i(2),$$

where the $u_i(1)$ are orthonormal, the $v_i(2)$ are orthonormal, the sum of the absolute squares of the expansion coefficients c_i is unity, and the sum contains at least two terms with non-zero coefficients. By constructing self-adjoint operators S and T of which the u_i and the v_i are eigenstates with distinct eigenvalues, one obtains a violation of Outcome Independence.

The quantum mechanical predictions do not violate Parameter Independence if the Hamiltonian of the composite system can be written in the form

(18) $H_{tot} = H_1 + H_2,$

where H_1 is the Hamiltonian of particle 1 alone (in the environment to which it is exposed), and H_2 is the Hamiltonian of particle 2 alone (in the environment to which it is exposed), with no interaction Hamiltonian, and with no influence of particle 1 upon the environment of particle 2 and conversely. If the composite system $1+2$ is prepared at the initial time 0 in the state $\phi(0)$, then the state $\phi(t)$ at a later time t is determined by the Hamiltonian of Eq. (18) and also $\phi(0)$. It is straightforward to prove [4] that the expectation value of any self-adjoint operator S on the space of states of particle 1 is independent of H_2, and the expectation value of any self-adjoint operator T on the space of states of particle 2 is independent of H_1. Now the choice of a parameter a of the analyzer of particle 1 is

3. See, for example, von Neumann (1955), pp. 431–4.
4. Eberhard (1977); Ghirardi, Rimini, and Weber (1980); and Page (1982).

effectively the choice of the Hamiltonian of particle 1, and likewise concerning the choice of parameter b. Parameter Independence follows. This general argument may be made more intuitive by considering the special case of a pair of photons with ϕ of Eq. (7) as the state at time 0. At time t photon 1 impinges upon a polarization filter with one of two orientations of its transmission axis: (i) $a = 0$, or (ii) $a = \pi/4$. In either case, the filter upon which photon 2 will impinge will be taken to have its transmission axis along the x direction, i.e., parameter b is 0. We calculate the probability that photon 2 will pass through the filter in each of the two cases.

(i) Photon 1 has probability $\frac{1}{2}$ of passing through the filter with $a = 0$, in view of Eq. (7), and if it does so the term $u_x(1)u_x(2)$ is picked out of the superposition, so that the conditional probability that photon 2 will pass through its filter is 1. Photon 1 also has probability $\frac{1}{2}$ of not passing, in which case the term $u_y(1)u_y(2)$ is picked out, and the conditional probability that photon 2 will pass its filter is 0. The net probability of passage of photon 2 is $\frac{1}{2} \cdot 1 + \frac{1}{2} \cdot 0 = \frac{1}{2}$.

(ii) It is useful to rewrite Eq. (7) in the equivalent form

(19) $\phi = 2^{-1/2}[u_{x'}(1)u_{x'}(2) + u_{y'}(1)u_{y'}(2)],$

where x' is the direction in the x-y plane making an angle $\pi/4$ to both the x and y directions, and y' is perpendicular to x' in the x-y plane. (The equivalence of Eq. (19) to Eq. (7) follows from Eqs. (10) and (11) and their counterparts for photon 2.) There is probability $\frac{1}{2}$ that photon 1 will pass through its filter, picking out the term $u_{x'}(1)u_{x'}(2)$, in which case the conditional probability that photon 2 will pass through its filter is $\cos^2 \pi/4$. There is also probability $\frac{1}{2}$ that photon 1 will not pass through its filter, in which case the term $u_{y'}(1)u_{y'}(2)$ is picked out, and the conditional probability that photon 2 will pass through its filter is $\cos^2 3\pi/4$. The net probability of passage of photon 2 is $\frac{1}{2} \cdot \frac{1}{2} + \frac{1}{2} \cdot \frac{1}{2} = \frac{1}{2}$. The equality of the net probabilities of passage in cases (i) and (ii) illustrates Parameter Independence.

Since the standard quantum mechanical treatment of polarization correlation assigns the same quantum state to all photon pairs of the ensemble of interest (either the ϕ of Eq. (7) or an appropriate variant of it), there is no question of a quantum mechanical violation of the third premiss utilized in deriving Bell's Inequality (i.e., that the distribution over the complete states is independent of the parameters a and b).

It is very interesting now to consider the relations between violations of Parameter Independence and Outcome Independence and relativistic locality. Suppose that a violation of Parameter Independence occurred in the situation schematized by Figure 1 because for some k and m

(20) $p^1(m/k, a, b') \neq p^1(m/k, a, b'').$

Then one binary unit of information can be transmitted from the location of the second analyzer to the location of the first analyzer by making the choice between b' and b'' at the former location, in the following way. Aspect shows that the choice between b' and b'' can be made extremely quickly. We can also suppose (as a thought experiment) that a large number of pairs of particles $1+2$ are prepared in a time which is short compared to the time needed to choose between b' and b'', and also that the complete state of each of these pairs is k. Then the difference in probability in Inequality (20) will with near certainty, by the law of large numbers, produce a clear difference between the statistics of occurrence of the value s_m conditional upon the two choices of the parameter b. Hence with near certainty an observer of the outcomes s_m can infer whether the choice made at the other analyzer was b' or b''. By the hypothetical arrangement, this binary unit of information is transmitted at superluminal speed between the two analyzers. Hence in principle a violation of the relativistic upper limit upon the speed of a signal can be obtained by exploiting failure of Parameter Independence in a situation where the analyses of the two particles are events with space-like separation. If there is a violation of Outcome Independence, a binary unit of information can also be transmitted, but it is easy to see that the transmission is slower than the speed of light. Suppose that for some k and m

(21) $$p^1(m/k, a, b, n') \neq p^1(m/k, a, b, n'').$$

Again prepare a large number of pairs $1+2$ in the state k, and for each pair analyze particle 2 with the same parameter setting b. While the analysis is being performed, particle 1 is to be placed "on hold," e.g., by being kept in a circular light guide. An antecedent agreement is made that particle 1 will be released only if the result of analysis of particle 2 is $t_{n'}$ or $t_{n''}$ but not both, and that a uniform decision will be made for all the pairs $1+2$. An observer of the statistics of s_m can then infer with near certainty whether the choice has been made to release particles 1 of which the partners are analyzed with result $t_{n'}$ or with result $t_{n''}$, since the difference in probability in Inequality (21) will, by the law of large numbers, produce a difference in the statistics with near certainty. The transmission of a binary unit of information in this way, however, will be subluminal, because the analysis of particle 2 must be completed, then the result of the analysis must be transmitted to the ring where particle 1 is "on hold," then particle 1 must be released, then it must propagate towards its analyzer, and finally it must be analyzed. Clearly, this complex process takes longer than a straight radar signal between the two analyzers.

In an experimental arrangement like that of Aspect et al. a violation of either Parameter Independence or of Outcome Independence produces some tension with the theory of relativity. The violation of Parameter Independence seems to be the more serious of the two, because it entails the possibility in principle of superluminal signalling. The fact that quantum mechanics does not violate Parameter Independence but does violate Outcome Independence is most remarkable on two counts: it does show that quantum mechanical entanglement can be responsible for a kind of causal relation between two events with space-like separation, but also that quantum mechanics can "coexist peacefully" with relativity theory because of the impossibility of exploiting entanglement for the purpose of superluminal communication. By using the locution "peaceful coexistence" I do not wish to convey the impression that there is nothing problematic in the state of affairs which has been exhibited. A deeper analysis is certainly desirable. It is possible that a deepened understanding of space–time structure will be required in order to clarify quantum mechanical nonlocality. Or it may be that the concept of "event," which has been borrowed from pre-quantum physics, will have to be radically modified. However, I have not yet seen a promising development of either of these suggestions.

Because the implications of Bell's theorem and of the experiments which it inspired are philosopically momentous, it is important to pay attention to the loopholes in the experimental reasoning.

The first loophole is due to the periodicity of the switches which Aspect et al. employed to choose between a' and a'' and between b' and b''. The switches are not randomly turned off and on, but rather operate periodically with a total period of 20 nanoseconds. Even though relativity theory does not permit a direct causal connection between contemporaneous settings of the switches (where "contemporaneous" must of course be understood relative to some definite frame of reference, such as that of the laboratory), the periodicity may enable clever demons located in one analyzer to *infer* the contemporaneous setting of the other switch and to regulate the outcome of analysis of the particle accordingly. The attribution of such a process of inductive reasoning to the demons would not violate relativity theory. In order to block this loophole it would be necessary to operate the switches stochastically. It has been suggested by Clauser, for example, that each should be controlled by the arrival of starlight gathered by a telescope pointed to a distant galaxy. Blocking the periodicity loophole seems to be experimentally feasible in principle, but it would greatly complicate an experiment that is already difficult and delicate. It remains to be seen whether any experimenter is sufficiently motivated to make the great effort that would be required. See also Zeilinger (1986).

The second loophole is due to the fact that actual particle detectors are not 100% efficient. In the foregoing discussion of Figure 1 it was tacitly assumed that if the outcome of analyzing particle 1 is s_m (e.g., the particle passes into channel m), then this fact can be known with certainty because the particle detectors are ideally efficient; and likewise concerning the analysis of particle 2. In the polarization correlation tests of Bell's Inequality the photodetectors were less than 20% efficient, and therefore fewer than 4% of the photon pairs that jointly pass through their respective filters are actually detected. It is not inconceivable that the passage rates satisfy Bell's Inequality, but that the counting rates agree with the predictions of quantum mechanics (in disaccord with the Inequality) because of peculiarities in the way that the complete states k determine the probability of detection. There are, in fact, several models[5] which preserve Parameter Independence and Outcome Independence and nevertheless yield counting rates in agreement with the predictions of quantum mechanics. Inefficiency of the particle detectors is crucial for these models.

The following argument shows that the detection loophole can be blocked in a polarization correlation experiment if technology improves and photodetectors of efficiency greater than 0.841 are constructed. The foregoing realization of the schema in Figure 1 can be modified by taking the analyzers not to be polarization filters, which allow photons polarized along the transmission axis to pass but absorb those polarized in the perpendicular direction, but rather Wollaston prisms, which allow the first set of photons to emerge in one ray (the "ordinary ray") and the second to emerge in another (the "extraordinary ray"). Then three outcomes of analysis of photon 1 can be distinguished: detection in the ordinary ray, detection in the extraordinary ray, and non-detection; and the values of s_m assigned to these three outcomes can be conventionally taken to be 1, −1, and 0. The outcomes t_n of analysis of photon 2 will likewise have three values 1, −1, and 0, with analogous interpretations. If Parameter and Outcome Independence are satisfied, and the distribution w is independent of the parameters, then Bell's Inequality (Inequality (4)) follows. The expression for the quantum mechanical expectation of the product of s_m and t_n is equal to the expectation value of the operator product $S_a T_b$, which was given by Eq. (16), multiplied by the probability of joint detection of a pair of photons emerging from the respective Wollaston prisms. If, for simplicity, we assume that the four photodetectors intercepting the ordinary and extraordinary rays from the two Wollaston prisms have the

---

5. Clauser and Horne (1974), and Marshall, Santos, and Selleri (1983). The earlier model of Pearle (1970) achieves agreement with quantum mechanics only if the probability that a pair will be detected once it has passed through the pair of filters has a rather special form $g(b-a)$, and the constant function (specifically with value η^2, as discussed above) is not of his required form; hence his model does not achieve all that those of Clauser and Horne and of Marshall et al. have established.

same efficiency η, then the probability of joint detection is η^2. Hence the expectation value of present interest is

(22) $$E_\phi^{\text{det}}(a, b) = \eta^2 \cos 2(b - a).$$

For the choice of angles made before Eq. (17) we have

(23) $$E_\phi^{\text{det}}(a', b') + E_\phi^{\text{det}}(a', b'') + E_\phi^{\text{det}}(a'', b') - E_\phi^{\text{det}}(a'', b'') = 2.828\eta^2.$$

Disaccord with Bell's Inequality results provided that

(24) $$\eta > 0.841.$$

Mermin and Schwarz (1982) and Garg and Mermin (1987) have shown that the detection loophole can be blocked if a less stringent constraint is placed upon the efficiency of the photodetectors, namely,

(25) $$\eta > 0.828,$$

but their arguments are more complex than the simple one just given and depend upon some additional (but empirically testable) symmetry assumptions. A molecular experiment designed to satisfy Inequality (25) was proposed by Lo and Shimony (1981) but has not been performed. An experiment satisfying this Inequality will probably be performed in the near future by E. Fry and collaborators (1992).

Finally, I wish to point out that even though I have used polarization correlation to discuss Bell's Inequality, there is nothing about the Inequality that is intrinsically restricted to polarization experiments. That should be obvious, in fact, from the generality of Figure 1 and of the proof given of Bell's Inequality. For example, one can use pairs of photons with entangled momentum states to test the Inequality by means of two-photon interferometry (Horne, Shimony, and Zeilinger 1989, 1990, and Rarity and Tapster 1990).

REFERENCES

Aspect, A., 1983, Thèse, Université de Paris–Sud (unpublished).
Aspect, A., Dalibard, J., and Roger, G., 1982, *Physical Review Letters,* 49: 1804.
Bell, J. S. 1964, *Physics,* 1: 195. Reprinted in Bell (1987).
Bell, J. S., 1971, in "Foundations of Quantum Mechanics," B. d'Espagnat, ed., Academic Press, New York, 171. Reprinted in Bell (1987).
Bell, J. S., 1987, "Speakable and Unspeakable in Quantum Mechanics," Cambridge University Press, Cambridge, England.
Clauser, J. F., Horne, M. A., Shimony, A., and Holt, R. A., 1969, *Physical Review Letters,* 26: 880.
Clauser, J. F. and Horne, M. A., 1974, *Physical Review,* D10: 526.
Clauser, J. F. and Shimony, A., 1978, *Reports on Progress in Physics,* 41: 1881.

Eberhard, P., 1977, *Nuovo Cimento,* 38B: 75.
Fry, E., 1992, unpublished manuscript, Physics Department, Texas A & M University.
Garg, A. and Mermin, N. D., 1987, *Physical Review,* D35: 3831.
Ghirardi, G. C., Rimini, A., and Weber, T., 1980, *Lettere al Nuovo Cimento,* 27: 293.
Horne, M. A., Shimony, A., and Zeilinger, A., 1989, *Physical Review Letters,* 62: 2209.
Horne, M. A., Shimony, A., and Zeilinger, A., 1990, in "Sixty-two Years of Uncertainty," A. Miller, ed., Plenum Press, New York, 113.
Jarrett, J., 1984, *Noûs,* 18: 569.
Lo, T. K. and Shimony. A., 1981, *Physical Review,* A23: 3003.
Marshall, T. W., Santos, E., and Selleri, F., 1983, *Physics Letters,* 98A: 5.
Mermin, N. D. and Schwarz, G. M., 1982, *Foundations of Physics,* 12: 101.
Neumann, J. v., 1955, "Mathematical Foundations of Quantum Mechanics," Princeton University Press, Princeton, N.J.
Ou, Z. Y. and Mandel, L., 1988, *Physical Review Letters,* 61: 50.
Page, D., 1982, *Physics Letters,* 91A: 57.
Pearle, P., 1970, *Physical Review,* D2: 1418.
Rarity, J. G. and Tapster, P. R., 1990, *Physical Review Letters,* 64: 2495.
Redhead, M., 1987, "Incompleteness, Nonlocality, and Realism," Clarendon Press, Oxford, England.
Shih, Y. H. and Alley, C. O., 1988, *Physical Review Letters,* 61: 2921.
Zeilinger, A., 1986, *Physics Letters A,* 118: 1.

COMMENT

The demonstration of Bell's Inequality presented in this paper belongs to a family of demonstrations – including the pioneering one of Bell (1964) – which assumes that there is a well-defined space of complete states of the particle pair and a well-defined probability distribution over this space when an experimental procedure for specifying an ensemble of pairs is given. H. P. Stapp has presented arguments for Bell's Inequality dispensing with these assumptions in a number of publications, including "S-matrix Interpretation of Quantum Theory," *Physical Review* D3: 1303 (1971), and "Quantum Nonlocality and the Description of Nature," in J. Cushing and E. McMullin (eds.), *Philosophical Consequences of Quantum Theory,* Notre Dame, IN: University of Notre Dame Press, pp. 154–74 (1989). There have been some serious criticisms, however, of Stapp's use of counterfactuals in his arguments, among them J. F. Clauser and A. Shimony, "Bell's Theorem: Experimental Tests and Implications," *Reports on Progress in Physics* 41: 1881 (1978), especially pp. 1898–1900, and B. d'Espagnat, "Nonseparability and the Tentative Description of Nature," *Physics Reports* 110: 202–64 (1984), especially pp. 224–26.

9
Contextual hidden variables theories and Bell's Inequalities

Noncontextual hidden variables theories, assigning simultaneous values to all quantum mechanical observables, are inconsistent by theorems of Gleason and others. These theorems do not exclude contextual hidden variables theories, in which a complete state assigns values to physical quantities only relative to contexts. However, any contextual theory obeying a certain factorisability condition implies one of Bell's Inequalities, thereby precluding complete agreement with quantum mechanical predictions. The present paper distinguishes two kinds of contextual theories, 'algebraic' and 'environmental', and investigates when factorisability is reasonable. Some statements by Fine about the philosophical significance of Bell's Inequalities are then assessed.

I. INTRODUCTION

In most of the discussions of Bell's Inequalities little notice is taken of the historical circumstances of their discovery. J. S. Bell was told by J. Jauch of the work of A. Gleason [1957] which showed that simultaneous values cannot be assigned to all observables of a quantum mechanical system in a way that respects their algebraic structure (except in the simple case in which the relevant Hilbert space has dimension less than three). Gleason's theorem thus precludes a type of hidden variables theory that has come to be called 'noncontextual'. The question of agreement between the statistical predictions of quantum mechanics and those of a noncontextual hidden variables theory does not even arise, since such a theory cannot be consistently formulated. Bell noticed [1966], however, that there is a large family of hidden variables theories, which have come to be called 'contextual', that are not precluded by Gleason's and related theorems. He found, in fact, that the hidden variables theories which had been developed in the literature in some detail were contextual, even though their authors had not articulated the distinction. Bell noticed further that one of these theories, that of Bohm [1952], postulated a peculiar kind of nonlocality or action at a distance in order to recover the quantum mechanical predictions concerning correlated, spatially separated systems of the

The work originally appeared in *British Journal for the Philosophy of Science* 35 (1984), pp. 25–45. Reprinted by permission of the publisher.

kind studied by Einstein, Podolsky, and Rosen [1935]. Bell asked whether such nonlocality is a necessary condition for the recovery of the statistical predictions of quantum mechanics. His positive answer to this question [1964] is called 'Bell's Theorem'. (The paper of 1966 was written earlier than that of 1964 but because of an editorial mishap was published later.) Bell proved the theorem by showing that a certain factorisability condition upon a hidden variables theory implies that certain statistical correlations between two spatially separated systems always satisfy an inequality which he formulated, while quantum mechanics implies that this inequality is sometimes violated. The factorisability condition is reasonable when the two systems are separated by typical laboratory distances (a few metres), and especially reasonable if the operations performed upon the two systems are events with space-like separation. Bell concluded that no physically acceptable contextual hidden variables theory could agree completely with the statistical predictions of quantum mechanics. Since the experiments inspired by his work overwhelmingly support quantum mechanics, it follows that no contextual hidden variables theory is viable unless it postulates a kind of nonlocality. A survey article by J. Clauser and A. Shimony [1978] gives details of the history of this topic up to its publication date, but it must be supplemented by the very important paper of Aspect, Dalibard, and Roger [1982].

One of the purposes of this paper has been achieved in the first paragraph, by pointing out that Bell's Theorem was the result of a question concerning the physical adequacy of contextual hidden variables theories. A further purpose is to examine the concept of contextual hidden variables theory in more detail than is common in the literature. It is a complex concept, and at least two very different types can be discriminated – to be called 'algebraic' and 'environmental'. The conceptual study will help to clarify the content and the philosophical significance of Bell's Theorem and the Inequalities involved in it.

In two recent papers A. Fine [1982a,b] presented some theorems concerning the relation between Bell's Inequalities and joint probability distributions, and on the basis of these theorems he drew some conclusions about the philosophical significance of the Inequalities. The theorems are correct and interesting, but I wish to challenge the statements about philosophical significance. Specifically, I believe that these statements neglect the importance of contextual hidden variables theories and of locality.

2. NONCONTEXTUAL HIDDEN VARIABLES THEORIES

The set $\mathcal{L}$ of 'eventualities' or 'propositions' concerning a physical system S is a convenient entity to study, because its mathematical structure is considerably simpler than that of the set of observables of S. I prefer the term 'eventuality', which was introduced by H. Stein [1972], because it does not invite the conflation of questions about the structure of $\mathcal{L}$ with questions of deductive logic. Each eventuality is a bivalent possibility – either true or false – if it has a definite value, but the value may be indefinite in certain states of S. $\mathcal{L}$ is assumed to be partially ordered with respect to a relation $\leq$, where $a \leq b$ intuitively means that whenever a is true, b also is. There is an orthocomplementation operation $'$ on $\mathcal{L}$, such that $(a')' = a$ and $a \leq b$ implies $b' \leq a'$. The intuitive interpretation is that a' is false if a is true, and a' is true if a is false, though the possibility is open that neither has a definite truth value. In the standard Hilbert space formulation of quantum mechanics $\mathcal{L}$ is assumed to be isomorphic to the lattice $\mathcal{L}_H$ of projectors (or equivalently of closed linear subspaces) of an appropriate Hilbert space $\mathcal{H}$ (*cf.* Mackey [1963]). Unless otherwise noted, this assumption will be made throughout the present paper. This isomorphism guarantees that the g.l.b. $\bigwedge a_i$ and the l.u.b. $\bigvee a_i$ ($i = 1, 2, \ldots$) with respect to the relation $\leq$ are well-defined, though of course the isomorphism to $\mathcal{L}_H$ is not a necessary condition for these σ-lattice properties to hold.

There is a standard procedure for defining observables in terms of the lattice $\mathcal{L}_H$, and a theorem of J. von Neumann establishes an isomorphism between the algebra of observables and the algebra of self-adjoint operators on $\mathcal{H}$ (see Varadarajan [1968], pp. 143–4). The eventualities can, in fact, be taken as a subclass of observables – *viz.*, bivalent observables with (conventionally) +1 representing truth and 0 representing falsity. The self-adjoint operators corresponding to the eventualities in accordance with von Neumann's theorem are the projectors, which of course have no other eigenvalues than 1 and 0. If e is an eventuality, E_e will denote the projector corresponding to it. For the purposes of this paper there will be no need to discuss further any observables other than eventualities.

G. Mackey [1963] introduced the important concept of a measure on $\mathcal{L}_H$, that is, a mapping $p \colon \mathcal{L}_H \to [0, 1]$ such that (i) $p(1) = 1$, where 1 is the identity operator (and hence the projector that has the whole of $\mathcal{H}$ as its range), and (ii) if the a_i are such that $a_i \leq a_j'$ whenever i and j are distinct ($i, j = 1, 2, \ldots$), then

$$p(\bigvee a_i) = \sum p(a_i).$$

Clearly, Mackey's concept is similar in some ways to that of a classical probability measure, except that it is adapted to the lattice structure of $\mathcal{L}_H$, which in general is not Boolean (as the lattice of subsets of a phase space is). Quantum mechanics supplies some exemplary measures on $\mathcal{L}_H$. Let $\psi \in \mathcal{H}$ and have norm 1. If p_ψ is defined for all $e \in \mathcal{L}_H$ by the equation

$$p_\psi(e) = (\psi, E_e \psi),$$

it can easily be shown to satisfy conditions (i) and (ii) of Mackey's definition. Convex combinations of denumerably many measures of the form p_ψ also satisfy these conditions.

Gleason proved the fundamental theorem that if the dimension of $\mathcal{K}$ is greater than 2 and p is a measure, then there is a finite or denumerable sequence of normalized vectors ψ_i and of non-negative real numbers w_i summing to 1, such that for all $e \in \mathcal{L}_H$

$$p(e) = \Sigma\, w_i p_{\psi_i}(e).$$

A corollary is the non-existence of a p which assigns only 1 or 0 to each $e \in \mathcal{L}_H$ (sometimes called a 'dispersion-free measure'). The corollary was proved directly, without proceeding via Gleason's theorem, by Bell [1966], S. Kochen and E. Specker [1967], and F. Belinfante [1973]. These proofs show in fact that there are finite sublattices of $\mathcal{L}_H$ on which dispersion-free measures cannot exist.

Gleason's theorem and its corollary doomed the program of noncontextual hidden variables theories. This program envisaged a space Λ of complete states of S, each member λ of which determines a dispersion-free measure on $\mathcal{L}_H$. It envisaged, furthermore, a σ-algebra $\Sigma(\Lambda)$ of subsets of Λ and classical probability functions ρ such that each triple $\langle \Lambda, \Sigma(\Lambda), \rho \rangle$ constitutes a classical probability space. And finally it envisaged that for each normalized $\psi \in \mathcal{K}$ there is a classical probability measure function ρ_ψ on $\Sigma(\Lambda)$ such that for all $e \in \mathcal{L}_H$

(1) $$p_\psi(e) = \int_\Lambda p_\lambda(e)\, d\rho_\psi.$$

This program was actually achieved (in the sense of a mathematical model) by Bell [1966], Kochen and Specker [1967], and Pearle [1965], when the dimension of $\mathcal{K}$ is 2. It fails when the dimension of $\mathcal{K}$ is greater than 2 because of Gleason's theorem and its corollary. The failure, of course, occurs in the first step of the program, for there are no dispersion-free states p_λ, hence no space Λ of complete states having the desired properties; and therefore one cannot even pose the question of whether there is a ρ_ψ over $\Sigma(\Lambda)$ satisfying Eq. (1).

3. CONTEXTUAL HIDDEN VARIABLES THEORIES

Contextual hidden variables theories are not excluded by these negative results. This kind of hidden variables theory was exemplified by the models of L. de Broglie [1926], D. Bohm [1952], and N. Wiener and A. Siegel [1953], but the significance of the difference from noncontextual theories was first enunciated by Bell [1966]. Again a space Λ of complete states and a σ-algebra $\Sigma(\Lambda)$ of subsets of Λ are envisaged. Now, however, p_λ is conceived to be a function of two variables e and C, where e is a bivalent possibility (perhaps an eventuality of S, but other options will later be noted), and C is the 'context' in which the value of e is determined. Bell defends the idea of contextual hidden variables theories by saying, 'The result of an observation may reasonably depend not only on the state of the system (including hidden variables) but also on the disposition of the apparatus' [1966, p. 451]. The character of the contexts, however, is not the same in all expositions.

Gudder [1970] developed in considerable mathematical detail a contextual hidden variables theory in which the context C is taken to be some subset of the lattice $\mathcal{L}_H$. It is reasonable to call this type of theory 'algebraic'. Specifically, Gudder proposes to let C be a maximal Boolean sub-σ-algebra of $\mathcal{L}_H$, and $p_\lambda(e, C)$ is well defined if and only if $e \in C$. When $\mathcal{K}$ has dimension greater than 2, an eventuality e can belong to more than one maximal Boolean sub-σ-algebra, and therefore the specification of C is not redundant. It is assumed that $p_\lambda(e, C)$ is either 1 or 0 (hence the measure is dispersion-free), and Mackey's conditions (i) and (ii) are satisfied within each C. If e belongs both to C and C', the possibility is left open that $p_\lambda(e, C) \neq p_\lambda(e, C')$. Not only do Gleason's theorem and its corollary not preclude the existence of such p_λ, but the remainder of the hidden variables program can be carried out: for each normalised $\psi \in \mathcal{K}$ there is a classical probability measure ρ_ψ such that

(1') $$p_\psi(e) = \int_\Lambda p_\lambda(e, C)\, \mathrm{d}\rho_\psi.$$

Hence Gudder has provided a mathematical existence proof of a contextual hidden variables theory which agrees with the statistical predictions of quantum mechanics, whatever the quantum state of S may be.

Physically, however, Gudder's proposal is not fully satisfactory, since rarely does an actual apparatus single out a maximal Boolean sub-σ-algebra along with the eventualities which are to be observed. If ideal apparatus should do so, then Eq. (1') assures agreement with quantum mechanics, but it is not obvious that agreement would be achieved when the apparatus is non-ideal. The same difficulty would arise concerning other contextual hidden variables theories of the algebraic type.

It is desirable, therefore, to consider another type of contextual hidden variables theory, which may appropriately be called 'environmental', in which the context C is a state of the physical environment (including apparatus) with which the system S interacts. (Heywood and Redhead [1983] also use the term 'environmental', but in a somewhat different sense.) In principle C could be the complete state of the environment.

Then according to a deterministic theory C together with the complete state λ of the system S determine the value of any relevant bivalent possibility e. In practice, of course, the environment is very imprecisely known and characterized. Consequently, there is motivation to let C be a 'coarse-grained state' of the environment, such as is specified by a macroscopic description of the apparatus. It is reasonable to assume that the complete state of S, together with a coarse-grained state of the environment, suffice to determine a probability $p_\lambda(e, C)$ that the bivalent possibility e will have a designated value (say $+1$). Good experimental design in microphysical investigations is, in fact, intended to establish situations in which such probabilities are well defined: situations which exhibit sensitivity to the microscopic features of the system S of interest, but sensitivity also to the macroscopic features of the apparatus. There is no *a priori* guarantee that a hidden variables theory will achieve these two types of sensitivity, so that $p_\lambda(e, C)$ will be well defined. One that does not do so, however, does not seem to be capable of making statistical predictions about the outcomes of actual experiments and therefore would be of little interest to working scientists.

4. BELL'S THEOREM

The question may now be raised: can a contextual hidden variables theory of the environmental type account for all the experimental data that quantum mechanics explains? The question is not sharply formulated, because the class of theories is ill-defined. I suppose, however, that the question could be sharpened in a natural way, and the answer would then be positive: the environmental theory would be linked to an algebraic theory, and the agreement of the latter with quantum mechanics could be established along the lines indicated by Bell and Gudder. Bell, rather elliptically, suggests such a procedure at the end of his article of 1966. He does not pursue details on this point, since a question of principle intrudes. He asks effectively (not using this language) whether a contextual hidden variables theory of the environmental type is able to recover the statistical predictions of quantum mechanics without introducing physically undesirable features. His first inequality [1964] and other inequalities inspired by his work (which I shall call collectively 'Bell's Inequalities') provide a strong negative answer to this question of principle. This result is especially remarkable in view of the looseness of the characterisation of theories of environmental type, and it is possible only because the inequalities can be demonstrated with great generality. The negative answer removes most of the physical motivation to characterise this type of hidden variables theory with precision and to investigate them in detail.

The inequality which is most convenient for the purposes of this paper is that of Clauser and Horne [1974], but I shall state the conditions under which it holds in a new way in order to emphasise generality.

Theorem. *Let a be a parameter which can take on the values a', a", and b be a parameter which can take on the values b', b". Let $\langle \Lambda, \Sigma(\Lambda), \rho \rangle$ be a probability space: i.e., Λ is a nonempty set, $\Sigma(\Lambda)$ is a σ-algebra of subsets of Λ, and ρ is a probability measure on $\Sigma(\Lambda)$, such that for each value of a and b there are random variables $p_\lambda^1(a)$, $p_\lambda^2(b)$, and $p_\lambda(a, b)$ with values between 0 and 1 for each $\lambda \in \Lambda$. Suppose that*

(2) $$p_\lambda(a, b) = p_\lambda^1(a) p_\lambda^2(b).$$

Then

(3) $$-1 \le p(a', b') + p(a', b'') + p(a'', b') - p(a'', b'') - p^1(a') - p^2(b')$$
        $$\le 0,$$

where

(4a) $$p^1(a) = \int_\Lambda p_\lambda^1(a) \, d\rho,$$

(4b) $$p^2(b) = \int_\Lambda p_\lambda^2(b) \, d\rho,$$

(4c) $$p(a, b) = \int_\Lambda p^2(a, b) \, d\rho.$$

Comments. (*i*) Inequality (3) will be called 'the Bell/CH Inequality'. The proof of the theorem is given by Clauser and Horne.

(*ii*) The statement given of the theorem is purely mathematical, but it is obviously amenable to physical interpretation. Specifically, the set Λ can be interpreted as the space of complete states of a system S. The parameters a and b can label bivalent possibilities A_a and B_b (with values 1 and 0, for example), and $p_\lambda^1(a)$ and $p_\lambda^2(b)$ are respectively the probabilities that A_a and B_b will have the value 1 if the complete state is λ, while $p_\lambda(a, b)$ is the probability that A_a and B_b both will be 1 if the complete state is λ. ρ is the probability distribution over $\Sigma(\Lambda)$ associated with some procedure for preparing an ensemble of replicas of S.

(*iii*) There remain various options for interpreting A_a and B_b. (a) One obvious option is to interpret these as eventualities of S: e.g., if S is a photon propagating in the $\hat{z}$-direction, A_a can be the eventuality that S is linearly polarized along a line that makes the angle a relative to a fixed $\hat{x}$-direction, and B_b has a similar meaning for a photon propagating in the $-\hat{z}$-direction. $p_\lambda^1(a) = 1$ means that A_a is true when λ is the complete state of S, $p_\lambda^2(b) = 1$ means that B_b is true in this state, and $p_\lambda(a, b) = 1$

means that the conjunction of A_a and B_b is true in this state; in each instance a value 0 means falsity. If $p_\lambda^1(a)$ and $p_\lambda^2(b)$ can only take on the values 1 and 0, and do so in a way that is compatible with Mackey's conditions (i) and (ii) on measures, then one has a fragment of a noncontextual hidden variables theory – not the whole of such a theory unless every eventuality in $\mathcal{L}_H$ is identifiable with one of the A_a and B_b. (b) Another option is to let a be an ordered pair (e, C), where C is a maximal Boolean sub-σ-algebra of $\mathcal{L}_H$ and $e \in C$. If $p_\lambda^1(a)$, etc. can assume only the values 1 and 0 and satisfy Mackey's conditions (i) and (ii) within each C, then we have a fragment of a contextual hidden variables theory of algebraic type. (c) A third option is that A_a and B_b refer to bivalent possibilities under prescribed experimental conditions. For instance, there could be an experimental arrangement in which two photons, propagating respectively in the $\hat{z}$ and $-\hat{z}$ directions, impinge perpendicularly upon sheets of polaroid, which are oriented with transmission axes making arbitrary angles to the fixed $\hat{x}$-axis. A_a can be interpreted as the bivalent possibility that the first photon will be transmitted if the first sheet of polaroid is oriented at the angle a, with $+1$ signifying transmission and 0 signifying non-transmission; and B_b has an analogous interpretation. Less elliptically, specifying the angle a is tantamount to a choice of a pair (e, C), where e is a bivalent possibility and C is a state (normally macroscopically specified) of part of the environment; and similarly with b. Since the type of environment and type of binary possibility are antecedently fixed in discussions of correlation experiments, the only variables needed to complete the macroscopic description of the experiment are the angles of orientation of the analysers. We have here a fragment of a contextual hidden variables theory of environmental type. Since the contexts C are not normally complete states of the environment, one does not expect $p_\lambda^1(a)$, $p_\lambda^2(b)$, and $p_\lambda(a, b)$ to be restricted to 1 and 0, but only to have values lying in the interval $[0, 1]$.

(*iv*) Eq. (2) is referred to by Clauser and Horne, following Bell, as the 'locality condition'. In the interest of generality, however, it is desirable to avoid a locution which suggests the locality of relativity theory, and therefore I shall follow Garuccio and Selleri [1978] and Fine [1982a,b] in calling it the 'factorisability condition'. Whether it is a reasonable condition to impose depends upon the options concerning A_a and B_b, and also upon our knowledge of the physical world. There are two very different cases in which it is clearly a reasonable condition. (a) In a noncontextual hidden variables theory all eventualities are assigned definite truth values in a complete state, and all observables possess definite values simultaneously. If both A_a and B_b are true, so is their conjunction; and then the interpretation of $p_\lambda^1(a)$, $p_\lambda^2(b)$, and $p_\lambda(a, b)$ automatically ensures that the factorisability condition is satisfied. This argument does not depend upon the nature of S (*e.g.*, whether or not S consists of two spatially separated parts S_1 and S_2), or on the character of the eventualities A_a and B_b (whether the first refers to part S_1 and the second to part S_2, whether the projectors corresponding to A_a for different values of a are commuting, etc.). The fact that the factorisability condition is satisfied in so general a way in a noncontextual hidden variables theory provides special motivation for avoiding the weighted term 'locality condition'. (b) Under the right conditions Eq. (2) will hold for a contextual hidden variables theory of environmental type. The conditions are the ones commonly stated in deriving Bell's Inequalities. The system S consists of two spatially separated parts S_1 and S_2. The relevant parts of the environment are two bivalent analysers, each with a channel labelled $+$ and a channel labelled $-$, one analyser for S_1 and the other for S_2. Each analyser has an adjustable feature (typically an orientation) parametrised by a and b respectively. The analysers are well separated on a laboratory scale (typically by several meters), with precautions taken that the choice of parameter of each analyser not influence the analysis performed by the other analyser. Let $p_\lambda^1(a)$ be the probability that S_1 goes into the $+$ channel of its analyser if the parameter has value a and the complete state of S is λ; let $p_\lambda^2(b)$ be the probability that S_2 will go into the $+$ channel of its analyser if the parameter has value b and the complete state of S is λ; and let $p_\lambda(a, b)$ be the probability that both S_1 and S_2 will go into their respective $+$ channels, given a, b, and λ. Under these conditions it is reasonable that $p_\lambda(a)$ be independent of b, as the notation indicates, and that $p_\lambda(b)$ be independent of a. Also, since λ is the complete state of $S_1 + S_2$ all correlations between the behaviour of S_1 and that of S_2 should be carried by λ. If λ were not given, then the passage of S_1 into its $+$ channel would convey some (usually probabilistic) information relevant to the behaviour of S_2, and conversely, but this is not the case if λ is given and if there is no direct causal connection between the two parts of the experiment. It follows that the factorisability condition, Eq. (2), is reasonable. Of course, some physical influence could be exerted by the choice of b upon the behaviour of S_1 and by the choice of a upon the behaviour of S_2, in spite of precautions to the contrary. In order to preclude unnoticed causal influence, it suffices (provided that relativity theory is true) for the compound event which consists of the choice of a and the passage of S_1 into a channel to have space-like separation from the corresponding compound event concerning S_2. The terminology 'locality condition' for Eq. (2) is certainly appropriate when this additional precaution about space-like separation is taken; but I think that in a commonsense way it is even appropriate without this additional precaution, provided that care is taken to prevent manifest causal influence between the two sides of the experiment. For a discussion of these points see the 1981 publications of Fine and Shimony.

The term 'Bell's Theorem' means more than Bell's Inequalities. It also comprises the fact that there are some situations in which the quantum mechanical predictions violate Ineq. (3) (and other Bell's Inequalities), so that hidden variables theories satisfying the factorisability condition cannot completely agree with the statistical predictions of quantum mechanics. This disaccord is, of course, very interesting as it stands. But of even greater interest is the fact that there are situations in which the experimental evidence violates Ineq. (3) or others of Bell's Inequalities. The experimental situation is chosen to be one in which the quantum mechanical prediction violates the Inequalities, for it is anticipated that these predictions will be correct. That expectation has been overwhelmingly fulfilled. The most impressive experiment is that of Aspect, Dalibard, and Roger [1982], which obtained results violating one of Bell's Inequalities, but agreeing with quantum mechanics, when the two compound selection-analysis events have space-like separation.

5. PHILOSOPHICAL SIGNIFICANCE

What can one conclude, with all due regard for fallibility?

First, no noncontextual hidden variables theory can be completely correct. This is so, because the factorisability condition, Eq. (2), is automatically satisfied in the noncontextual case.

Second, no contextual hidden variables theory of the environmental type can be completely correct, unless Eq. (2) is abandoned. And the experimental situation of Aspect *et al.* does not permit the abandonment of Eq. (2) without a radical change of world view. Whether the requisite change is a revision of relativistic space–time structure is too large a question to be discussed here. I have elsewhere [1978] suggested the possibility of 'peaceful coexistence' between the quantum mechanical nonlocality exhibited in the correlation experiments and relativistic space–time structure, and hope to explore the issue further. Even 'peaceful coexistence', however, requires a radical change of our conception of the way in which events occur in the theatre of space–time.

I say nothing about contextual hidden variables theories of the algebraic type, just because their connection with experiment is difficult to discuss at the level of abstraction on which they have been proposed. Silence on this point does not mean, however, that they offer a viable loophole from the results of the correlation experiments. If detailed connection is established between a theory of algebraic type and experiments, that theory will be subject to disconfirmation by the correlation experiments unless it violates the factorisability condition.

Since the predictions of quantum mechanics have been confirmed and those of its most promising alternatives have been disconfirmed, we have stronger reasons than ever for believing that the following peculiar implications of quantum mechanics will remain permanently as established parts of physical theory: (1) In any state of a physical system S there are some eventualities which have indefinite truth values. (2) If an operation is performed which forces an eventuality with indefinite truth value to achieve definiteness (whether this happens only by measurement or also by some other means is one of the outstanding problems of the foundations of quantum mechanics) the outcome is a matter of chance. (3) There are 'entangled systems' (in Schrödinger's phrase) which have the property that they constitute a composite system in a pure state, while neither of them separately is in a pure state.

If a noncontextual hidden variables theory had been viable, then some aspects of the world view of classical physics would have achieved a new lease of life: definiteness of truth values of all eventualities, elimination of chance at the fundamental level, no entanglement.

If a contextual hidden variables theory of environmental type had been viable, then a world view definitely different from that of quantum mechanics, and yet different in some respects from that of classical physics would have been suggested. In this world view a system S has a definite complete state, and so does its environment. The outcome of an experiment performed upon S is determined cooperatively by these two complete states. Hence, this kind of theory does not have the fundamental stochastic character of quantum mechanics. (A further generalisation, which would permit the outcome to be a matter of chance even when the state of the entire universe is given, will not be considered here, since our concern is with the possibility of world views which differ sharply from that of quantum mechanics.) The entanglement of systems can be avoided if the complete state of each subsystem of a composite system is definite even when the quantum state (which from the standpoint of a hidden variables theory contains only partial information) cannot be factorised into definite quantum states for each subsystem. Instead of entanglement, which is a kinematical feature of a composite system in quantum mechanics, there would be causal interaction between components, which is a dynamical matter. The outcome of an experiment is a cooperative effect, just as it is conceived to be in classical physics. Classically, however, the system S possesses a definite array of properties, and the interaction of S with measuring apparatus produces an effect which is indexical of one or more of these properties – perfectly so if the apparatus is ideal, imperfectly with actual apparatus. A contextual hidden variables theory of environmental type differs from classical physics by regarding as relational many features which in classical physics, and also in noncontextual hidden variables theories, are thought to be intrinsic to S. Of course, exactly which features are fixed by the complete state λ of S, and which are fixed by λ together with the complete state of the environment, cannot be told without giving the details of the hidden variables theory. The contextual hidden variables theories which are most relevant to the correlation experiments performed so far are those which conceive the components of linear polarisation of photons, or the components of spin of certain atoms, to depend both upon the state of the particle involved and upon the state of the environment, and not upon the former by itself.

The philosophical significance of Bell's Inequalities, in my opinion, is that they permit a near decisive test of those world views which are contrary to that of quantum mechanics. Bell's work made possible, therefore, some near decisive results in experimental metaphysics. These results certainly do not imply that quantum mechanics as it is now formulated will never be displaced or ameliorated by any physical theory. After all, only a handful of quantum mechanical predictions are tested in the correlation experiments. This handful should not be under-estimated, however, since they concern salient, sensitive points – points at which a clash of world view occurs. The confirmation of the quantum mechanical predictions at these points and the disconfirmation of one or another of Bell's Inequalities provide strong grounds for confidence about future physical theories: any theory which displaces but improves upon quantum theory will preserve the indefiniteness of eventualities, the fundamental role of chance, the entanglement of systems, and a kind of nonlocality.

The whole vast body of experimental results which have probed and confirmed quantum mechanics can in fact be regarded as contributions to experimental metaphysics, and the mathematical analysis of the structure of quantum mechanics by Gleason [1957], Kochen and Specker [1967], and others has helped to make explicit its metaphysical content. Notably, the structure of the lattice of eventualities for all but the simplest quantum systems (all for which the appropriate Hilbert space has dimension greater than 2) precludes a noncontextual hidden variables theory, hence the possession of definite values simultaneously by all observables. This is a profound metaphysical result, which antedated Bell's Inequalities. It must be emphasised, however, that the experimental evidence for the structure of the lattice of eventualities of physical systems is fragmentary and for the most part indirect. The only case in which there is abundant direct evidence is that of photon polarisation, and this is unsatisfactory for the purpose at hand, since the quantum mechanical treatment of polarisation of a single photon requires only a two-dimensional Hilbert space. A systematic set of experiments on spin-1 atoms would suffice to establish the structure of the lattice of eventualities in the simplest relevant case, but these experiments have never been performed. Stern–Gerlach methods would suffice to establish the structure of a certain interesting sublattice of eventualities for the spin-1 system (Hultgren and Shimony [1976]), but even these have not been carried out; and the structure of the full lattice of eventualities for the spin-1 system requires methods that are much more elaborate than those of Stern–Gerlach (Swift and Wright [1980]). Consequently, a die-hard defender of noncontextual hidden variables theories might express scepticism about the premiss of Gleason and of Kochen and Specker, etc., that the lattice of eventualities is isomorphic to the lattice of projectors on a Hilbert space. This (rather perverse) defence of noncontextual hidden variables theories is defeated by the experimental violation of Bell's Inequalities. Many fewer measurements are needed for checking one of the Inequalities than for establishing the structure of the lattice of eventualities, and therefore Bell's result is a very helpful instrument even for ends which in principle could be achieved in other ways. And for the purpose of disconfirming contextual hidden variables theories which conform to relativistic restrictions upon causality, we have at present no other means than Bell's Inequalities. (A large family of local contextual hidden variables theories could in principle be refuted by using the theorem in Section 3 of Heywood and Redhead [1983] or a related unpublished result of Kochen. In order to use their results for this purpose it would be necessary to exhibit certain correlations between a pair of systems, each of which is known experimentally to have a lattice of eventualities $\mathcal{L}_H$ with dim $\mathcal{H} > 2$. Such an experiment would be much more intricate than those based upon Bell's Theorem.)

6. COMMENTS ON FINE'S PAPERS

Fine [1982a,b] proves several new theorems concerning hidden variables, joint probability, and Bell's Inequalities, and on the basis of these theorems he makes some strong philosophical claims, which I wish to assess.

In order to state Fine's main theorem, the notation introduced in Section 4 must be extended in a natural way: $p^1(\bar{a})$ is the probability (when a specified method of preparing the system of interest is given) that the bivalent possibility A_a has the value 0, and similarly for $p^2(\bar{b})$; $p(a, \bar{b})$ is the probability that A_a has value 1 and B_b has value 0, and similarly for $p(\bar{a}, b)$ and $p(\bar{a}, \bar{b})$. The parameter a can take on values a' and a'' (and perhaps others not of immediate interest), while b can take on values b' and b'' (and perhaps others). Fine inquires about the possibility of extending the set of single and double probabilities $p^1(a)$, $p^1(\bar{a})$, $p^2(b)$, $p^2(\bar{b})$, $p(a, b)$, etc. to a set of compatible quadruple probabilities, where 'compatibility' has the natural meaning:

(5) $$p(a', b') = p(a', a'', b', b'') + p(a', a'', b', \bar{b}'')$$
$$+ p(a', \bar{a}'', b', b'') + p(a', \bar{a}'', b', \bar{b}''),$$

and similarly for $p(a', b'')$, $p(a', \bar{b}')$, etc. He proves that a necessary and sufficient condition for this extension to be possible is that the singles and doubles satisfy the Bell/CH Inequality (3). This theorem is proved both in his [1982a] and his [1982b], and I shall refer to it as Fine's Theorem 3, which is its number in the latter, more comprehensive paper.

Theorem 3 is the central theorem upon which he bases the following conclusions about the philosophical significance of Bell's Inequalities: 'our investigations suggest that what the different hidden variables programs have in common, and the common source of their difficulties, is the provision of joint distributions in those cases where quantum mechanics denies them' [1928b, p. 1309], and '. . . hidden variables and the Bell inequalities are all about . . . imposing requirements to make well defined precisely those probability distributions for noncommuting observables whose rejection is the very essence of quantum mechanics' [1982a, p. 294].

I am very sceptical of Fine's strong claims. He has, in my opinion, neglected the diversity of hidden variables theories and the diversity of the aims of different programmes. I shall try to indicate briefly that the results of research on hidden variables theories are too intricate to be characterized accurately by his condensed statements. Furthermore, I doubt that his own theorems, though correct mathematically, help much to elucidate the philosophical significance of Bell's Theorem.

One class of hidden variables programmes starts with the premiss that quantum mechanics gives a correct characterisation of physical systems. Since there are various inequivalent formulations of quantum mechanics, this premiss can be interpreted in many ways, but for the present purposes we can restrict attention to Mackey's axiomatization [1963]. Upon the assumption that quantum mechanics is correct one can either have a programme of showing that the quantum mechanical formalism is susceptible of a hidden variables interpretation or a programme of demonstrating the contrary, and each of these programmes ramifies according to the meaning of 'hidden variables interpretation'. If, in particular, 'hidden variables interpretation' is understood in the noncontextual sense, then the main result was already mentioned in Section 2: no noncontextual hidden variables interpretation is possible if the dimension of $\mathcal{H}$ is greater than 2, because no dispersion-free measure $p_\lambda(e)$ can be imposed upon the lattice $\mathcal{L}_H$. If a dispersion-free measure did exist, then it would be straightforward to define the joint probability of an arbitrary set of eventualities $e_1, \ldots, e_n$, corresponding to projectors which may not pairwise commute: namely

$$(6) \qquad p_\lambda(e_1, \ldots, e_n) = 1 \quad \text{if } p_\lambda(e_i) = 1 \text{ for each } i = 1, \ldots, n,$$
$$= 0 \quad \text{otherwise.}$$

If, furthermore, the values of the parameter λ constitute a space Λ with a σ-algebra of subsets $\Sigma(\Lambda)$, then other joint probability distributions could be obtained by integrating $p_\lambda(e_1, \ldots, e_n)$ with arbitrary probability measures over $\Sigma(\Lambda)$. The non-existence of even a single dispersion-free measure p_λ (proved by Gleason [1957], Bell [1966], Kochen and Specker [1967], and Belinfante [1973]) is thus an obstacle to the construction of joint probability distributions for arbitrary eventualities. Fine's conclusion, quoted above, about 'probability distributions for noncommuting observables whose rejection is the very essence of quantum mechanics' applies well enough to the non-existence results of Gleason *et al.*, without pretending to provide an exact summary of their results.

When the dimension of $\mathcal{H}$ is 2, Fine's conclusion about probability distributions for noncommuting observables does not apply without serious qualification. In this case, dispersion-free states p_λ can be constructed on $\mathcal{L}_H$, and Λ and $\Sigma(\Lambda)$ can be constructed so that for every quantum state ψ there is a probability measure ρ_ψ over $\Sigma(\Lambda)$ such that for every $e \in \mathcal{L}_H$

$$(7) \qquad p_\psi(e) = \int_\Lambda p_\lambda(e) \, \mathrm{d}p_\psi.$$

When $p_\psi(e_1, \ldots, e_n)$ is defined as in Eq. (6), then

$$(8) \qquad p_\psi(e_1, \ldots, e_n) = \int_\Lambda p_\lambda(e_1, \ldots, e_n) \, \mathrm{d}\rho_\psi$$

is a joint probability of $e_1, \ldots, e_n$, whether or not the corresponding projectors commute, and the usual $p_\psi(e_i)$ can be recovered from Eq. (8) by marginalisation. At first glance, the case of dim $\mathcal{H} = 2$ appears to provide a counter-example to Theorem 7 of Fine [1982b], namely: observables $A_1, \ldots, A_n$ satisfy the joint distribution condition if and only if all pairs commute. A counter-example does not materialise, however, because Fine uses 'joint distribution condition' in a strong sense, *viz.*, 'corresponding to every n-place Borel function f, there is an observable of the system with operator $f(A_1, \ldots, A_n)$, and corresponding to every state Ψ of the system there is probability measure $\mu_{\Psi, A_1, \ldots, A_n}$ on the Borel set of R^n that returns the quantum single distributions P_{A_i} as marginals, such that

$$\mu_{\Psi, A_1, \ldots, A_n}(f^{-1}(S)) = P_{f(A_1, \ldots, A_n)}(S)$$

for every state Ψ and Borel set S of reals'. The demand in this definition for an observable with operator $f(A_1, \ldots, A_n)$ cannot be met, even when a joint probability distribution is defined by Eq. (8). This fact can be illustrated simply by letting $n = 2$ and A_1, A_2 be noncommuting projectors representing eventualities e_1, e_2. There exists no self-adjoint operator A such that

$$p_\psi(e_1, e_2) = P_A^\psi(\{1\}),$$

where the expression on the right means the quantum mechanical probability, given ψ, that A will have its value in the (unit) set $\{1\}$; specifically, A cannot be chosen to be $A_1 A_2$ or $A_2 A_1$, because neither of these operators is self-adjoint. Although Fine's Theorem 7 is correct, Eq. (8) provides a reasonable way to introduce joint probability distributions in the case dim $\mathcal{H} = 2$, by a procedure less restrictive than his. Another way of putting the matter is that if dim $\mathcal{H} = 2$ then the lattice $\mathcal{L}_H$ can be embedded in a Boolean lattice $\mathcal{B}$ (see Kochen and Specker 1967). Not surprisingly, the greatest lower bound (glb) of A_1 and A_2 is not the same in $\mathcal{B}$ as in $\mathcal{L}_H$. In $\mathcal{B}$ the glb is an element which intuitively is the ordinary conjunction of the eventualities represented by A_1 and A_2, whereas in $\mathcal{L}_H$ the glb is the null operator, which represents the impossible eventuality. The embeddability of $\mathcal{L}_H$ in $\mathcal{B}$ underlies the existence of joint probability distributions for arbitrary eventualities in the case of dim $\mathcal{H} = 2$.

Something remains to be said about joint distributions of arbitrary eventualities even in the case of dim $\mathcal{H} > 2$, in spite of the non-embeddability of $\mathcal{L}_H$ in a Boolean lattice and the non-existence of a dispersion-free measure, and hence the non-applicability of Eq. (8). Suppose that we focus upon a fragment $\mathcal{F} \subset \mathcal{L}_H$, not necessarily a sublattice, which is singled out for some reason such as the design of an experiment, and suppose also that it is embeddable in a Boolean lattice $\bar{\mathcal{F}}$. Then probability distributions can be constructed on $\bar{\mathcal{F}}$, which automatically provide joint probability distributions for any set $e_1, \ldots, e_n$ of eventualities in $\mathcal{F}$. One may inquire, however, whether a probability distribution on $\bar{\mathcal{F}}$ exists which is compatible with a *pre-assigned* distribution on $\mathcal{F}$. Fine's Theorem 3 gives a mathematically pleasing answer to this question in a special case. Let $\mathcal{F}$ consist of the eight eventualities $a', a'', b', b'', a' \wedge b', a' \wedge b''$, $a'' \wedge b', a'' \wedge b''$, and $\bar{\mathcal{F}}$ be the Boolean lattice constructed from a', a'', b', b'' by conjunctions and negations. Fine's Theorem 3 then says that a necessary and sufficient condition for the existence of probability distributions on $\bar{\mathcal{F}}$ compatible with a preassigned distribution p on $\mathcal{F}$ is that p satisfies the Bell/CH Inequality. Garg and Mermin [1982] have derived results applying to other fragments $\mathcal{F} \subset \mathcal{L}_H$, but the question concerning an arbitrary fragment still remains open. It should be mentioned, as a qualification of Fine's philosophical conclusion quoted earlier, that his Theorem 3 does not yield a straightforward connection between commutativity and the existence of joint distributions. If a', a'', b', b'' are eventualities represented by commuting projectors, then indeed for any $\psi \in \mathcal{H}$ the probability measure p_ψ satisfies the Bell/CH Inequality, and therefore joint distributions compatible with p_ψ are guaranteed. If, however, the projectors representing a' and a'' or those representing b' and b'' fail to commute, then p_ψ may or may not satisfy the Bell/CH Inequality, for the quantum mechanical probabilities violate the Bell/CH Inequality only for special cases.

The philosophical significance of Fine's Theorem 3 can be assessed by considering both cases in which the experimental data confirm the Bell/CH Inequality and those in which they disconfirm it. (*i*) If the data confirm the Inequality, then Fine's Theorem 3 permits the fragment $\mathcal{F}$ to be interpreted within a fragmentary noncontextual hidden variables theory, which, as we have seen earlier, retains in broad outlines the world view of classical physics. This is a new result, which does not seem to be accessible without Fine's theorem. However, this retention of a classical world view is only local, for the theorems of Gleason *et al.* show that definite truth values cannot be assigned to all eventualities of $\mathcal{L}_H$ if dim $\mathcal{H} > 2$, and hence a global classical world view is impossible if the formalism of quantum mechanics is correct. (*ii*) If the experimental data disconfirm the Bell/CH Inequality then not even for the fragment $\mathcal{F}$ can a noncontextual hidden variables theory be consistently formulated. This result does not require Fine's theorem, for it is just below the surface in the earlier literature on Bell's Theorem: a noncontextual hidden variables theory automatically guarantees the factorisability condition (as pointed out in Section 4), from which the Bell/CH Inequality follows.

Let us now consider programmes which keep the premiss that quantum mechanics is correct, but admit contextual hidden variables theories. Such theories, in contrast to noncontextual hidden variables theories, do not explicitly aim at joint distributions for non-commuting observables. This contrast is particularly clear for contextual theories of algebraic type, but it also holds for those of environmental type. If e_1 and e_2 are eventualities represented quantum mechanically by non-commuting projectors, then there does not exist a single Boolean sub-σ-algebra to which both e_1 and e_2 belong. Hence, even though a complete state λ of S assigns a definite truth value to e_1 when a context (maximal Boolean sub-σ-algebra) to which it belongs is singled out, and likewise λ assigns a definite truth value to e_2 when a context to which it belongs is singled out, there is no one context in which definite truth values are assigned to both e_1 and e_2. Therefore, it does not even make sense to pose the question: what is the probability that e_1 and e_2 are jointly true? Of course the same strictures apply when more than two eventualities $e_1, \ldots, e_n$ not all belonging to a single maximal Boolean sub-σ-algebra are considered. If the contextual hidden variables theory is of environmental type, then e_1 and e_2 must be interpreted as bivalent physical possibilities, and C_1 and C_2 as states of the respective environments in which these possibilities are actualized. If C_1 and C_2 are mutually exclusive (*e.g.*, they describe different orientations of a polaroid sheet), then it makes no sense to speak of e_1 and e_2 jointly having the value 1, or jointly being true, and *ipso facto* no sense to speak of the probability of this joint outcome.

Indeed, the restrictions on joint distributions imposed by contextual hidden variables theories are very much like those of quantum mechanics itself – and exactly the same in the case of Gudder's version of a contextual theory of algebraic type. This concordance is not surprising when one recalls that Bell, by a judo-like manoeuvre, cited Bohr in order to vindicate a family of hidden variables theories in which the values of observables depend not only upon the state of the system but also upon the context [1966, p. 452]. The concordance is so great, in fact, that one may wonder whether there is really any difference between contextual hidden variables theories and quantum mechanics as it is ordinarily interpreted. This question has already been answered at the end of Section 5, but it is worthwhile reiterating the answer. A contextual hidden variables theory envisages a space Λ of complete states, such that the specification of $\lambda \in \Lambda$ together with the complete context (algebraic or environmental) uniquely determines the values of all bivalent physical possibilities that are compatible with that context. Hence chance, which is a fundamental element of quantum mechanics, is denied in a contextual hidden variables theory (at least in the deterministic case to which we have restricted our attention). Probability may enter a contextual hidden variables theory, but only in the way that it enters classical statistical mechanics: via probability distributions over $\Sigma(\Lambda)$. Whether this probability should be interpreted as personal, or logical, or propensity probability, or in some other way, is not important at this point. What is important is that the deterministic contextual hidden variables theories do not admit chance at a fundamental level, but claim rather that the state of the universe (system plus total environment) suffices to determine all outcomes. Also, the contextual hidden variables theories do not admit the entanglement of systems at the level of complete states, and interpret quantum mechanical entanglement only epistemically.

It has been suggested by Fine (private correspondence) that joint probability distributions of non-commuting observables can be introduced in contextual hidden variables theories by speaking counterfactually. Suppose, as considered in Section 4, that a' is the ordered pair (e', C'), where

C' is a maximal Boolean sub-σ-algebra containing e', and a'' is the analogous ordered pair (e'', C''). Then $a'(\lambda) = 1$ means that if the complete state of S is λ and the context C' is realized, then e' is true; and $a''(\lambda) = 1$ has an analogous meaning. Then $(a' \wedge a'')(\lambda) = 1$ could be interpreted as saying that if C' were the context then e' would be true, and if C'' were the context e'' would be true, all upon supposition that the complete state of S is λ. The lattice of propositions introduced in this way, which may be called $\mathcal{L}_C$, is Boolean and admits joint distributions determined by all probability measures ρ over $\Sigma(\Lambda)$. Of course, the price paid for introducing joint distributions in this way is that the new lattice $\mathcal{L}_C$ has an algebraic structure which disagrees globally with $\mathcal{L}_H$, though it agrees locally, *i.e.,* on the Boolean sublattices of $\mathcal{L}_H$. The Boolean structure of $\mathcal{L}_C$ suffices to guarantee the existence of joint distributions without any consideration of the Bell/CH Inequality or Fine's Theorem 3.

Finally, we consider hidden variables programmes which make no commitment to the quantum mechanical formalism, but inquire whether the experimental evidence can be accounted for by a contextual hidden variables theory of environmental type. Much of what was said above about joint probability distributions in contextual theories of algebraic type applies also to those of environmental type. Two different analyses can be envisaged concerning the joint probabilities of bivalent possibilities which are associated with mutually exclusive environmental contexts. One analysis denies that joint probability makes sense when there is no single experimental arrangement (or other environmental context) in which the bivalent possibilities all have definite truth values. The other analysis prescribes speaking counterfactually, so that the conjunction of a' and a'' would be well-defined even if a' and a'' refer to mutually exclusive contexts. The conjunction of a' and a'' consists in asserting that e' would be true were context C' realized and also that e'' would be true if context C'' were realized. When conjunction is thus defined it makes sense to talk about the probability that the conjunction is true, and of course this analysis can be extended to quadruples a', a'', b', b'', etc. There is some good sense in each of these two analyses, and I do not wish to adjudicate between them in this paper, because to do so would require an extended investigation of counterfactuals.

Without adjudication, however, one can see how each of the two analyses would assess the applicability of Fine's Theorem 3. (*i*) From the standpoint of the first analysis, nothing of interest is lost if quadruple probabilities $p(a', a'', b', b'')$, etc. are undefinable when the Bell/CH Inequality is false, since the conjunction of a', a'', b', b'' is considered to be meaningless. On the other hand, if the Bell/CH Inequality is experimentally true, so that the quadruple probabilities exist according to Fine's Theorem 3, then the first analysis would merely say that this is a formal result without physical significance. The only significant probabilities, according to the first analysis, are those of a', a'', b', and b'' singly, and of the doubles $a' \wedge b'$, $a' \wedge b''$, $a'' \wedge b'$, and $a'' \wedge b''$, since each of these unequivocally refers to a single realisable context. (*ii*) According to the second analysis the joint probabilities of a' and a'', of b' and b'', and of the quadruples a', a'', b', b'', etc. are always defined, via counterfactuals, regardless of the mutual exclusiveness of the contexts. But now Fine's Theorem 3, which says that a necessary and sufficient condition for the existence of compatible quadruples is the satisfaction of the Bell/CH Inequality, becomes irrelevant. The quadruple probabilities exist whether or not the Inequality holds. The reason for this irrelevance is that the existence of compatible quadruple probabilities, when conjunctions are understood counterfactually, does not imply the factorisability condition.

Bell's Inequalities remain unequivocally relevant to the question of whether the empirical evidence can be accounted for by a contextual hidden variables theory of environmental type. Whenever the experimental arrangement and the theoretical presuppositions (notably relativity theory) imply the factorisability condition (or an appropriate variant thereof), then Bell's Inequalities provide the crucial link between theory and experiment. Specifically, in Section 4 it was argued that the factorisability condition is reasonable in the experimental arrangement of Aspect, Dalibard, and Roger [1982], if relativity theory is assumed. However, Fine [1981] presented a case against the factorisability condition even under these circumstances. His papers [1982a,b] can be read as supplying a new argument against factorisability, even though he is not explicit on this point. He could (and perhaps implicitly does) argue as follows: Factorisability seems harmless and reasonable in the situation constructed by Aspect *et al.* But we know that the Bell/CH Inequality implies the existence of joint probabilities of a', a'', b', b'', etc. Hence, implicit in the factorisability condition is an extremely strong and far from harmless assumption about the existence of joint probabilities.

According to the first of the analyses considered above, the quadruple probabilities which are inferred from factorisability have no physical significance, and their existence is just as harmless as their non-existence would be. According to the second analysis (which Fine endorses in correspondence), the existence of quadruple probabilities is not surprising after all. We can indeed ask about the probability that both e' would be true if its context C' were realised and e'' would be true if its context C'' were realised, and similarly for other pairs, triples, and quadruples. The existence of joint probabilities in this sense does not depend upon the truth of the factorisability condition, but only upon having a contextual hidden variables theory which is employed counterfactually. One must be very careful, however, to specify the environmental contexts accurately. In the polarisation correlation experiment of Aspect *et al.* and others, a typical context C is a state of a part of the environment of a pair of photons, specifying at least the angles of orientation of both polarisation analysers and perhaps other things as well. The Bell/CH Inequality does not follow from the existence of joint (double, triple, quadruple) probabilities when the contexts are understood in this way. An extra premiss is needed: namely, a context sufficient for fixing the probability of a specifies only the orientation of the first polarisation analyser, so that the orientation of the second polarisation analyser is irrelevant, and likewise for the probability of b. Whether this additional premiss is reasonable from the standpoint of relativistic non-quantum physics can and should be debated (as in the 1981 papers of Fine and Shimony), but this question is untouched by considerations of the existence of joint probabilities in a contextual hidden variables theory.

7. CONCLUSIONS

Cataloguing possible world views is a formidable enterprise, and cataloguing those which are compatible with a certain body of experimental evidence is yet more formidable. Some of the world views which have been proposed and discussed in the literature on the foundations of quantum mechanics have not even been mentioned in this paper. Among those which have been considered, however, the evidence has narrowed the choices – though again due regard must be given to the possibility of experimental error. The world view associated with noncontextual hidden variables theories is overwhelmingly disconfirmed, whether or not one pays attention to relativistic locality. That associated with contextual hidden variables theories is overwhelmingly disconfirmed if relativistic locality is assumed. Bell's Theorem is a useful instrument for strengthening the first of these judgments, and at present it is indispensable for achieving the latter. That, in my opinion, is the philosophical significance of Bell's Inequalities and Theorem.

Acknowledgments. I wish to thank Arthur Fine and Michael Redhead for comments which were valuable in revising this paper.

REFERENCES

Aspect, A., Dalibard, J., and Roger, G. [1982]: 'Experimental test of Bell's Inequalities using variable analyzers', *Physical Review Letters,* **49,** pp. 1804–1807.
Belinfante, F. [1973]: *A Survey of Hidden-Variables Theories.* Pergamon Press.

Bell, J. S. [1964]: 'On the Einstein Podolsky Rosen Paradox', *Physics*, 1, pp. 195–200.

Bell, J. S. [1966]: 'On the problem of hidden variables in quantum mechanics', *Review of Modern Physics*, 38, pp. 447–52.

Bohm, D. [1952]: 'A suggested interpretation of the quantum theory in terms of "hidden variables"', *Physical Review*, 85, pp. 166–79, pp. 180–93.

Broglie, L. de [1926]: 'Sur la possibilité de relier les phénomenès d'interférence et de diffraction à la théorie des quanta de lumière', *Comptes Rendus*, 183, pp. 447–8.

Clauser, J. F. and Horne, M. A. [1974]: 'Experimental consequences of objective local theories', *Physical Review D*, 10, pp. 526–35.

Clauser, J. F. and Shimony. A. [1978]: 'Bell's theorem: experimental tests and implications', *Reports on Progress in Physics*, 41, 1881–927.

Einstein, A., Podolsky, B., and Rosen, N. [1935]: 'Can quantum-mechanical description of physical reality be considered complete?', *Physical Review* 47, pp. 777–80.

Fine, A. [1981], 'Correlations and physical locality', *PSA 1980*, vol. 2, ed. A. Asquith and R. Giere (Philosophy of Science Association, E. Lansing, Michigan), pp. 535–62.

Fine, A. [1982a]: 'Hidden variables, joint probability, and the Bell Inequalities', *Physical Review Letters*, 48, pp. 291–5.

Fine, A. [1982b]: 'Joint distributions, quantum correlations, and commuting observables', *Journal of Mathematical Physics*, 23, pp. 1306–10.

Garg, A. and Mermin, D. [1982]: 'Comment on "Hidden variables, joint probability, and the Bell Inequalities"', *Physical Review Letters*, 49, p. 242.

Garuccio, A. and Selleri, F. [1978]: 'On the equivalence of deterministic and probabilistic local theories', *Lettere al Nuovo Cimento*, 23, pp. 555–8.

Gleason, A. [1957]: 'Measures on the closed subspaces of a Hilbert space', *Journal of Mathematics and Mechanics*, 6, pp. 885–93.

Gudder, S. [1970]: 'On hidden-variable theories', *Journal of Mathematical Physics*, 11, pp. 431–6.

Heywood, P. and Redhead, M. [1983]: 'Nonlocality and the Kochen–Specker paradox', *Foundations of Physics*, 13, pp. 481–99.

Hultgren, B. and Shimony, A. [1976]: 'The lattice of verifiable propositions of the spin-1 system', *Journal of Mathematical Physics*, 18, pp. 381–94.

Kochen, S. and Specker, E. [1967]: 'The problem of hidden variables in quantum mechanics', *Journal of Mathematics and Mechanics*, 17, pp. 59–88.

Mackey, G. [1963]: *Mathematical Foundations of Quantum Mechanics*. Benjamin.

Pearle, P. [1965]: 'A framework for hidden variable theories', preprint, Harvard University (unpublished).

Shimony, A. [1978]: 'Metaphysical problems in the foundations of quantum mechanics', *International Philosophical Quarterly*, 18, pp. 3–17.

Shimony, A. [1981]: 'Critique of the papers of Fine and Suppes', *PSA 1980*, vol. 2, ed. A. Asquith and R. Giere (Philosophy of Science Association, E. Lansing, Michigan), pp. 572–80.

Stein, H. [1972]: 'On the conceptual structure of quantum mechanics', *Paradigms and Paradoxes,* ed. R. Colodny (U. of Pittsburgh Press, Pittsburgh), pp. 367–438.

Swift, A. and Wright, R. [1980]: 'Generalized Stern–Gerlach experiments and the observability of arbitrary spin operators', *Journal of Mathematical Physics,* **21**, pp. 77–82.

Varadarajan, V. [1968]: *The Geometry of Quantum Theory.* Van Nostrand.

Wiener, N. and Siegel, A. [1953]: 'A new form for the statistical postulate of quantum mechanics', *Physical Review,* **91**, pp. 1551–60.

COMMENT

In Section 5 an argument by Kochen and independently by Heywood and Redhead was mentioned, showing that no local contextual hidden variables can agree with all of the nonstatistical assertions of quantum mechanics concerning a pair of spin-1 systems prepared in the singlet state. It seems not to have been much noticed that they had thus succeeded in proving Bell's theorem (which essentially states that any local hidden-variables theory, even if contextual, must disagree with some of the observable predictions of quantum mechanics), and had done so without an inequality of the usual type and without considering any statistical predictions. The neglect was partly due to the fact that the argument relies on the theorem (due to Gleason, Bell, and Kochen and Specker) that a noncontextual hidden-variables theory cannot be constructed for a single spin-1 system, and this theorem is far from intuitive and simple. It therefore was surprising (even to the few people who were aware of the result of Kochen and of Heywood and Redhead) that a very intuitive proof of Bell's theorem without inequalities or reference to statistics was given by D. Greenberger, M. Horne, and A. Zeilinger, somewhat inaccurately entitled "Going Beyond Bell's Theorem," in *Bell's Theorem, Quantum Theory, and Conceptions of the Universe,* edited by M. Kafatos (Dordrecht: Kluwer Academic, 1989), pp. 73–76. Variants, mathematical details, and generalizations of their proof were given by N. D. Mermin (1990, *American Journal of Physics* 58: 731); D. Greenberger, M. Horne, A. Shimony, and A. Zeilinger (1990, *American Journal of Physics* 58: 1131); and R. Clifton, M. Redhead, and J. Butterfield (1991, *Foundations of Physics* 21: 149). I shall present a somewhat simplified version of an argument in the next to last of these papers, which is doubly instructive since it dispenses with considerations of spin or polarization that have dominated discussions of Bell's theorem.

Consider the three-photon arrangement of Figure 1. The three photons are supposed to be produced by the down-conversion process from a single photon of a beam propagating in a direction perpendicular to the paper

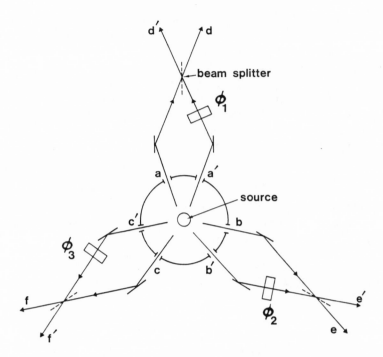

and pumping a nonlinear crystal. A cylindrical shield surrounds the source, pierced by three pairs of pinholes, with photon 1 going into beams a or a', photon 2 going into beams b or b', and photon 3 going into beams c or c' (though from the viewpoint of quantum mechanics the "or" does not connote a definite choice). Momentum conservation implies that the six beams all are tilted along the direction of the pumping beam, but for simplicity of representation all six are flattened onto the plane of the paper. The quantum state of the triple of photons is assumed to be

(1) $$|\psi\rangle = \frac{1}{\sqrt{2}}[|a\rangle_1|b\rangle_2|c\rangle_3 + |a'\rangle_1|b'\rangle_2|c'\rangle_3].$$

(More accurately, the quantum state is obtained by symmetrizing the right-hand side of Eq. (1), since photons are bosons, but nothing in the following calculations is essentially changed by symmetrization.) The ket $|a\rangle_1$ represents a state in which photon 1 enters path a, $|a'\rangle_1$ represents a state in which photon 1 enters path a', etc. Phase plates causing controllable phase shifts ϕ_1, ϕ_2, ϕ_3 are placed respectively in paths a', b', and c'. The beams in paths a and a' are brought together at a half-silvered mirror whose output feeds detectors d and d'; similarly, b and b' are brought together to feed detectors e and e', and c and c' are brought together to feed detectors f and f'. The numerical value $+1$ is assigned to the firing of the unprimed detectors d, e, and f, and the value -1 is assigned to the firing of d', e', and f'. The experimental arrangement is assumed to be ideal, so that any photon entering one of the paths is detected by one of the detectors fed by the appropriate half-silvered mirror. A straightforward interferometric calculation (Greenberger et al. 1990, p. 1141) shows that the probability of triple outcomes A, B, C having product ABC equal to $+1$ or -1 is

(2) $P_\psi(ABC = \pm 1 | \phi_1, \phi_2, \phi_3) = \frac{1}{2}[1 \pm \sin(\phi_1 + \phi_2 + \phi_3)]$,

where A is the outcome $+1$ or -1 for detection of photon 1, and B and C likewise for photons 2 and 3; the sign on the right-hand side is plus if ABC is $+1$ and minus if ABC is -1. In particular,

(3a) $P_\psi(ABC = 1 | \phi_1, \phi_2, \phi_3) = 1$,

(3b) $P_\psi(ABC = -1 | \phi_1, \phi_2, \phi_3) = 0$,   if $\phi_1 + \phi_2 + \phi_3 = \pi/2 \pmod{2\pi}$,

and

(4a) $P_\psi(ABC = 1 | \phi_1, \phi_2, \phi_3) = 0$,

(4b) $P_\psi(ABC = -1 | \phi_1, \phi_2, \phi_3) = 1$,   if $\phi_1 + \phi_2 + \phi_3 = 3\pi/2 \pmod{2\pi}$.

Thus the entangled three-photon state of Eq. (1) permits some predictions to be made with certainty, in the sense of probability unity.

The certainties expressed in Eqs. (3a,b) and (4a,b) permit an EPR (Einstein–Podolsky–Rosen) argument to go through (see Greenberger et al. 1990, pp. 1134–36). If EPR's famous sufficient condition for the existence of an element of reality is accepted, together with their assumption of no action-at-a-distance, then one deduces that each triple of photons entering the interferometric arrangement is in a complete state λ, belonging to a space Λ of complete states, with the following properties: For each ϕ_1 the outcome A is specified for all but a set of measure zero of complete states to be definitely $+1$ or definitely -1, and hence it can be expressed as $A_\lambda(\phi_1)$, and with the same reservations B can be expressed as $B_\lambda(\phi_2)$ and C as $C_\lambda(\phi_3)$; and furthermore

(5a) $A_\lambda(\phi_1)B_\lambda(\phi_2)C_\lambda(\phi_3) = 1$ for all but a set of measure zero, for each specified triple ϕ_1, ϕ_2, ϕ_3 summing to $\pi/2 \pmod{2\pi}$,

and

(5b) $A_\lambda(\phi_1)B_\lambda(\phi_2)C_\lambda(\phi_3) = -1$ for all but a set of measure zero, for each specified triple ϕ_1, ϕ_2, ϕ_3 summing to $3\pi/2 \pmod{2\pi}$, in order to agree with the quantum-mechanical predictions of Eqs. (3a,b) and (4a,b).

Alternatively, one could bypass the EPR argument and simply postulate a contextual hidden-variables theory for the triple of photons, permitting the outcomes A, B, and C each to depend upon all three of the phase shifts ϕ_1, ϕ_2, ϕ_3. If one then imposes the locality condition that each photon is insensitive to the phase plates inserted along paths it cannot possibly take, then A can only depend upon ϕ_1, B upon ϕ_2, and C upon ϕ_3. In other words, the imposition of locality has the effect of restricting the latitude of the contextual hidden-variables theory and yielding a noncontextual hidden-variables theory, just as in the argument of Kochen and of Heywood and Redhead. Agreement with the predictions of quantum mechanics would again require Eqs. (5a,b).

No matter how Eqs. (5a,b) are derived, they are easily seen to be inconsistent. They imply by substitution that

(6a) $$A_\lambda(\pi/2)B_\lambda(0)C_\lambda(0) = 1,$$

(6b) $$A_\lambda(0)B_\lambda(\pi/2)C_\lambda(0) = 1,$$

(6c) $$A_\lambda(0)B_\lambda(0)C_\lambda(\pi/2) = 1,$$

for all but a set of measure zero, and taking the product of Eqs. (6a,b,c) one obtains

(7) $$A_\lambda(\pi/2)B_\lambda(\pi/2)C_\lambda(\pi/2) = 1$$

for all but a set of measure zero, in contradiction with Eq. (5b). This inconsistency is Bell's theorem, derived without any inequalities or consideration of statistical predictions of quantum mechanics.

10
Controllable and uncontrollable
non-locality*

As suggested by Jarrett, a condition from which a Bell-type Inequality is derivable can be expressed as the conjunction of two conditions. In appropriate circumstances the violation of either of these conditions appears to be non-local from the standpoint of relativity theory, but in one case the non-locality is controllable, hence usable for sending a superluminal signal, and in the other it is uncontrollable. A straightforward calculation shows that the non-locality implicit in the entangled quantum state of spatially separated particles is uncontrollable. Some questions are raised about the non-local hidden variables programs of Bohm and collaborators and of Vigier and collaborators.

I. INTRODUCTION

All of the Bell-type inequalities are derived from some variant of a condition which Bell[1] himself (somewhat controversially) called "locality." It will be convenient in this paper to restrict attention to the following variant,

(1.1) $\Pi_\lambda(d_L, x_L; d_R, x_R) = \Pi_\lambda(d_L, x_L; 0, 0) \cdot \Pi_\lambda(0, 0; d_R, x_R)$,

which is similar to Jarrett's "strong locality." Here L and R are the left and right analyzers (as seen from some observer's perspective), and d_L and d_R are the values of controllable parameters of the analyzers; the special value 0 of $d_L(d_R)$ signifies the absence of the analyzer from the path of the left or right propagating particle. The outcome variables x_L and x_R can take on the values 1, −1 and 0, with 1 representing passage of the left or right propagating particle through its respective analyzer, −1 representing impingement but non-passage, and 0 representing non-impingement. λ is the complete state of a pair of particles generated together from some source and propagating in the left and right directions respectively. $\Pi_\lambda(d_L, x_L; d_R, x_R)$ is the probability of joint outcomes x_L and x_R upon assumption that λ is the complete state of the particle pair

This work originally appeared in S. Kamefuchi et al. (eds.), *Foundations of Quantum Mechanics in the Light of New Technology,* Tokyo: The Physical Society of Japan, 1984. Reprinted by permission of the publisher.

*This work was supported in part by a grant from the U.S. National Science Foundation (No. DMR 802-1057A01).

130

and d_L, d_R are the values of the controllable parameters. The notation x_L and x_R will occasionally be used to represent the events of passage or non-passage through the analyzers (or non-impingement). Context easily determines the intended meanings of the symbols.

Whether eq. (1.1) is a reasonable condition depends upon the presuppositions of physical theory and also upon the details of the experimental arrangement. In particular, if the special theory of relativity and the "realistic" viewpoint of Einstein, Podolsky, and Rosen[3] are presupposed, and if the experimental arrangement of Aspect, Dalibard, and Roger[4] is employed (in which the choices of d_L and d_R are events with space-like separation), then eq. (1.1) can be reasonably regarded as a consequence of relativistic locality. The disconfirmation of a Bell-type inequality by Aspect *et al.*, and hence of eq. (1.1), is therefore generally regarded as a disproof of the above-mentioned physical presuppositions. In the author's opinion this conclusion is roughly correct, despite the criticisms which have been leveled against it.[5] Since the conceptual issues are momentous and subtle, however, further scrutiny is desirable. Considerable insight is afforded by a recent analysis of Jarrett,[2] who pointed out that eq. (1.1) can be expressed as the conjunction of two logically independent conditions, which will be stated in Section 2. The violation of either of these conditions in an experimental arrangement like that of Aspect *et al.* would *prima facie* be in disaccord with relativistic locality, but in different ways: the violation of the first would constitute a case of controllable non-locality (as Jarrett points out), whereas the violation of the second could be interpreted as uncontrollable non-locality (which Jarrett does not say). There is experimental evidence only for the latter. Furthermore, the non-locality implicit in quantum mechanics must be recognized to be of the uncontrollable variety, that cannot be used for the purpose of sending a signal faster than light. In this sense there may be "peaceful coexistence" between quantum mechanics and relativity theory. There is an open question, on the other hand, whether some of the recently proposed non-local hidden variables theories, to be discussed in Section 4, can peacefully coexist with relativity theory.

2. DECOMPOSITION OF STRONG LOCALITY

Jarrett[2] pointed out that the strong locality condition of eq. (1.1) is equivalent to the conjunction of the following conditions I and II:

(2.1a)   I. $\Pi_\lambda(d_L, x_L; 0, 0) = \sum_{X=\pm 1} \Pi_\lambda(d_L, x_L; d_R, X)$, for all $d_R \neq 0$,

(2.1b) $\Pi_\lambda(0, 0; d_R, x_R) = \sum_{X=\pm 1} \Pi_\lambda(d_L, X; d_R, x_R)$, for all $d_L \neq 0$,

(2.2)   II. $\Pi_\lambda(d_L, x_L; d_R, x_R) = \sum_{X=\pm 1} \Pi_\lambda(d_L, x_L; d_R, X)$

$$\cdot \sum_{Y=\pm 1} \Pi_\lambda(d_L, Y; d_R, x_R).$$

Condition I (eqs. 2.1a,b) is cognate to a condition which Jarrett calls "locality," and Condition II (eq. 2.2) to one that he calls "completeness."[1] The latter term seems inappropriate, however, since it fails to be satisfied by the quantum mechanical probabilities, and nevertheless the quantum mechanical description may be "complete" in the sense of saying all that there is to say about a system. It will be argued below, furthermore, that a violation of Condition II is in disaccord with relativistic locality in certain situations. For these reasons, it seems best to depart from Jarrett's terminology and denote the conditions merely by Roman numerals.

Jarrett showed that if Condition I holds, then it is impossible for an experimenter located at R to send a bit of information to some one located at L by choosing between $d_R = 0$ and some pre-arranged non-zero value of d_R; and similarly with R and L interchanged. He also showed

---

1. Actually, wherever the notation $\Pi_\lambda(d_L, x_L; d_R, x_R)$ is used above Jarrett writes

$$\Pi_\lambda^{\mu_R, \mu_L}(d_L, x_L; d_R, x_R),$$

where μ_L is all the information in excess of d_L needed to specify the complete state of the analyzer L in all microscopic detail, and similarly for μ_R. The probability considered in this paper, without any reference to μ_L and μ_R, can be thought of as an average over Jarrett's probabilities, using a probability distribution over the possible values of the excess information, where this probability distribution is presumably determined by the macroscopic description of the experiment. It is non-trivial to assume that $\Pi_\lambda(d_L, x_L; d_R, x_R)$ is well-defined, but the assumption is unavoidable when one wishes to draw conclusions from experiments, since macroscopic apparatus cannot be microscopically controlled. Furthermore, the occurrence of stable statistics in repeated runs in a well-performed experiment lends some support to the assumption. Jarrett's notation avoids the need for this assumption, since his probabilities are contingent upon specified complete states of L and R. But his notation gives rise to a serious problem: whether μ_L and μ_R can be chosen independently of each other. The microstate of a macroscopic piece of apparatus is an enormously complicated and detailed compendium of information. Consequently, even if the controllable parameters d_L and d_R can be chosen independently, according to any reasonable standard of independence (such as reliance upon two distinct stochastic devices), it seems very likely that events in the past which lead to the detailed choice of μ_L will leave some small trace upon μ_R, and conversely. Consequently, Jarrett's equation which is cognate to eq. (2.1a),

$$\Pi_\lambda^{\mu_L, \mu_R^0}(d_L, x_L; 0, 0) = \sum_{x \pm 1} \Pi_\lambda^{\mu_L, \mu_R}(d_L, x_L; d_R, X),$$

and likewise his other equations, may be physically empty, because of the impossibility or at least extreme improbability of having the same μ_L both when R is described by μ_R and when it is described by μ_R^0. A similar objection was raised by Shimony, Clauser and Horne[6] against an argument by Bell[7] which turned upon assumptions concerning microstates of space–time regions.

that if Condition I fails, then a bit of information can be transmitted in this way, using an ensemble of particle pairs all prepared in the complete state λ as the medium of communication. The statistics of passage through one analyzer will depend upon the controllable preparation of the other analyzer. And if the events of choosing $d_L(d_R)$ have space-like separation, then the bit of information is transmitted superluminally. Since there is no physical obstacle to the choice of $d_R = 0$ or the choice of $d_R \neq 0$, the failure of Condition I can reasonably be called *controllable non-locality*. (It should be noted, however, that there is some idealization in using the term "controllable", since in practice it may be very difficult to prepare an ensemble of particle pairs all in the same complete state λ.)

Suppose now that Condition I holds but Condition II fails. Then there must be correlations between the outcomes x_L and x_R that are not implicit in the state λ: it must sometimes happen that the probability of the L-outcome being x_L, given λ together with x_R, is different from the probability of this outcome given λ alone, or else that such a difference between conditioned and unconditioned probabilities will occur with L and R interchanged. But how is such a difference of probabilities to be understood if λ is the complete state of the particle pair, as has been assumed? If the outcomes x_L and x_R were events with time-like or light-like separation, there would be no obstacle to a causal explanation: the earlier of the two events would affect the propensity of the latter. If, however, x_L and x_R have space-like separation, then a causal connection between them would be an action at distance. And even if one abstains from using a causal locution in this situation, the very fact that the correlation cannot be attributed to the complete state λ constitutes a kind of nonlocality[2] (which perhaps is *sui generis* and appropriately named "passion at a distance"). The violation of Condition II, however, does not permit superluminal communication between R and L as long as Condition I is satisfied. An experimenter at R, for example, cannot affect the statistics of outcomes x_L by selective measurements. Since λ is presumed to be complete, there is no way for the experimenter to select, prior to the measurement, a sub-

---

2. The conclusion of non-locality can be avoided by theorizing that correlations are established in the overlap region O of the backward light cones B_L and B_R of the outcome events x_L and x_R. Suppose, for example, that the value of $x_R(x_L)$ is sensitive to the microstate of the R (L) analyzer, and that correlations between microstates of R and L are established in the region O. Then the value of x_R provides some information about the microstate of R, hence indirectly about the microstate of L, and hence indirectly about the probability of x_L. No ontic non-locality would then be needed in order to understand the violation of Condition II. Upon reflection, however, this hypothesis appears to be desperate, since the macroscopic analyzers L and R are inevitably subject to perturbations which make their microstates uncontrollable, and hence hardly reliable enough for carrying out the conspiracy which the hypothesis requires.

ensemble with statistics different from that of the entire ensemble. Even though the propensity of an outcome x_L in a single instance is somehow sensitive to the outcome x_R, the statistics actually collected by the experimenter at L will be the lumped results of the propensity of x_L contingent upon $x_R = 1$ and of that contingent upon $x_R = -1$, *i.e.,*

$$\sum_{x_R = \pm 1} \Pi_\lambda(d_L, x_L; d_R, x_R).$$

According to Condition I, however, this sum equals $\Pi_\lambda(d_L, x_L; 0, 0)$, and therefore no information has been transmitted from R to L. It is reasonable, then, to call the violation of Condition II *uncontrollable nonlocality.*

3. THE UNCONTROLLABILITY OF QUANTUM NON-LOCALITY

Quantum mechanics supplies abundant probability evaluations in conflict with Condition II, for example the probabilities of spin results for a pair of spin-$\frac{1}{2}$ particles in the singlet state. There seems to be widespread knowledge among experts that quantum mechanics does not violate Condition I, though to the author's knowledge the only explicit proof in the literature is a paper by Ghirardi, Rimini, and Weber.[8] The following is a variant of their argument. Consider a pair of particles r, l prepared in a quantum state Ψ, which can always be written in the Schmidt representation[9] as

(3.1) $$\Psi = \sum c_i u_i \otimes v_i,$$

where the u_i are orthonormal vectors in the Hilbert space $\mathcal{K}_r$ associated with particle r and the v_i are orthonormal vectors in the Hilbert space $\mathcal{K}_l$ associated with particle l, and where the $|c_i|^2$ sum to 1. If there is more than one value of the index i with $c_i \neq 0$, then Ψ cannot be written in the product form $u \otimes v$, with $u \in \mathcal{K}_r$ and $v \in \mathcal{K}_l$, and Ψ is said to be an "entangled" state of the composite system $r + l$. Because of this entanglement there is a definite sense in which an operation performed upon $r(l)$ can change the state of $l(r)$. For example, if the u_i are eigenfunctions of a self-adjoint operator F on $\mathcal{K}_r$, such that

(3.2) $$Fu_i = f_i u_i,$$

with non-degenerate eigenvalues f_i, then an ideal repeatable measurement of F would ensure that immediately after the result f_k the state of $r + l$ is $u_k \otimes v_k$, which shows that the state of l has indeed been affected by the operation performed upon r. If the measurement upon r is performed at a space–time point which lies upon the world-line of r but not upon the world-line of l, the result just described appears to be non-local.

In order to assess carefully what has just happened it is useful to write the total hamiltonian of $r+l$ together with that of the apparatus A which interacts with r; it is assumed that neither A nor r interacts with l during the time interval under consideration. The total hamiltonian of $A+r+l$ can be written in the form

$$(3.3) \qquad H_{\text{tot}} = \bar{H}_r \otimes 1 + 1 \otimes H_l,$$

where H_l is the hamiltonian of l alone and is an operator on $\mathfrak{IC}_l$, while $\bar{H}_r$ is the hamiltonian of $A+r$ and hence operates upon the Hilbert space $\mathfrak{IC}_r = \mathfrak{IC}_A \otimes \mathfrak{IC}_r$. Of course, $\bar{H}_r$ can be further analyzed into the hamiltonians of A and r in isolation together with their interaction hamiltonian, but nothing will be gained by expressing this analysis explicitly. Suppose now that at an initial time t_0, when the quantum state of $r+l$ is represented by the Ψ of eq. (3.1), the quantum state of A is represented by $w \in \mathfrak{IC}_A$. Then the quantum state of $A+r+l$ can be expressed as

$$(3.4) \qquad \bar{\Psi}(t_0) = \sum c_i \bar{u}_i \otimes v_i,$$

where $\bar{u}_i = w \otimes u_i$ is a vector in $\mathfrak{IC}_r$. (If we began with the apparatus not in a pure quantum state but in a mixture of w^j, with non-negative real weights summing to 1, then we should have to describe $A+r+l$ by a corresponding mixture with the same weights, but our conclusion would not be affected by this complication.) The time evolution operators for $A+r+l$ are

$$(3.5) \qquad U(t) = e^{iH_{\text{tot}}t} = e^{i\bar{H}_r t} \otimes e^{iH_l t},$$

as can be checked by series expansion. From eqs. (3.4) and (3.5) we obtain a vector representing the state of $A+r+l$ at an arbitrary time t, provided that A, r and l undergo no other interactions than the ones indicated between t_0 and t:

$$(3.6) \qquad \bar{\Psi}(t) = U(t-t_0)\bar{\Psi}(t_0)$$
$$= \sum c_i e^{i\bar{H}_r(t-t_0)} \bar{u}_i \otimes e^{iH_l(t-t_0)} v_i.$$

If the operation which is performed by arranging the interaction of A and r is to convey a message *via* l, it is necessary that the expectation value of some observable G of l alone be sensitive to the choice of A. That observable is represented in $\mathfrak{IC}_r \otimes \mathfrak{IC}_l$ by the operator $1 \otimes G$, and hence

$$(3.7) \qquad \langle G \rangle_{\bar{\Psi}(t)} = (\bar{\Psi}(t), 1 \otimes G\bar{\Psi}(t)) = \sum |c_i|^2 (v_i', Gv_i'),$$

where

$$v_i' = e^{iH_l(t-t_0)} v_i.$$

Clearly the expectation value of eq. (3.7) is independent of the choice of the apparatus A and of its interaction with r. Consequently, if quantum mechanics is correct there is no way to affect the statistics of measurements upon replicas of l by local operations performed upon replicas of r, even though each replica of $r+l$ is described at an initial time by the entangled state of eq. (3.1). The quantum mechanical predictions conform to Condition I of Section 2, even in those situations in which they violate Condition II. We conclude that the non-locality of quantum mechanics is uncontrollable.

4. NON-LOCAL HIDDEN VARIABLES THEORIES

Although the experiments inspired by Bell's Theorem, especially that of Aspect *et al.*, have thrown grave doubt upon the whole family of local hidden variables theories, a motivation remains for speculating about hidden variables, even at the price of concessions to non-locality. If the quantum mechanical description is complete, then the quantum mechanical probabilities cannot be interpreted only epistemically, and the quantum state must be regarded as a network of potentialities, actuality being reserved for those observables of which the state in question is an eigenstate. But potentiality is an empty concept unless there is a process of actualizing potentialities, and it is hard to see how a dynamics based upon the time-dependent Schrödinger equation can supply such a process.[10] Hidden variables theories, by contrast, escape this problem completely by interpreting the quantum state epistemically, as a compendium of statistical information, and not ontically as a network of potentialities.

Non-local hidden variables theories have been suggested programmatically by Bohm and associates[11] and by Vigier and associates.[12] No attempt will be made here to analyze their programs in detail, but a few comments and questions may be useful for future discussions. Both programs postulate a fundamental material field (called "the random ether" by Vigier), which is governed by deterministic non-linear differential equations. There can be direct interaction between elements of this field with space-like separation, and hence genuine relativistic non-locality occurs. On the other hand, the instabilities of the field are such that ordered configurations of the field cannot be propagated superluminally. (The contrast between the rapid disorderly motion of a Brownian particle and its slow drift is one of the inspirations of these programs.)

A fundamental difficulty is to explain how the postulated turbulent medium is able to account for the experimentally observed precise correlation of the polarizations of well-separated photons. Even though we can acknowledge the general proposition that any statistical physical theory is able to account for the emergence of some stability out of disorder, for example because of conservation laws or laws of large numbers, we nevertheless want to know in detail how the polarization correlations are stably maintained.

A second question hangs upon the first. The programs under discussion wish to retain relativistic locality on a phenomenological level, by ensuring the impossibility of superluminal signalling. But it is not clear that a medium sufficient to establish polarization correlations could not, in principle, be used to signal faster than light. Can it be shown rigorously that the postulated non-locality of the hidden variables theories is uncontrollable rather than controllable?

A third question concerns the predictions of the programs of Bohm, Vigier, *et al.* in a proposed modification of the experiment of Aspect, Dalibard, and Roger.[4] Suppose that in the interval after the commutators of that experiment have been actuated, but before the polarization analysis of the photons has been completed, a strong burst of laser light is propagated transverse to but intersecting the paths of the propagating photons. (If the experiment is actually performed, care will have to be taken to prevent scattered photons from this burst from causing accidental coincidence counts, and to protect the source of the photon pairs from perturbation.) Because of the non-linearity of the fundamental material medium which has been postulated, this burst would be expected to generate excitations, which could conceivably interfere with the non-local propagation that is responsible for polarization correlations. The question is: what do the programs under discussion actually predict concerning the effect of such a laser burst? If none, why not? If some effect, might there not be a relatively easy modification of apparatus now in existence for the purpose of testing these versions of non-local hidden variables theories?

REFERENCES

[1] J. S. Bell: Physics **1** (1964) 195.

[2] J. Jarrett: *Bell's Theorem, Quantum Mechanics and Local Realism* (Ph.D. Thesis, Univ. of Chicago, 1983, unpublished).

[3] A. Einstein, B. Podolsky and N. Rosen: Phys. Rev. **47** (1935) 777.

[4] A. Aspect, J. Dalibard and G. Roger: Phys. Rev. Lett. **49** (1982) 1804.

[5] A. Fine: *PSA 1980,* ed. A. Asquith and R. Giere (Philosophy of Science Association, E. Lansing, Michigan, 1981) Vol. 2, p. 535; comments by A. Shimony, *ibid.*, p. 572.

[6] A. Shimony, J. Clauser and M. Horne: Epistemological Letters **13** (1976) 1. Reprinted in Dialectica **39** (1985) 97.

[7] J. S. Bell: Epistemological Letters **9** (1976) 11. Reprinted in Dialectica **39** (1985) 86.

[8] G. C. Ghirardi, A. Rimini and T. Weber: Lett. Nuovo Cim. **27** (1980) 293. Two

other arguments have been brought to my attention: P. Eberhard: Nuovo Cim. **46B** (1978) 392; D. Page: Phys. Lett. **91A** (1982) 57.

[9] J. von Neumann: *Mathematical Foundations of Quantum Mechanics* (Princeton University Press, Princeton, 1955) p. 434.

[10] B. d'Espagnat: *Conceptual Foundations of Quantum Mechanics* (W. A. Benjamin, Reading, Massachusetts, 1976) Part 4.

[11] D. Bohm: *Wholeness and the Implicate Order* (Routledge and Kegan Paul, London, 1980) Chap. 4; D. Bohm and B. Hiley: Found. Physics **5** (1975) 93.

[12] J.-P. Vigier: Astr. Nachr. **303** (1982) 55; N. Cufaro-Petroni and J.-P. Vigier: Phys. Lett. A **81** (1981) 12; P. Droz-Vincent: Phys. Rev. D **19** (1979) 702; A. Garuccio, V. A. Rapisarda and J.-P. Vigier: Lett. Nuovo Cim. **32** (1981) 451.

*　　　　　*　　　　　*

A. J. Leggett: Presumably, for the proposed test of Vigier's theory, you would have to get him to specify the magnitude of the effect of a transverse laser pulse in advance, since standard quantum mechanics would predict the same effect (presumably too small to measure at present).

A. Shimony: Yes, I would add that a failure to derive an order of magnitude prediction could be a serious reason for skepticism about his theory.

A. Zeilinger: In your suggestion to upset by laser pulses the subquantal medium, is it really necessary to introduce the disturbance of the beam paths of the EPR apparatus?

A. Shimony: I am relying upon my knowledge of wave propagation in other media to infer (plausibly) that a disturbance at the beam paths would have some effect on causal connections between the separated particles. However, we must await Vigier's comments about the implications of his theory on this point. His subquantal ether may be very different from other media.

COMMENT

Elsewhere I use the terminology "parameter independence" for Condition I (Eqs. 2.1a,b), which Jarrett calls "locality"; and "outcome independence" for Condition II (Eq. 2.2), which he calls "completeness."

The statement in the last paragraph of Section 2 that the violation of Condition II does not permit superluminal communication is correct, but the reason given is not. It is possible for an experimenter at R to perform certain operations which, because of the violation, would permit a message to be sent to an experimenter at L – by a procedure exhibited in Section 4 of "Events and Processes in the Quantum World," the next paper in this collection. One easily sees, however, that the message thus transmitted is not superluminal.

As David Albert and Sheldon Goldstein (private communications) have pointed out, my argument that the violation of Parameter Independence implies the possibility of superluminal communication depends upon the feasibility of preparing an ensemble in the complete state λ mentioned in Inequality (19). The non–local hidden-variables theory of David Bohm (1952, *Physical Review* 85: 166 and 180) evades superluminal communication by postulating a physical limitation upon the controlled preparation of the ensemble. I am skeptical of this evasion, because it seems to mingle an anthropocentric consideration with fundamental physical considerations of causality.

11
Events and processes in the quantum world

I. PROSPECTIVE

The concern of this paper will be different from that of most studies of quantum gravity. There will be no discussion of the problem of quantizing the gravitational field equations, hence nothing about the modifications which quantization requires of space–time structure at distances of the order of the Planck length ($\sim 10^{-33}$ cm). The emphasis will be rather upon the implications of quantum mechanics for certain general properties of events and processes which occur in the theatre of space–time, specifically with *non-locality* and *nonlinearity*.

Quantum mechanics is undoubtedly a non-local theory when it treats correlated spatially separated systems. When one examines closely the character of this non-locality, however, one does not find reasons for modifying the causal structure of space–time as described by special or general relativity theory, but rather reasons for refining the concept of an event. The occurrence of definite outcomes of measurements implies that there are processes in nature governed by nonlinear laws. Nonlinearity is very peculiar from the standpoint of quantum mechanics, since it is contrary to the linearity of the time-dependent Schrödinger equation, which is commonly assumed to govern the dynamics of any isolated physical system. Hence it is necessary to inquire whether a rational treatment of the occurrence of outcomes is possible without some modifications of current quantum mechanics. Whatever decision is reached on this question, however, it does not appear to bear directly on space–time structure, but rather on the general character of processes.

One of the attractions of emphasizing events and processes, rather than space–time structure, is the availability of relevant experimental evidence at the atomic level, in contrast to the frustrating inaccessibility of evidence at the level of the Planck length. Rejoicing over this fact must be tempered, however, by the formidable conceptual difficulties of inter-

This work originally appeared in R. Penrose and C. Isham (eds.), *Quantum Concepts in Space and Time,* Oxford University Press, 1986. Reprinted by permission of the publisher.

preting experimental results. The main purpose of this chapter is to present these difficulties to an audience of experts on the interrelations of quantum mechanics and space–time theory, in the hope of eliciting new ideas which will free us from our current impasse.

2. SOME FEATURES OF THE QUANTUM MECHANICAL WORLD-VIEW

When the formalism of quantum mechanics is supplemented by a few principles of interpretation, there are striking consequences concerning the nature of the physical world. These consequences by themselves do not constitute a philosophy – neither a theory of knowledge, nor a metaphysics, nor a methodology, and obviously not a value theory – but they are so fundamental that a comprehensive philosophy has an obligation to take them into account. Accordingly, it seems appropriate to refer collectively to these consequences as 'the quantum mechanical world-view'.

One consequence is the possibility of a pure state in which a physical variable has an *indefinite* value. Pure states are represented by normalized vectors of an appropriate Hilbert space $\mathcal{3C}$, or more accurately, they are in one-to-one correspondence with the rays or one-dimensional subspaces of $\mathcal{3C}$. If $\mathcal{F}$ is a physical variable (usually called an 'observable', but this locution should be avoided if one accepts the principle of avoiding anthropocentrism in interpreting the formalism) which has the value f_1 in the state represented by the normalized vector u_1 and the value f_2 ($\neq f_1$) in the state represented by the normalized vector u_2, then in the state represented by the vector u,

(1) $$u = c_1 u_1 + c_2 u_2, \quad c_1 \neq 0, \quad c_2 \neq 0, \quad |c_1|^2 + |c_2|^2 = 1,$$

the physical variable $\mathcal{F}$ does not have a definite value. Furthermore, if we add to the formalism the principle of interpretation that the pure state represents a maximal specification of the system itself, rather than a compendium of someone's knowledge of the system, then *the indefiniteness of $\mathcal{F}$ is objective and not a matter of incomplete knowledge*. In this way, the superposition principle, which is exemplified in eqn (1), has a striking philosophical implication. (It should be added that even if superselection principles hold, restricting the domain of applicability of the superposition principle, there will still remain an abundance of instances of eqn (1), and its philosophical implication remains intact.)

The coefficients c_1 and c_2 in eqn (1) determine the probabilities of the outcomes f_1, f_2 respectively, upon condition that a measurement of the physical variable $\mathcal{F}$ is performed. Specifically, $|c_1|^2$ is the probability of the outcome f_1, and $|c_2|^2$ is the probability of the outcome f_2. Given the principle of interpretation that a pure quantum state represents a maximal specification of the system itself, we cannot regard these probabilities as epistemic – i.e., as the reasonable degrees of belief concerning the values of a quantity which has a definite though unknown value prior to measurement. The probabilities of classical statistical mechanics can reasonably be understood in this way, but not the quantum mechanical probabilities. The latter must be regarded as ontic probabilities for the objectively indefinite physical variable to take on one or another value when it becomes definite; they are implicit in the physical state of affairs rather than in someone's knowledge.

The combination of indefiniteness of value with definite probabilities of possible outcomes can be compactly referred to as *potentiality*, a term suggested by Heisenberg (1958, p. 185). When a physical variable which initially is merely potential acquires a definite value, it can be said to be *actualized*. So far, the only processes we have mentioned in which potentialities are actualized are measurements, but in a non-anthropocentric view of physical theory the measurement process is only a special case of the interaction of systems, of special interest to scientists because knowledge is thereby obtained, but not fundamental from the standpoint of physical reality itself. It remains an open problem, however (discussed further in Section 5), to determine under what circumstances the actualization of potentialities generically takes place.

The pure quantum states of a composite system $1+2$ are represented by normalized vectors in the tensor product Hilbert space

(2) $$\mathcal{K} = \mathcal{K}_1 \otimes \mathcal{K}_2,$$

where $\mathcal{K}_i$ is the Hilbert space associated with component i $(i = 1, 2)$. If u_1, u_2 are orthogonal normalized vectors in $\mathcal{K}_1$ and v_1 and v_2 are orthogonal normalized vectors in $\mathcal{K}_2$, then both $u_1 \otimes v_1$ and $u_2 \otimes v_2$ are normalized vectors in $\mathcal{K}$, representing possible pure states of $1+2$. Application of the superposition principle yields immediately that

(3) $$\Psi = \frac{1}{\sqrt{2}}(u_1 \otimes v_1 + u_2 \otimes v_2)$$

is a normalized vector $\mathcal{K}$ and represents a pure state of $1+2$. But a little algebraic analysis shows that

(4) $$\forall w \in \mathcal{K}_1, \quad z \in \mathcal{K}_2, \quad w \otimes z \neq \Psi.$$

In Schrödinger's locution, Ψ is an 'entangled' state – one in which neither system 1 by itself nor system 2 by itself can be said to be in a pure state; neither can be fully specified without reference to the other. It must be emphasized that the concept of entanglement is inseparable from the role of potentiality in quantum mechanics. Suppose that u_1 and u_2 are related to the physical variable $\mathfrak{F}$ as above, and v_1 and v_2 represent states in which a physical variable $\mathfrak{G}$ of 2 has respective values g_1 and g_2 ($g_1 \neq g_2$). When the state of $1+2$ is represented by Ψ, then both $\mathfrak{F}$ and $\mathfrak{G}$ are merely potential, but in an interlocked manner: $\mathfrak{F}$ and $\mathfrak{G}$ can be actualized so as to have respective values f_1 and g_1, or so as to have respective values f_2 and g_2, and in fact the probability of each of these joint outcomes is $\frac{1}{2}$; but they cannot be actualized either as f_1, g_2 or as f_2, g_1.

If 1 and 2 are spatially well separated, then entanglement implies a peculiar kind of quantum non-locality, the character of which is the primary concern of Section 4.

It may be objected at this point that we have proceeded recklessly in our interpretation of the quantum mechanical formalism. By supplementing the formalism with the principle that a pure quantum state is a maximal specification of the system itself, and by insisting upon the strict avoidance of anthropocentrism, have we not inserted an unhealthy dose of metaphysics into a scientific theory? There are two evident avenues for escaping from the metaphysical consequences which have been catalogued above. One is a hidden variables theory, which maintains that a complete specification of a physical system requires the supplementation of the quantum state, the supplementary information being commonly known as 'hidden variables'. If this avenue is taken, then quantum indefiniteness can be understood epistemically, as incomplete knowledge of the details concerning the system. Quantum mechanical probabilities can then be interpreted in the same way as the probabilities of classical statistical mechanics when some physical constraints are imposed (e.g. equilibrium with specified reservoirs) but some details are unknown. Finally, quantum mechanical entanglement can also be understood epistemically, as the information that one system supplies about the other in virtue of correlations established by their mode of joint preparation. *Prima facie,* this avenue of escaping from the metaphysical peculiarities inferred above is very attractive indeed: it is non-anthropocentric, it is 'realistic' concerning the ontological status of physical systems, and it continues an intellectual tradition which was immensely successful in classical physics. We can easily understand why this avenue was recommended by Einstein and de Broglie, in spite of their full realization of the power of quantum mechanics as a statistical theory and in spite of their own great contributions to it. The significance of Bell's theorem and the experiments based upon it is that this very attractive avenue of interpretation must be abandoned, unless modifications are made which deprive it of its attractiveness. In Section 3 a summary will be given of the impasse which Bell's theorem has exhibited in the program of hidden variables theories.

The other avenue for escaping from the peculiar metaphysics of quantum mechanics is to relax the taboo against anthropocentrism. If *any* physical theory – and quantum mechanics in particular – is regarded as nothing more than an instrument for summarizing human experience and anticipating more experience in the future, then the peculiar features of quantum mechanics cannot be attributed to the things in themselves. There is a long philosophical tradition behind this point of view, from Berkeley through Kant to Mach, and many of the founding fathers of quantum mechanics accepted it either deliberately, as a thoughtfully considered epistemological commitment, or opportunistically, as a way of avoiding painful metaphysical difficulties. There is a large literature on the cogency of positivistic and instrumentalistic interpretations of scientific theories, and I shall refer to Suppe's (1974) survey in order to avoid reviewing familiar arguments. As a surrogate for an argument I shall mention a paraphrase of a statement of Einstein which Wigner gave in an unpublished lecture (Varenna, 1970): 'Do you really believe that the *sun* is nothing more than our perceptions?'

3. BELL'S THEOREM AND LOCAL HIDDEN VARIABLES THEORIES

The program of a hidden variables interpretation of quantum mechanics is logically viable, provided that sufficient latitude is allowed in the interpretation. Specifically, it is essential that the interpretation be 'contextual', in the sense that the value which is assigned by a complete state λ to a physical variable $\mathfrak{F}$ must depend also upon the 'context' of the measurement of $\mathfrak{F}$ (see Bell 1966; Gudder 1970; Shimony 1984*a*). Contextualism permits the program of hidden variables interpretations to escape the well known impossibility theorems of Gleason (1957), of Kochen and Specker (1967), and of Bell himself (1966).

Logical viability is not good enough, however. An acceptable hidden variables theory must also satisfy physical desiderata. In particular, unless revolutionary evidence is presented in favour of changing the relativistic conception of space–time structure, an acceptable hidden variables theory must conform to this structure. So far the profound differences between the special and the general theories of relativity have not impinged significantly upon the analysis of hidden variables theories. The demand upon hidden variables theories which has played the central role is common to the special and general theories: namely, there is no direct causal connection between points with space-like separation. This thesis, which will be called simply 'locality', does not actually occur as a premise of Bell's theorem (1964), but it did provide the heuristics for his crucial premise. For convenience of analysis I shall present neither Bell's original theorem nor any of his later variants, but rather shall give an argument inspired by his work. The argument consists of a modification of a theorem of Clauser and Horne (1974), a modification of a theorem of Jarrett (1984), and some informal discussion.

Theorem 1. *Let* $\langle \Lambda, \Sigma(\Lambda), \rho \rangle$ *be a classical probability space, i.e.* Λ *is a non-empty set,* $\Sigma(\Lambda)$ *a σ-algebra of subsets of* Λ, *and* ρ *a probability measure on* $\Sigma(\Lambda)$. *Let a and b be parameters, for each value of which and for each* $\lambda \in \Lambda$ *there are random variables* $p_\lambda^1(a)$, $p_\lambda^2(b)$, *and* $p_\lambda(a, b)$ *with values in* $[0, 1]$. *Suppose further that the factorizability condition*

(5) $$p_\lambda(a, b) = p_\lambda^1(a) p_\lambda^2(b)$$

is satisfied. Then

(6) $$-1 \le p(a', b') + p(a', b'') + p(a'', b') - p(a'', b'') - p^1(a') - p^2(b')$$
$$\le 0,$$

where

(7a) $$p^1(a) = \int_\Lambda p_\lambda^1(a) \, d\rho,$$

(7b) $$p^2(b) = \int_\Lambda p_\lambda^2(b) \, d\rho,$$

(7c) $$p(a, b) = \int_\Lambda p_\lambda(a, b) \, d\rho.$$

Although Theorem 1 is somewhat more abstract than that stated by Clauser and Horne (1974), it is proved in essentially the same way.

There are various possible interpretations of the terms of Theorem 1 (see Shimony (1984a), Section 4). In the most interesting interpretation, λ is the complete state of a pair of particles propagating in different directions from a common origin and subjected to tests chosen at will by experimenters, with a and b being parameters characterizing the tests to which the first and second particles respectively are subjected. For example, if particles 1 and 2 are photons, then a may be the angle which the transmission axis of a polarization analyser for 1 makes with a fixed axis perpendicular to the propagation direction; and similarly for b. Among the values of the parameter a we shall include ∞, which means that particle 1 is not subjected to a polarization test – i.e. no polarization analyser is placed in its path; a similar interpretation is given to $b = \infty$. We shall restrict our attention to situations in which the only possible outcomes when $a \neq \infty$ are $x_a = \pm 1$, and the only possible outcomes when $b \neq \infty$ are $x_b = \pm 1$. For example, $x_a = 1$ could mean that particle 1 passes through its analyser when the parameter has the value a, and $x_a = -1$ could mean that particle 1 impinges upon the analyser with this value of the parameter but fails to pass through it; and $x_b = \pm 1$ is similarly interpreted. When $a = \infty$, i.e. particle 1 is not subjected to a polarization test, then the trivial outcome is designated by $x_a = \infty$; and likewise if $b = \infty$. It will be assumed that for each $\lambda \in \Lambda$ and each choice of the parameters a, b there is a probability function $p_\lambda(x_a, x_b \mid a, b)$ which satisfies the following conditions:

(8a) $\qquad 0 \leq p_\lambda(x_a, x_b \mid a, b) \leq 1 \quad$ for $\begin{array}{l} x_a = \pm, \infty \\ x_b = \pm, \infty; \end{array}$

(8b) $\qquad$ if $a \neq \infty$, $b \neq \infty$, then $\displaystyle\sum_{X = \pm 1} \sum_{Y = \pm 1} p_\lambda(X, Y \mid a, b) = 1;$

(8c) $\qquad$ if $a \neq \infty$, then $\displaystyle\sum_{X = \pm 1} p_\lambda(X, \infty \mid a, \infty) = 1;$

(8d) $\qquad$ if $b \neq \infty$, then $\displaystyle\sum_{Y = \pm 1} p_\lambda(\infty, Y \mid \infty, b) = 1;$

(8e) $\qquad p_\lambda(\infty, \infty \mid \infty, \infty) = 1.$

It will be useful to define some marginal and conditional probability functions in terms of $p_\lambda(x_a, x_b \mid a, b)$:

(9a) $\qquad p_\lambda^1(x_a \mid a, b) = \displaystyle\sum_X p_\lambda(x_a, X \mid a, b),$

where $X = \pm 1$ if $b \neq \infty$, $X = \infty$ if $b = \infty$;

(9b) $\qquad p_\lambda^2(x_b \mid a, b) = \displaystyle\sum_{X = \pm 1} p_\lambda(X, x_b \mid a, b)$

where $X = \pm 1$ if $a \neq \infty$, $X = \infty$ if $a = \infty$;

(10a) $\qquad p_\lambda^1(x_a \mid a, b, x_b) = \dfrac{p_\lambda(x_a, x_b \mid a, b)}{p_\lambda^2(x_b \mid a, b)};$

(10b) $\qquad p_\lambda^2(x_b \mid a, b, x_a) = \dfrac{p_\lambda(x_a, x_b \mid a, b)}{p_\lambda^1(x_a \mid a, b)}.$

Two kinds of independence conditions can now be compactly defined.

Parameter Independence. For all λ, a, b,

(11a) $\qquad p_\lambda^1(x_a \mid a, b) = p_\lambda^1(x_a \mid a, \infty),$

(11b) $\qquad p_\lambda^2(x_b \mid a, b) = p_\lambda^2(x_b \mid \infty, b).$

(According to this condition the probability of the outcome concerning a single particle of the pair is independent of the choice of the parameter of the test performed upon the other particle.)

Outcome Independence

(12a) $$p^1_\lambda(x_a \,|\, a, b) = p^1_\lambda(x_a \,|\, a, b, x_b).$$

(12b) $$p^2_\lambda(x_b \,|\, a, b) = p^2_\lambda(x_b \,|\, a, b, x_a).$$

We can now prove a theorem which is a modification of the decomposition theorem of Jarrett (1984), with his 'Completeness Condition' replaced by our condition of Outcome Independence.

Theorem 2. *If the conditions of Parameter and Outcome Independence are both satisfied, then*

(13) $$p_\lambda(x_a, x_b \,|\, a, b) = p_\lambda(x_a, \infty \,|\, a, \infty) \cdot p_\lambda(\infty, x_b \,|\, \infty, b).$$

Proof.

$$p_\lambda(x_a, x_b \,|\, a, b) = p^1_\lambda(x_a \,|\, a, b, x_b) \cdot p^2_\lambda(x_b \,|\, a, b)$$

$$= \frac{p^2_\lambda(x_b \,|\, a, b, x_a) p^1_\lambda(x_a \,|\, a, b)}{p^2_\lambda(x_b \,|\, a, b)} p^2_\lambda(x_b \,|\, a, b)$$

$$= p^1_\lambda(x_a \,|\, a, b) p^2_\lambda(x_b \,|\, a, b) = p^1_\lambda(x_a \,|\, a, \infty) p^2_\lambda(x_b \,|\, \infty, b)$$

$$= p_\lambda(x_a, \infty \,|\, a, \infty) p_\lambda(\infty, x_b \,|\, \infty, b).$$

If the conditions of Parameter and Outcome Independence are satisfied, then the conclusion of Theorem 2 constitutes the crucial premise of Theorem 1, once some additional notational identifications are made. It suffices to define

(14a) $$p_\lambda(a, b) = p_\lambda(1, 1 \,|\, a, b),$$

(14b) $$p^1_\lambda(a) = p_\lambda(1, \infty \,|\, a, \infty),$$

(14c) $$p^2_\lambda(b) = p_\lambda(\infty, 1 \,|\, \infty, b).$$

With these identifications, inequality (6) follows. What is most remarkable is the generality of the inequality: it holds regardless of the character of the probability space $\langle \Lambda, \Sigma(\Lambda), \rho \rangle$. Inequality (6) does, however, depend upon the factorizability condition (5), which in itself is neither obvious nor obligatory, but Theorem 2 shows that there are conditions which imply factorizability. Furthermore, if the event which consists of the choice of parameter a together with the consequent test performed on particle 1 has space-like separation from the event which consists of the choice of parameter b together with the test performed on particle 2, then relativistic locality ensures that these two events are not directly connected causally. Consequently, it is plausible that both Parameter and

Outcome Independence should hold. This crucial final step in the argument will, however, be re-examined in Section 4, since the concept of locality is subtle.

If ideal apparatus is assumed, it is easy to give examples of quantum states of a particle pair for which inequality (6) is violated by the quantum mechanical predictions. For example, let the quantum polarization state of a pair of photons propagating respectively in the $\hat{z}$ and $-\hat{z}$ directions be

(15) $$\Psi = \frac{1}{\sqrt{2}}[u_x(1)\otimes u_x(2)+u_y(1)\otimes u_y(2)],$$

where $u_x(i), u_y(i)$ represent states of polarization of photon i $(i=1,2)$ along the $\hat{x}$ and $\hat{y}$ axes respectively. The quantum mechanical prediction based upon Ψ for the joint passage of 1 and 2 through analysers with transmission axes making angles a, b respectively with the x-axis is easily shown to be

(16) $$p_\Psi(a, b)= \tfrac{1}{2}\cos^2(a-b),$$

and the probability of the passage of a single photon is

(17) $$p_\Psi^1(a)=p_\Psi^2(b)=\tfrac{1}{2}.$$

If a', a'', b', b'' are respectively chosen to be $22\tfrac{1}{2}°$, $0°$, $45°$, $67\tfrac{1}{2}°$, then

$$p_\Psi(a', b')= p_\Psi(a', b'')= p_\Psi(a'', b')=0.4268,$$

and

$$p_\Psi(a'', b'')=0.0732.$$

Hence

(18) $$p_\Psi(a', b')+p_\Psi(a', b'')+p_\Psi(a'', b')$$
$$-p_\Psi(a'', b'')-p_\Psi^1(a')-p_\Psi^2(b')=0.2072,$$

in violation of equality (6). Hence no hidden variables theory satisfying relativistic locality (as constructed above) can agree with all the predictions of quantum mechanics. This result is one of the forms of Bell's theorem.

Even when the assumption of ideal apparatus is dropped it is possible to exhibit situations in which the quantum mechanical predictions violate inequality (6). As a result, experimental confrontations between quantum mechanics and all hidden variables theories satisfying the factorizability condition (5) are possible. The confrontation is sharpened in the experiment of Aspect *et al.* (1982), in which the polarization analysis events on the two photons have space-like separation, which by the argument above leads almost inevitably to the factorizability condition. This spectacular experiment yielded results in agreement with quantum mechanics and in disagreement with inequality (6). As in all actual experiments, a determined advocate of a hypothesis can find loopholes to explain the disconfirming evidence, but in this case the loopholes are unattractive and unpromising. The evidence is overwhelmingly against the family of local hidden variables theories.

4. ASPECTS OF NON-LOCALITY

The experimental violation of inequality (6) indicates that a re-examination of Parameter and Outcome Independence is needed. Is each of them required by relativistic locality? Which of them is violated by quantum mechanics?

It is easy to see that a violation of Parameter Independence would be the basis of a violation of relativistic locality, as Jarrett (1984) pointed out. Suppose that for some specified a, b

(19) $$p_\lambda^1(x_b \mid a, b) \neq p_\lambda^1(x_b \mid \infty, b),$$

and let an ensemble of pairs of particles be prepared, each pair in the complete state λ, propagating within a short time interval from a common origin. Let Experimenter I interact with particles 1 of the ensemble in a region R_1 and Experimenter II with particles 2 of the ensemble in a region R_2, and suppose that R_1 and R_2 have space-like separation. The protocol is also laid down that Experimenter I will either choose the value a for each of the analysers upon which the particles 1 impinge or else the value ∞ (i.e. the analysers will be removed from the paths of the 1 particles). Experimenter II chooses the value b for the analysers upon which the particles 2 impinge and observes the resulting statistics. Because of Inequality (19) the statistics observed by Experimenter II will depend upon the choice made by Experimenter I, and by making the ensemble large enough the information about the decision can be transmitted from I to II with a probability as close to unity as desired. Furthermore, the whole process can be automated and conscious observers dispensed with, so that an explosion can be triggered in R_2 (with a probability as close to unity as desired) by the automated selection in R_1 of one of the two parameter values a or ∞. The only way to deny that a violation of relativistic locality occurs under the conditions just supposed is to insist upon certainty as a criterion for transmission of a message, and that demand does not seem reasonable.

The quantum mechanical predictions concerning ensembles of pairs of particles do not violate Parameter Independence, provided that non-locality is not explicitly built into the interaction Hamiltonian of the particle pair. Specifically, it is impossible to capitalize upon the entanglement of the quantum state of a two-particle system for the purpose of sending a message to an observer of one of the particles by performing an operation upon the other particle. This is a theorem which has been proven in great generality by Eberhard (1978), Ghirardi *et al.* (1980), and Page (1982), and I shall not give a proof here.

There seems to be no way to use a violation of Outcome Independence for the purpose of sending a message faster than light. Consider again the two Experimenters I and II observing respectively the 1 and 2 particles of an ensemble of pairs, and suppose that for specified values a, b of the adjustable parameters

(20) $$p_\lambda^2(1 \mid a, b) \neq p_\lambda^2(1 \mid a, b, 1),$$

from which it follows that

(21) $$p_\lambda^2(1 \mid a, b, 1) \neq p_\lambda^2(1 \mid a, b, -1).$$

Experimenter I may try to capitalize upon inequality (21) to send a message to Experimenter II by blocking the transmission of all 2 particles which have partners (i.e., 1 particles) for which the outcome is j ($j = 1, -1$). Since the statistics of $x_b = 1$ differ, in view of inequality (21), according to the choice of j, a message can be transmitted with a probability as close to unity as desired by sufficiently increasing the size of the ensemble. But this message cannot be transmitted any faster than the flight of the 2 particles themselves, and therefore it cannot be superluminal. I see no way of varying the experimental arrangement so as to achieve superluminal transmission. Consequently, a violation of Outcome Independence does not constitute a counter-example to relativistic locality in the same sense as a violation of Parameter Independence. Shimony (1984*b*) called the kind of non-locality implicit in a violation of Outcome Independence 'uncontrollable non-locality'. I now think that this terminology is misleading, since inequality (21) does permit a controlled message to be transmitted, though not superluminally.

Situations can be readily exhibited in which the quantum mechanical predictions violate Outcome Independence. Consider again the Ψ of eqn (15), and suppose that it represents the complete polarization state of the composite system $1 + 2$. Let the parameters a and b be chosen equal, for specificity 0 (in other words, both polarization analysers are oriented with transmission axes along $\hat{x}$). Then, if ideal analysers are assumed,

(22) $$p_\Psi^1(1 \mid 0, 0) = \tfrac{1}{2},$$

(23) $$p_\Psi^1(1 \mid 0, 0, 1) = 1,$$

thus violating Outcome Independence.

We have now answered both of the questions with which Section 4 commenced, and the answers mesh in a most interesting manner. Quantum mechanics violates Outcome Independence, which cannot be used to send a superluminal message; and it conforms to Parameter Independence, a violation of which would permit superluminal communication. We may summarize by adapting a well known political slogan: there is 'peaceful coexistence' between quantum mechanics and relativity theory. This result, incidentally, throws light upon a question that must have occurred to some readers: if quantum mechanics is non-local in some sense, why should we feel obligated to restrict our attention to local hidden variables theories? Why not escape from most of the peculiar metaphysics of quantum mechanics by paying the price of accepting a small part of this peculiarity, namely non-locality? There are, in fact, some serious proposals, notably by Vigier and collaborators (Vigier 1982) for escaping from Bell's Theorem by building non-locality into their hidden variables theories, specifically, by postulating shock waves of superluminal velocity in a hypothetical 'quantum ether'. It is far from clear, however, that there is peaceful coexistence between relativity theory and such hidden variables theories. (See Shimony (1984b) for a challenge on this point.) Consequently, local hidden variables theories may be possible only if relativity theory is replaced by a radically different theory of space–time structure, whereas the non-locality of quantum mechanics may leave relativistic space–time structure intact, but only change our conception of an event in space–time.

In spite of the discriminations made above, the idea of peaceful coexistence between relativistic space–time structure and quantum mechanics remains puzzling, and a further examination is needed of the concept of an event. Let us return once more to the state of photon pair $1+2$ represented by Ψ of eqn (15), which we shall again assume to constitute a complete specification of the polarization properties of the pair. Let $A_x(i)$ be the physical variable which takes on the value 1 if photon i ($i = 1, 2$) is polarized along the direction $\hat{x}$ and the value -1 if it is polarized along the direction $\hat{y}$ (which is perpendicular both to x and to directions $\hat{z}$ and $-\hat{z}$ of propagation of the two photons); and let $A_{x'}(i)$ have an analogous meaning concerning another direction $\hat{x}'$ in the plane perpendicular to $\hat{z}$. We note – as we have not done so far – that Ψ is cylindrically symmetrical and hence can be written not only in the form of eqn (15), but also as

(24) $$\Psi = \frac{1}{\sqrt{2}}[u_{x'}(1) \otimes u_{x'}(2) + u_{y'}(1) \otimes u_{y'}(2)].$$

Now let us suppose that $A_x(1)$ and $A_{x'}(2)$ are measured with ideal apparatus, and that the acts of measurement have space-like separation.

Furthermore, let Σ be an inertial frame of reference in which the measurement of $A_x(1)$ occurs earlier than that of $A_{x'}(2)$, and Σ' an inertial frame in which the reverse time order holds. It is reasonable (though, as we shall see, not obligatory) to describe the sequence of relevant events from the standpoint of Σ to be the following: $t_0 < t_1 < t_2$ and

at t_0: $A_x(1), A_x(2), A_{x'}(1), A_{x'}(2)$ are all indefinite;
at t_1: $A_x(1)$ and $A_x(2)$ are definite, but not $A_{x'}(1)$ and $A_{x'}(2)$ (because the first measurement reduces the superposition Ψ, yielding either $u_x(1) \otimes u_x(2)$ or $u_y(1) \otimes u_y(2)$);
at t_2: $A_x(1)$ and $A_{x'}(2)$ are definite, but $A_x(2)$ and $A_{x'}(1)$ are indefinite (because the second measurement reduces $u_y(2)$ so as to yield either $u_{x'}(2)$ or $u_{y'}(2)$ but leaves $u_x(1)$ intact).

The sequence of events from the standpoint of Σ' is the following (with reasons suppressed, since they are analogous to the ones given for Σ): $t_0' < t_1' < t_2'$ and

at t_0': $A_x(1), A_x(2), A_{x'}(1), A_{x'}(2)$ are all indefinite;
at t_1': $A_{x'}(2)$ and $A_{x'}(1)$ are definite, but $A_x(2)$ and $A_x(1)$ are indefinite;
at t_2': $A_{x'}(2)$ and $A_x(1)$ are definite, but $A_{x'}(1)$ and $A_x(2)$ are indefinite.

The criterion for definiteness of a physical variable A at a given time is that the quantum state at that time is an eigenstate of A, as commonly assumed in the interpretation of the quantum mechanical formalism. In addition, a plausible premise has been tacitly employed, even though it is not essential to the interpretation of the formalism: namely, that the definiteness or indefiniteness of a physical variable constitutes an event.

When the sequences from the standpoints of Σ and Σ' are compared there are some agreements and some disagreements. There is agreement concerning the initial events (at t_0 and t_0' respectively), and also concerning the final events (at t_2 and t_2' respectively). There is disagreement concerning the intermediate events. Consequently, the two accounts of *processes* from initial to final sets of events are in disaccord. But it is important to note that the process is a theoretical construction. $A_x(2)$ is not measured by an observer in Σ but only inferred from the measured value of $A_x(1)$ in virtue of the correlation expressed by Ψ; and likewise $A_{x'}(1)$ is inferred rather than measured directly by an observer in Σ'. As to causality, there is also some agreement and some disagreement between the descriptions in Σ and Σ'. There is no pair of events such that an observer in Σ identifies one as the cause and the other as the effect, while an observer in Σ' makes the opposite identification. However, an observer in Σ identifies the measurement of $A_x(1)$ as the cause of the transition of $A_x(2)$ from being indefinite to being definite, whereas an observer in Σ' does not recognize this causal sequence; and likewise an observer in Σ' identifies the measurement of $A_{x'}(2)$ as the cause of the transition of $A_{x'}(1)$ from being indefinite to being definite, whereas an observer in Σ does not do so. The thesis of peaceful coexistence presupposes a conceptually coherent reconciliation of the descriptions from the standpoints of Σ and Σ'. Even more desirable, in the spirit of the geometrical formulation of space–time theory, would be a coordinate-free account.

I do not believe that we are able, in our present state of understanding of the foundations of quantum mechanics, to meet these desiderata. The reason is that the puzzling events in question consist of transitions from indefiniteness to definiteness of physical variables, in other words, of the actualization of potentialities. Although the concept of potentiality is crucial to the world-view of quantum mechanics (as argued in Section 2), the theory is notoriously uninformative about the process of actualization. More will be said about this problem in Section 5, though without a firm conclusion. For the present, however, I wish to present a conjecture which has some coherence even in the absence of a full account of the actualization of potentialities.

The conjecture is that *the reduction of a superposition, however that is achieved, does not* ipso facto *constitute the actualization of a physical variable of which the reduced state is an eigenstate.* When this conjecture is applied to the previously considered sequence of events in Σ the consequence is that $A_x(2)$ is not definite at t_1; and when it is applied to the sequence of events in Σ' the consequence is that $A_{x'}(1)$ is not definite at t_1'. These statements are made in full cognizance of the fact that at t_1 the state of the composite system $1+2$, from the standpoint of Σ, is represented either by $u_x(1) \otimes u_x(2)$ or by $u_y(1) \otimes u_y(2)$, and in either state the probability that $a_x(2)$ has the value 1 is either unity or zero. That is to say, an observer who is aware of the result of the measurement made upon $A_x(1)$ would know with certainty what would be found if $A_x(2)$ were measured – provided, of course, that there were no intervening disturbance of system 2, such as the measurement of $A_{x'}(2)$. My conjecture, however, says that certainty with regard to what would be found concerning $A_x(2)$ is not tantamount to the actualization of $A_x(2)$, or equivalently, to the occurrence of the event that $A_x(2)$ is definite. According to the conjecture, the actualization of a potentiality must not be conceived as a limiting case of probability – as probability 1 or 0. Instead, actuality and potentiality are radically different modalities of reality. These modalities are obviously related, since the probabilities associated with a quantum mechanical potentiality govern the propensities of actualization, and in the extreme case of probabilities 1 and 0 they positively exclude one or the other possible outcome. Nevertheless, excluding one outcome does not constitute actualization of the other outcome.

The conjecture just presented removes the discrepancy between the description of intermediate states from the standpoints of Σ and Σ'. According to the conjecture, from neither standpoint is $A_x(2)$ or $A_{x'}(1)$ definite. Of course, there is a discrepancy of the time ordering of the measurements of $A_x(1)$ and $A_{x'}(2)$, but this is the familiar frame-dependence of temporal ordering of events with space-like separation, and quantum mechanics introduces nothing new on this point.

My conjecture seems to be at odds with the well known criterion for the existence of an element of physical reality formulated by Einstein *et al.* (1935): 'If, without in any way disturbing a system, we can predict with certainty (i.e., with probability equal to unity) the value of a physical quantity, then there exists an element of physical reality corresponding to this physical quantity'. The metaphysical content of the criterion of Einstein *et al.* is minimal. Beyond assuming that there are systems and properties that have existence independent of human experience, they seem to be making no metaphysical commitment. The primary import of their criterion is epistemological: it gives a sufficient condition for asserting that the value of a physical variable is objectively definite. My conjecture, however, has a considerable metaphysical content, for it distinguishes between two modalities of reality, and it modifies the criterion of Einstein *et al.* by abstaining from the identification of the 'existence of an element of physical reality' with existence in the modality of actuality. It is well known that Bohr (1935) offered an answer to the argument of Einstein *et al.* for the incompleteness of quantum mechanics. However, my conjecture is only vaguely cognate to Bohr's answer. Bohr's fundamental dichotomy is between results of experiments, describable in the language of classical physics, and the quantum mechanical wave function, which he took to be an algorithm or intellectual instrument for calculating the probabilities of experimental outcomes. By contrast, the conjecture presented above is quite metaphysical and attributes a mode of reality to the quantum state, and a different one to actualizations, and it makes no commitment that the latter occur only in the context of deliberate experimentation by human beings.

5. NONLINEARITY AND THE ACTUALIZATION OF POTENTIALITIES

The quantum mechanical law of dynamical development for an isolated system is the time-dependent Schrödinger equation, which can be written in integrated form as

(25) $$\psi(t) = U(t)\psi(0),$$

where

(26) $$U(t) = e^{-iHt/\hbar}$$

for an appropriate self-adjoint operator H. For each t, $U(t)$ is a unitary operator on the Hilbert space to which the vectors $\psi(t)$ belong. For our present purposes, the most important characteristic of $U(t)$ is linearity, which has the effect of propagating superpositions. When eqn (25) is applied to a composite system consisting of a microscopic system together with a measuring apparatus, it has the consequence – under suitable initial conditions – of leading to a final state in which the indexical apparatus variable (i.e. the one from which the value of the microscopic variable of interest is inferred) is indefinite: a pointer needle does not point in any definite direction. This peculiarity of the measurement process shows that quantum mechanics does not have, in any obvious way, the conceptual tools for explaining how potentialities are actualized. There is a large and sophisticated literature on attempts to solve the problem of the actualization of potentialities strictly within the formalism of quantum mechanics, but in my opinion none of these attempts succeed. I shall not give even a brief review of this literature, partly because of shortage of space and partly because of the availability of excellent surveys (e.g. d'Espagnat 1976).

The problem of actualizing potentialities is, of course, completely circumvented by a hidden variables theory, and indeed this is one of the major motivations for investigating such theories. But we have already summarized the strongest evidence against hidden variables, and therefore must seek elsewhere. A very attractive program, which makes a minimal modification of quantum mechanics and avoids anthropocentric interpretations, is to keep the entire time-independent formalism of quantum mechanics (states, superpositions, physical variables, probabilities of outcomes, etc.) but to substitute a nonlinear differential equation for the linear time-dependent Schrödinger equation. It should be stated at the outset that there are obstacles to this program in the form of theorems (beautifully surveyed by Simon 1976) which derive the time-dependent Schrödinger equation from the time-independent part of the quantum mechanical formalism, provided that very mild and reasonable suppositions are made about temporal evolution. These obstacles are not insuperable, however, for these mild and reasonable suppositions may in fact be false, and alternatives to them are worth exploring.

A number of desiderata for a nonlinear quantum dynamics can be reasonably stated.

(i) Suppose that we have a situation in which phenomenologically it is known that actualization of potentialities occurs (e.g. a measurement situation), and suppose that initially the state of the system is represented by $\sum c_n u_n$, where each u_n is an eigenvector of the physical variable which is actualized. Then the nonlinear equation must govern a process

$$\sum c_n u_n \rightarrow u_k$$

for some k, with no residual contributions from u_i with $i \neq k$, in a finite time interval.

(ii) In such a situation, the probability of the transition to u_k is $|c_k|^2$.

(iii) The dynamical law should be general, not restricted to the measurement situation.

(iv) It must be possible to show in certain cases that the transition described in (i) is very rapid.

(v) There must be approximate agreement with the standard linear dynamics of quantum mechanics in the case of micro-systems.

Comments. Desideratum (i) retains the point of view that the quantum state is a complete description of the physical system – a thesis to which we have returned in view of the difficulties exhibited for hidden variables theories. Implicit, however, in desideratum (i) is a view of actualization: namely, that it is the limiting case of potentiality, in which the final state is an eigenstate of the physical variable of interest, or alternatively, in which the probability of one value of the physical variable is unity and that of all others is zero. Desideratum (i) is therefore in direct conflict with the conjecture concerning the relation between potentiality and actuality proposed for relativistic reasons in Section 4. The promise of success of a nonlinear variant of the Schrödinger equation which fulfils desideratum (i) would therefore seriously undercut my conjecture.

Desideratum (ii) is essential in order to regard the measuring process as a special case of the interaction between physical systems, which in turn is a necessary condition for the avoidance of anthropocentrism.

Desideratum (iv) is imposed by respect for phenomenology, for it is known that there are measurement processes in which the time interval between an initial state described by a superposition and a final definite result is very short indeed. Coincidence counting technology indicates that the time interval may be of the order of a few tens of picoseconds.

Desiderata (iii) and (v) are imposed by the excellent agreement of experimental results with quantum mechanical predictions in many experiments involving microscopic systems.

A serious investigation of a nonlinear variant of the time-dependent Schrödinger equation has been reported by Pearle (1969). Quite apart from considerations of agreement with experiment, Pearle's nonlinear equation is heuristically interesting. It is successful in achieving desiderata (i) and (ii), although I am sceptical that it, or any moderate variant of it, can achieve desiderata (iii) and (v).

The equation of Bialynicki-Birula and Mycielski (1976) was not designed for the purpose of solving the problem of the actualization of potentialities, and in fact it fails to fulfil the crucial desideratum (i). Nevertheless, it is noteworthy because of investigations which bear upon desideratum (v). The equation is

$$(27) \qquad \frac{\partial \psi}{\partial t} = H\psi - b \log|a\psi|^2 \psi,$$

where H is the usual quantum mechanical Hamiltonian operator, a is a constant introduced for dimensional reasons and of no great physical importance, and b is a non-negative quantity with dimension of energy which determines the magnitude of the nonlinearity. Neutron optics experiments have been performed for the purpose of setting upper bounds to b, and the results are very striking. A neutron interferometer experiment of Shull *et al.* (1980) showed that b has an upper bound of about 3×10^{-13} eV, and a neutron Fresnel diffraction experiment of Gähler *et al.* (1981) set a bound of about 3×10^{-15} eV. These are, to my knowledge, by far the best experimental confirmations of the linearity of the temporal evolution of the quantum state. Although these experiments tested only the Bialynicki-Birula–Mycielski equation, and not nonlinear equations generally, they indicate that nonlinear terms are likely to be required by experiment to be very small indeed.

It is premature to pronounce judgment upon the program of nonlinear variants of the Schrödinger equation. Nevertheless, my personal view is pessimistic. In particular, there is great tension between desiderata (iii) and (v). In order that the reduction of the superposition described in desideratum (i) should proceed very rapidly, the nonlinearity must be large, at least in certain measurement situations; but in order to account for the success of the time-dependent Schrödinger equation in the experiments of Shull *et al.* and Gähler *et al.* the nonlinearity must be small. The only way out of this impasse is to hypothesize that the magnitude of the nonlinearity depends crucially upon either the mass or the number of degrees of freedom of the system, or upon both; for the confirmation of the time-dependent Schrödinger equation has been achieved with microscopic systems (neutrons), whereas the rapid reduction of superpositions is required for a measuring apparatus, which is macroscopic. So far, no one has proposed a concrete implementation of this proposal. Consequently, I remain pessimistic that the reduction of a superposition, i.e. the actualization of a potentiality, can be correctly described by a continuous process in space–time, governed by a nonlinear differential equation.

6. CONCLUSIONS

Work during the last two decades has led us to a high plateau concerning the interpretation of quantum mechanics, the topography of which was described in Sections 2 and 3, and the first part of Section 4. These results can be stated with great confidence. By contrast, the statements in the latter part of the paper were made diffidently. It is only a conjecture that the actualization of a potentiality must be sharply distinguished from probabilities unity and zero. And it is only a conjecture that the reduction of superpositions is inexplicable by substituting some appropriate nonlinear differential equation for the time-dependent Schrödinger equation.

Nevertheless, it should be emphasized that these two conjectures mesh quite well. As noted in the first comment on the desiderata, the success of the program of a nonlinear variant of the Schrödinger equation would lend plausibility to the view that actualization is the limiting case of potentiality – the achievement of the extreme values of probability. Hence our brief survey of the poor prospects of a nonlinear differential equation governing the temporal evolution of the quantum state agrees well with the conjecture concerning actualization presented in Section 4.

Are we able to make any more positive statement about the conditions under which potentialities are actualized? I shall risk one more conjecture: that the actualization of a potentiality is discontinuous and stochastic. One obvious motivation for this conjecture is that desideratum (iv) is easily satisfied, for the stochastic process can be accomplished with arbitrarily great speed. Someone may comment at this point that after much agonizing I have returned to a rather conventional interpretation of quantum mechanics, for in textbook presentations and in all versions of the Copenhagen interpretation the outcome of a measurement is stochastic and discontinuous. I wish, however, to maintain a distance from this conventional interpretation, until and unless overwhelming considerations drive me to accept it. My reason, once again, is to avoid anthropocentrism. The actualization of potentialities seems to be such a fundamental transition in the physical world that I cannot seriously believe that it occurs only when experiments are performed by human beings. We simply are not that important in the scheme of things, and we should not try to undo the Copernican revolution. Consequently, we must be attentive to non-anthropocentric accounts of the actualization of potentialities, even if the formulations up to now are quite unsatisfactory. Károlyházy *et al.* (1982) have developed a theory in which the reduction of superpositions is linked to stochastic processes governing the space–time metric. Another possibility that is not exhausted is that the superposition principle fails for some kinds of macroscopic systems. Even if the fascinating efforts to exhibit macroscopic quantum coherence are successful, there still may be parts of the macroscopic domain which serve as the locus of actualization. Finally, there is mind, as London and Bauer pointed out in their work on measurement (1939), and as Wigner has been lecturing for three decades (e.g. 1961). If only human minds were the locus of actualizations of potentialities, then anthropocentrism would re-appear, but if there are non-human minds – as we have good reason to believe – might they not be as efficacious in the reduction of superpositions as our own?

My final conclusion is that the high plateau which we have reached in the interpretation of quantum mechanics is a darkling plain.

APPENDIX: THE EVERETT INTERPRETATION

Although I have not referred to the Everett interpretation (1957) in the body of this chapter, I briefly mentioned it in my original lecture, and then most of the discussion following the lecture was devoted to it. This is not the first time that I have observed a discussion of the Everett interpretation expanding to fill the available vessel.

Part of the appeal of the Everett interpretation is the metaphor of a trunk with many branches. The ramification occurs whenever, according to more orthodox interpretations, a superposition is reduced, but instead of a choice among many branches the Everett interpretation retains them all. The metaphor tacitly presupposes, however, a preferred basis, in which the vector representing the state of the system is expressed: the vector itself is the trunk, and the projections upon the basis vectors are the branches. But objectively there is no preferred basis. There is a continuum of possible bases, all on the same footing. But when one speaks in this more accurate way about the mathematics of the quantum state, the quasi-familiarity of the original metaphor is lost, and with it the appeal.

There have been some valiant attempts to re-establish the branching metaphor by singling out a basis – notably the 'pointer basis' of Zurek (1981) and the 'environmentally determined basis' of Deutsch (1981). I shall not attempt to explore the technicalities of these efforts, but shall remark upon a slippery feature which they share. Only if the entity which is responsible for the singling out of the basis – the pointer of the apparatus in the one case, the environment in the other – is endowed with a solid set of properties does the project even begin to have intuitive appeal. But if the fundamental ontological fact is merely that the universe is in a quantum state, this solidity evaporates. Only in some bases (relatively few, it would seem, whatever that precisely means) are there branches in which the pointer exists and in which the environment has its salient properties, and even for these bases most branches do not endow the apparatus or the environment with the requisite properties for singling out a basis. There is, consequently, a danger of a regress in the process of basis determination.

Even if a single branch system is accepted, the Everett interpretation appeals to our imaginations only because we endow each branch with the same kind and the same degree of actuality which we habitually (due to our subjectivity or our limitations) attribute to only one branch. Along the branch that I (subjectively) experience I may blunder, but along another branch, with equal actuality, I triumph gloriously. The Everett interpretation can be used in this way to mitigate sorrows, but this use is two-edged, for it equally well implies the speciousness of happiness. Moral questions apart, this procedure for attributing the same actuality to each branch results from the fundamental proposal of Everett to eliminate the distinction between potentiality and actuality. And, of course, if all branches are on the same footing, it is true that this distinction vanishes. But what is to be conflated with what? Making all branches equally actual appeals to our imaginations, nourished as they have been by drama and novels. But might it not be at least as appropriate, or more so, to make all branches equally potential, equally in limbo, equally remote from existential definiteness? Might not the elimination of actuality in favour of potentiality be demanded by the occurrence of interferences among branches, as provided by the quantum formalism?

The Everett interpretation is sometimes defended by an analogy to the Copernican theory: the true physical situation seems to be in contradiction with appearances because naive people fail to take into account their own physical states. I believe, however, that this analogy is faulty, and its weakness is instructive. The reconciliation of the apparent turning of the celestial sphere with the correct physical kinematics is achieved by an analysis of visual appearances in terms of the actual relative motions of the bodies observed, including that of the observer. But the Everett interpretation lacks an ingredient needed to implement this analogy: namely, a principle in terms of which one branch of the subject is endowed with subjective immediacy. Without such a principle there is only potentiality, as sketched in the preceding paragraph. But with such a principle the interpretation would lose its purity, and indeed there would be a reduction of a superposition based upon an interaction of the physical world with a mind.

My final objection to the Everett interpretation does not rest upon any anomalies or conceptual difficulties, but simply upon the consideration that it is an immense extrapolation of the linear dynamics of quantum mechanics from the microscopic domain which supplies experimental confirmation for this dynamics. We are so far from having evidence for the validity of this dynamics for macroscopic systems, for organisms, for creatures endowed with mentality, and for the space–time field that the extrapolation of this dynamics to the entire cosmos should be recognized as sheer conjecture. Some of the points which I have just made are amplified, and others are given, in Bell (1978) and Stein (1984).

REFERENCES

Aspect, A., Dalibard, J., and Roger, G. (1982). *Phys. Rev. Lett.* **49**, 1804.
Bell, J. S. (1964). *Physics* **1**, 195.
Bell, J. S. (1966). *Rev. Mod. Phys.* **38**, 447.
Bell, J. S. (1978). *Epistemol. Lett.* **20**, 1. Reprinted in revised form in *Speakable and unspeakable in quantum mechanics* (1987). Cambridge University Press, Cambridge, pp. 117–38.
Bialynicki-Birula, I. and Mycielski, J. (1976). *Ann. Phys. N.Y.* **100**, 62.
Bohr, N. (1935). *Phys. Rev.* **48**, 696.
Clauser, J. F. and Horne, M. A. (1974). *Phys. Rev.* D **10**, 526.
Deutsch, D. (1981). Quantum theory as a universal physical theory. University of Texas at Austin preprint.
Eberhard, P. (1978) *Nuovo Cimento* **46B**, 392.
Einstein, A., Podolsky, B., and Rosen, N. (1935). *Phys. Rev.* **47**, 777.
d'Espagnat, B. (1976). *Conceptual foundations of quantum mechanics* 2nd edn. Benjamin, Reading, Massachusetts.
Everett, H. (1957). *Rev. Mod. Phys.* **29**, 454.
Gähler, R., Klein, A. G., and Zeilinger, A. (1981). *Phys. Rev.* A **23**, 1611.
Ghirardi, G. C., Rimini, A., and Weber, T. (1980). *Lett. Nuovo Cimento* **27**, 293.
Gleason, A. (1957). *J. Math. Mech.* **6**, 885.
Gudder, S. (1970). *J. Math. Phys.* **11**, 431.
Heisenberg, W. (1958). *Physics and philosophy.* Harper and Row, New York.
Jarrett, J. (1984). *Noûs* **18**, 569.
Károlyházy, F., Frenkel, A., and Lukács, B. (1982). In *Physics as natural philosophy, essays in honor of László Tisza on his seventy-fifth birthday* (ed. A. Shimony and H. Feshbach). MIT Press, Cambridge, Massachusetts.
Kochen, S. and Specker, E. P. (1967). *J. Math. Mech.* **17**, 59.
London, F. and Bauer, E. (1939). *La théorie de l'observation en mécanique quantique* (Actualités scientifiques et industrielles 775). Hermann, Paris. Engl. transl. in *Quantum theory and measurement* (ed. J. A. Wheeler and W. Zurek) (1983). Princeton University Press, Princeton.
Page, D. (1982). *Phys. Lett.* **91A**, 57.
Pearle, P. (1969). *Phys. Rev.* D **13**, 857.
Shimony, A. (1984*a*). *Br. J. Phil. Sci.* **35**, 25. [Reprinted as Chapter 9 of this volume.]

Shimony, A. (1984*b*). In *Proc. Int. Symp. on Foundations of Quantum Mechanics in the Light of New Technology* (eds. S. Kamefuchi, H. Ezawa, Y. Murayama, M. Namiki, S. Nomura, Y. Ohnuki, and T. Yajima). Physical Society of Japan, Tokyo, pp. 25–30.

Shull, C. G., Atwood, D. K., Arthur, J., and Horne, M. A. (1980). *Phys. Rev. Lett.* **44**, 765.

Simon, B. (1976). In *Studies in mathematical physics: essays in honor of Valentine Bargmann* (ed. E. Lieb, B. Simon, and A. Wightman). Princeton University Press, Princeton, pp. 327–49.

Stein, H. (1984). *Noûs* **18**, 635.

Suppe, F. (ed.) (1974). *The structure of scientific theories*. University of Illinois Press, Urbana.

Vigier, J.-P. (1982). *Astr. Nachr.* **303**, 55.

Wigner, E. P. (1961). In *The scientist speculates* (ed. I. J. Good). Heinemann, London.

Zurek, H. (1981). *Phys. Rev. D* **24**, 1516.

12

An exchange on local beables

Comment* on Bell's theory

A. Shimony, M. A. Horne, and J. F. Clauser

Dr. Bell's paper, "The Theory of Local Beables",[1] performs a valuable service in clarifying two fundamental concepts: namely, *locality* and *physical reality*. His clarification leads him to a fundamental and highly reasonable assumption, expressed in equation (2) of Sect. 2. He then attempts in Sect. 4 to prove inequality (16) as a consequence of his equation (2). Unfortunately, we believe that his proof is not correct. A counter-example shows that (16) does not follow from (2) alone. Our objections are not given in a spirit of skepticism, since (16) does follow from other reasonable assumptions of locality and physical reality. These assumptions were discussed in an earlier paper[2] and will be reconsidered in this letter.

To illustrate the falsity of his claim we consider the following local beable situation. A person concocts a set of correlation experiment data. The data consist of four columns of numbers, indexed by event number j. Two of the columns contain the apparatus parameter settings, a_j and b_j, while the other two columns contain the experimental results, A_j and B_j. These data have been so contrived as to exhibit the correlation specified by quantum mechanics. The person sends the result columns (A_j and B_j) to an apparatus manufacturer; he sends the apparatus parameter settings to the secretaries of two physicists who will perform a correlation experiment using apparatus supplied by the manufacturer. The manufacturer preprograms the apparatus simply to display in sequence the results A_j (B_j) independently of what parameter setting is employed by physicist 1 (2). As physicist 1 (2) is about to record the result of the jth event, his secretary quietly whispers in his ear the suggestion that he set his apparatus parameter to the value a_j (b_j). The experimentalists thus record preprogrammed results and parameter settings *which are* consistent with

These two papers appeared in *Dialectica* 39 (1985), pp. 97–102 and pp. 107–9. Reprinted by permission of the publisher.
*Work done under the auspices of the U.S. Energy Research & Development contract No. W-7405-Eng-48.

1. J. S. Bell, Epistemological Letters 9, 11 (March 1976).
2. J. F. Clauser and M. A. Horne, Phys. Rev., *D10,* 526 (1974).

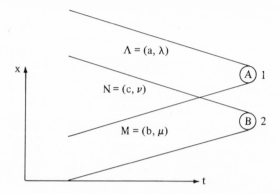

Figure 1. Space-time diagram of correlation experiment beables.

the quantum mechanical prediction. Thus, when they later compare their data, they find the resulting correlation is in violation of (16). Clearly, the violation occurs even though local beables alone were responsible for the results.

Now let us examine Bell's argument in some detail, in order to see what has gone amiss. We shall first recapitulate some of his notation. Recall that he is concerned with two beables A and B, localized respectively in space-time regions 1 and 2 which are space-like separated from each other. Denote by N the full set of beables contained in the region formed by the intersection of the backward light cones of 1 and 2. Denote by Λ the full set of beables in the remainder of the backward light cone of 1 and by M those in the remainder of the backward light cone of 2. (See Fig. 1.)

Bell uses the notation $\{C \mid D\}$ to mean the probability (or probability density, in case of a continuum of values for C) given D of the value C for some variable beable. (He denotes the variable by same letter C, an ambiguity in notation which causes no confusion.)

Bell's formulation of local causality, Eq. (2), is essentially the following: Let C be a variable beable localized in some space-time region, large or small. This region has a unique backward light cone; let D denote *all* the beables in this backward light cone. Then

$$\{C \mid D, E\} = \{C \mid D\}$$

holds for any beables E localized in space-time regions with a space-like separation from the region of C. We fully accept that this formulation is an unambiguous and concise statement of reasonable hypotheses about locality and reality. Furthermore, some of Bell's applications of Eq. (2) are certainly legitimate, specifically the replacements

$$\{A \mid \Lambda, M, N, B\} = \{A \mid \Lambda, N\}$$

$$\{B \mid \Lambda, M, N\} = \{B \mid M, N\}$$

when he proceeds from Eq. (5) to Eq. (6). We believe, however, that Bell is incorrect in drawing certain other consequences from Eq. (2).

He proceeds by dividing Λ into two parts a and λ; M into two parts b and μ; and N into two parts c and ν, where a, b, c, are controllable variables characterizing the experimental set-up and Λ, μ, ν are the respective residual parts. He then speaks of probability distributions of Λ, μ, ν and says that these may depend upon a, b, c, but unfortunately he does not explicitly state what depends on what, except for the remark: "Now applying again the locality hypothesis . . ., the distribution of λ and μ must be independent of b, - the latter being outside the relevant light cones". Let us now guess the dependences which Bell has in mind (subject, of course, to correction by him). By a standard rule of probability, we have

$$\{\lambda, \mu, \nu \mid a, b, c\} = \{\lambda \mid \mu, \nu, a, b, c\}\{\mu \mid \nu, a, b, c\}\{\nu \mid a, b, c\}.$$

We conjecture that Bell now wishes to make two separate appeals to Eq. (2) to obtain the following replacements:

(i) $\{\lambda \mid \mu, \nu, a, b, c\} = \{\lambda \mid \nu, a, c\}$;
(ii) $\{\mu \mid \nu, a, b, c\} = \{\mu \mid \nu, b, c\}$.

Both (i) and (ii) are consequences of Eq. (2) as we have reformulated it (more precisely, of slight extensions of it). For, even though the space–time region in which λ is located extends to negative infinity in time, ν, a, c are *all* the beables other than λ itself in the region of λ and in the backward light cone of this region, and μ and b *do* refer to beables with space-like separation from the λ region. Similar reasoning holds for assertion (ii).

Our guessing, however, is not finished. Bell's derivation of (16) apparently also requires a suitable assertion concerning the distribution of ν. Indeed, we see no way of filling out the outline of his argument[3] without using the following:

---

3. Even given (i), (ii), and (iii) there is a slip in Bell's argument for inequality (14) is incorrect, since the distribution for μ given b is different from the distribution for μ' given b'. However, given (i), (ii), and (iii), one could proceed as follows. Rewrite his Eq. 7 as

$$p(a, \lambda, b, \mu, c, \nu) = \sum_{A, B} AB\{A \mid a, \lambda, c, \nu\}\{B \mid b, \mu, c, \nu\}.$$

Multiplying both sides by $\{\lambda \mid \nu, a, c\}\{\mu \mid \nu, b, c\}$ and integrating we obtain

(iii) $\{v \mid a, b, c\} = \{v \mid c\}$.

It seems obvious that (iii) does not follow from Eq. (2), since the space–time regions containing a and b do not have space-like separation from the region of v. In fact, the forward light cone of the region containing v fills all of space–time. Could (iii) perhaps be a reasonable extension of Eq. (2)? We think not, at least not at the extreme level of generality that Bell seeks, since v is the complete specification of the region to which it refers (minus the one factor c) and consequently a specific value for v could hardly fail to influence the subsequent values of a and b. As a result of such influences, the probability distribution over the phase space of v values would in general be conditional upon the values of a and b. This dependence cannot be excluded without further argument. It seems to us that (iii) could be made reasonable only if the settings of a and b are the results of some spontaneous events, such as acts of free will of the experimenters (as Bell may have assumed tacitly in his derivation of (16) and explicitly in Sect. 8). This is a logical and metaphysical possibility, which we do not intend to exclude a priori. But since Bell's argument is intended to be general, it would not be legitimate for him to justify the assertion (iii) by relying upon a metaphysics which has not been proved and which may well be false.[4]

It should be noted that in the second paragraph of Sect. 6 of his letter, Bell expresses certain reservations about the decisiveness of experiments based on inequality (16). He emphasizes that, "it was supposed in (12) that the complete specification N of the overlap is the same for the various cases compared". These reservations are very close in spirit to the reservations which we have just now expressed against Bell's derivation of inequality (16) itself; thus, in a way he has anticipated our criticism.

We do not regard the flaws in Bell's argument as fatal to the enterprise of deriving an inequality which is valid for a reasonable class of local

$$p(a, b, c, v) = \bar{A}(a, c, v) \bar{B}(b, c, v),$$

where

$$\bar{A}(a, c, v) = \int \sum_A A\{A \mid a, \lambda, c, v\}\{\lambda \mid v, a, c\} d\lambda$$

$$\bar{B}(b, c, v) = \int \sum_B B\{B \mid b, \mu, c, v\}\{\mu \mid v, b, c\} d\mu.$$

Then Bell's argument (9)–(16) goes through, using (iii), without any friction at step (14).

4. The objection which we have raised against assertion (iii) holds equally well against an assumption made immediately after Eq. (4.2) in an earlier paper of Bell's [in *Proceedings of the International School of Physics "Enrico Fermi", Course IL, Varenna 1970* (Academic Press 1971), p. 178] and against the form of the joint probability distribution assumed for variables $\lambda, \lambda', \lambda''$ in a paper of Shimony [In *Logic, Methodology and Philosophy of Science,* P. Suppes, et al., eds., North-Holland Publishing Company, 1973, p. 567].

theories. We feel that such a derivation was given in the paper by Clauser and Horne (see note 2). In that paper two spatially separated analyzer-detector assemblies were considered along with a source of emissions located midway between the assemblies, with each emission presumably consisting of two objectively real components. It is to be noted that the set-up described is considerably more specific than that described by Bell, and therefore assumptions concerning it can be less general and more plausible than those needed by Bell. Let K denote the complete state of one of the emissions (denoted by λ in Ref. 1). Let $p_1(K, a)$ be the probability that assembly 1 registers a detection event, when K is the emission state and when an adjustable parameter of the assembly is chosen to be a. Let $p_2(K, b)$ be similarly defined for assembly 2. Finally, let $p_{12}(K, a, b)$ be the probability of a joint detection event given K, a, and b. Clauser and Horne make two suppositions:[5]

(1) $p_{12}(K, a, b) = p_1(K, a)p_2(K, b)$,

which they justify on grounds of locality and reality similar to Bell's.

(2) The distribution $\rho(K)$ of the emissions is independent of the settings a and b.

Supposition (2) of Clauser and Horne plays a role in their argument analogous to Bell's assertion (iii). The central question is whether the supposition (2) is more reasonable than (iii). Our contention is that it is, though we do not pretend to offer a definitive proof nor do we think that one can be given.

It is obvious to begin with that the assumption of Clauser and Horne is very much weaker than Bell's. In his notation, their assumption is

(2) $\{K \mid a, b, c\} = \{K \mid c\}$, which can be written out as
(2') $\int \{K \mid \nu\}\{\nu \mid a, b, c\} \, d\nu = \int \{K \mid \nu\}\{\nu \mid c\} \, d\nu$.

But this is just an integrated form of assertion (iii), $\{\nu \mid a, b, c\} = \{\nu \mid c\}$, with $\{K \mid \nu\}$ used as a weighting function in the integration. Of course, the assertion of the equality of two integrals is a much weaker statement than the equality of two integrands. But there is yet more to be said. It is well known in the statistical mechanics of extended systems that the normal dynamics of the system as well as external perturbations tend to wash out correlations between variables which are temporally or spatially well-separated, unless there are specific mechanisms of the system for maintaining these correlations. In the present case, a, b, and K are values associated

5. It is interesting to note that Bell's footnote 10 in his 1970 Varenna paper (see our previous footnote for reference) can be understood as an anticipation of a proof along the lines of that of Clauser and Horne.

with well separated events. Moreover, there is no mechanism that one can point to which sets up a correlation between the selection of parameter a (or of b) and the occurrence of an emission having state K. Therefore, even though the left hand side of (2′) contains a factor $\{K \mid \nu\}$ and another factor $\{\nu \mid a, b, c\}$, it is reasonable that the way in which the distribution of ν is influenced by a and b is irrelevant for the distribution of K.

Bell can, of course, reply that we do not know that the distribution of emissions K is insensitive to the values of a and b, or for that matter that there are no causal links between the act of selecting a and that of selecting b. After all, the backward light cones of those two acts do eventually overlap, and one can imagine one region which controls the decision of the two experimenters who chose a and b. We cannot deny such a possibility. But we feel that it is wrong on methodological grounds to worry seriously about it if no specific causal linkage is proposed. In any scientific experiment in which two or more variables are supposed to be randomly selected, one can always conjecture that some factor in the overlap of the backward light cones has controlled the presumably random choices. But, we maintain, skepticism of this sort will essentially dismiss all results of scientific experimentation. Unless we proceed under the assumption that hidden conspiracies of this sort do not occur, we have abandoned in advance the whole enterprise of discovering the laws of nature by experimentation.[6]

To sum up: the advantage of the Clauser–Horne approach over that of Bell's is not that it is supposition free. Rather, it is that the supposition needed is no stronger than one needs for experimental reasoning generically, and nevertheless just strong enough to yield the desired inequality.

Reply* to Bell

Abner Shimony

Bell's answer[1] to criticisms[2] by Shimony, Clauser, and Horne (CHS) of his earlier paper[3] has greatly diminished the distance between their respective positions. At risk of exhibiting what Freud called "narcissism of small differences" I shall comment on a crucial passage in Bell's answer. He says:

Consider the extreme case of a "random" generator which is in fact perfectly deterministic in nature – and, for simplicity, perfectly isolated. In such a device the

6. See A. Shimony, "Scientific Inference", in *The Nature and Function of Scientific Theories*, ed. R. Colodny (U. of Pittsburgh Press, Pittsburgh, 1971).

*Work supported by the National Science Foundation.

complete final state perfectly determines the complete initial state – nothing is forgotten. And yet for many purposes, such a device is precisely a "forgetting machine". A particular output is the result of combining so many factors, of such a lengthy and complicated dynamical chain, that it is quite extraordinarily sensitive to minute variations of any one of many initial conditions. It is the familiar paradox of classical statistical mechanics that such exquisite sensitivity to initial conditions is practically equivalent to complete forgetfulness of them. To illustrate the point, suppose that the choice between two possible outputs, corresponding to a and a', depended on the oddness or evenness of the digit in the millionth decimal place of some input variable. Then fixing a or a' indeed fixes something about the input – i.e., whether the millionth digit is odd or even. But this peculiar piece of information is unlikely to be the vital piece for any distinctively different purpose, i.e., it is otherwise rather useless. With a physical shuffling machine, we are unable to perform the analysis to the point of saying just what peculiar feature of the input is remembered in the output. But we can quite reasonably assume that it is not relevant for other purposes. In this sense the output of such device is indeed a sufficiently free variable for the purpose at hand. For this purpose the assumption (1) is then true enough, and the theorem follows.

This passage is excellent up to the last sentence. The last sentence, however, seems to me to be a non-sequitur, unless the phrase "true enough" is interpreted with extreme latitude. Suppose that in Bell's idealized example of a deterministic and completely isolated generator the set of hidden variables ν is such that a would be generated and a' not. Then for all b and c

$$\{\nu \mid a', b, c\} = 0,$$

where the expressions on the left hand side denotes the probability (or probability density) that the hidden variables have the value ν, on condition that the non-hidden variables are a', b and c. On the other hand, for at least some b and c

$$\{\nu \mid a, b, c\} \neq 0.$$

Hence Bell's assumption (1), which asserts the equality of $\{\nu \mid a, b, c\}$ and $\{\nu \mid a', b, c\}$, is false.

One can guess what Bell means by "true enough" from several sentences before the last in the passage quoted. Suppose K is some much less comprehensive feature of the generator than the complete set of hidden variables. Then from knowledge of K no inference can be drawn about whether a, a' or some other output will be generated. If so, then it is reasonable to assume that

$$\{K \mid a, b, c\} = \{K \mid a', b, c\}.$$

(Rigorous examples in which this assumption is true are provided by dynamical systems having the "mixing property".[4]) But this assumption is essentially the same as assumption (2) on p. 5 of Ref. 2. Thus the distance between the positions of Bell and of CHS seems to have converged to zero, but the latter can still claim to have articulated the common position with greater clarity.

There is an objection which Bell could have brought against CHS. They express a preference for the Clauser–Horne derivation[5] of Bell's inequalities over all other derivations, on grounds of generality and plausibility of assumptions. In my opinion, however, they should have recognized that Bell's derivation of 1971,[6] though very different in argument, proceeds from essentially the same assumptions. Bell assumes that the expectation value $\bar{A}$ depends only on the apparatus setting a and the hidden variables λ, and similarly for $\bar{B}$. Clauser and Horne assume that the probability p_1 of detection of emission 1 depends only upon a and λ, and similarly for the probability of detection of emission 2. When one recognizes the correct operational connections between the expectation values of observables and detection probabilities, the equivalence of these assumptions becomes clear.

A final remark concerns an entirely different point: namely, Bell's term "beable". The suffix "able" etymologically refers to a potentiality. An observable of a system is a property which can be observed, even though it may actually be the case that no one has observed it. But Bell's criterion for applying the term "beable" to things is that "they are there" (Ref. 3, p. 11); no potentiality is involved. The term "existent" would have been more accurate than his neologism. I hope that Gresham's law will not be confirmed in the present case.

REFERENCES

[1] J. S. Bell, Epistemological Letters *15,* Febr. 1977.
[2] A. Shimony, M. A. Horne, and J. F. Clauser, Epistemological Letters *13,* October 1976.
[3] J. S. Bell, Epistemological Letters *9,* March 1976.
[4] See, for example, J. Lebowitz and O. Penrose, Physics Today *26,* No. 2, 23 (1973).
[5] J. F. Clauser and M. A. Horne, Phys. Rev. D *10,* 526 (1974).
[6] J. S. Bell, in *Foundations of Quantum Mechanics:* Proceedings of the Int. School of Physics "Enrico Fermi", Course IL, Varenna 1970, ed. by B. d'Espagnat.

Note. All of Bell's papers listed here are reprinted in his book *Speakable and Unspeakable in Quantum Mechanics,* Cambridge University Press, Cambridge, 1987. Papers [1], [2], and [3] were also reprinted in Dialectica *39* (1985).

13
Physical and philosophical issues in the Bohr–Einstein debate

I. ASPECTS OF THE DEBATE

The debate between Niels Bohr and Albert Einstein concerning the interpretation of quantum mechanics extended from the fifth Solvay Conference in 1927 until the end of Einstein's life. The most dramatic exchange occurred in 1935, when Einstein, in collaboration with B. Podolsky and N. Rosen, published a paper in *Physical Review* entitled "Can quantum-mechanical description of physical reality be considered complete?",[1] concluding with a negative answer, and Bohr replied in the same journal, with a paper of the same title but giving a positive answer.[2] Their arguments were restated, in some respects with greater clarity, in the Library of Living Philosophers volume on Einstein in 1949.[3] The disagreements between Bohr and Einstein concerned not only the physical question expressed in the common title of their papers, but also philosophical questions about physical reality and human knowledge.

Much light was thrown upon the Bohr–Einstein debate by the theorem of J. S. Bell (1964)[4] and the experiments which it inspired.[5] Bell showed that any physical theory which applies to a pair of spatially separated

This work originally appeared in H. Feshbach, T. Matsui, and A. Oleson (eds.), *Niels Bohr: Physics and the World,* Chur, 1988. © Harwood Academic Publishers, GmbH. Reproduced with permission.

1. A. Einstein, B. Podolsky, and N. Rosen, "Can quantum-mechanical description of physical reality be considered complete?", Physical Review **47**, 777–780 (1935).
2. N. Bohr, "Can quantum-mechanical description of physical reality be considered complete?", Physical Review **48**, 696–702 (1935).
3. P. A. Schilpp, ed., *Albert Einstein: Philosopher-Scientist* (Evanston, IL: The Library of Living Philosophers, Inc., 1949).
4. J. S. Bell, "On the Einstein Podolsky Rosen paradox," Physics **1**, 195–200 (1964).
5. S. J. Freedman and J. F. Clauser, "Experimental test of local hidden variable theories," Physical Review Letters **28**, 938–941 (1972): E. S. Fry and R. C. Thompson, "Experimental test of local hidden-variable theories," Physical Review Letters **37**, 465–468 (1976); A. Aspect, J. Dalibard, and G. Roger, "Experimental test of inequalities using variable analyzers," Physical Review Letters **49**, 1804–1807 (1982). Other experiments are reviewed in J. Clauser and A. Shimony, "Bell's Theorem: experimental tests and implications," Reports on Progress in Physics **41**, 1881–1927 (1978).

systems (as considered in the thought experiment of Einstein, Podolsky, and Rosen) and satisfies a certain locality condition will, in certain circumstances, disagree statistically with quantum mechanics. Consequently, the supplementation of the quantum mechanical description envisaged by Einstein and his collaborators (usually referred to as a "hidden variables theory") must either violate the locality condition or clash with quantum mechanical predictions. Almost certainly this theorem would have surprised Einstein, who believed in locality in the relativistic sense but also treasured the predictive power of quantum mechanics as a statistical theory of the atomic domain. A series of experiments, culminating with that of A. Aspect, J. Dalibard, and G. Roger (1982), strongly supported quantum mechanics against the whole family of local hidden variables theories. In the experimental arrangement of Aspect et al., locality in the sense of relativity theory implies that Bell's locality condition is satisfied, and hence the results refuted the kind of hidden variables theory which Einstein seems to have had in mind. The physical question in the debate between Bohr and Einstein was thus answered in favor of Bohr. It should be said that there are some loopholes in the experiments performed so far, which have kept alive the hopes of dedicated advocates of local hidden variables theories. Most students of the subject, however, do not regard the exploitation of these loopholes to be promising.[6]

The lectures of Bell and Aspect in the present volume [*Niels Bohr: Physics and the World,* ed. H. Feshbach, T. Matsui, and A. Oleson (Chur: Harwood, 1988)] provide quite detailed accounts of the arguments of Einstein, Podolsky and Rosen, of Bell's theorem, and of the consequent experiments. Hence I shall not expand the condensed account given in the preceding paragraph. Instead, I shall be concerned primarily with the philosophical differences between Bohr and Einstein.

I shall maintain that the correctness of Bohr's answer to the question, "Can quantum-mechanical description of physical reality be considered complete?", does not by itself constitute a victory over Einstein in their philosophical disagreements. It will be argued in Section 2 that Einstein's physical realism can be suitably generalized to accommodate the results of the experiments inspired by Bell's theorem. The generalization consists in recognizing that a new modality of reality is implicit in quantum mechanical description, which may appropriately be called "potentiality," and in acknowledging a peculiar kind of quantum mechanical nonlocality. Einstein's fundamental philosophical thesis that the physical world has an

6. See, for example, the article of Clauser and Shimony in note 5, and also B. d'Espagnat, "Nonseparability and the tentative description of reality," Physics Reports **110**, 202–264 (1984). For a contrary opinion see T. Marshall and E. Santos, letter to Physics Today **38**, no. 11, 9–11 (1985) and references given there.

existence independent of human knowledge is preserved by this generalization, even though there is obviously a retrenchment from some of his characterizations of physical reality. Furthermore, it will be argued in Section 3 that the experimental results do not support the most radical of Bohr's philosophical innovations, and therefore other considerations are needed in order to assess his philosophy. I shall propose that Bohr's thought can be related to a certain philosophical tradition (including Hume and Kant) in which a theory of knowledge is worked out without commitment to a theory of existence. The placement of Bohr's thought in a well explored philosophical tradition helps to exhibit some serious lacunae in the realization of his philosophical program, and questions can be raised about the prospects of filling these lacunae.

2. NONLOCALITY, POTENTIALITY, AND EINSTEIN'S REALISM

As preparation for some critical remarks about Einstein's philosophical views, it will be very useful to analyze the concept of locality in some detail.

Figure 1 is an abstract representation of the kind of physical system studied by Bell, and earlier by Einstein, Podolsky, and Rosen. Two particles, labelled 1 and 2, propagate to the right and left respectively from a common source. At the moment of their departure from the source the *complete state* of the composite system $1+2$ is denoted by λ. For the present, no assumption is made that λ is the quantum mechanical state of $1+2$ or a state described by hidden variables. It is only assumed that λ provides a complete specification of the properties of $1+2$ when they leave the source. Particle 1 impinges upon an analyzer which has an adjustable parameter a, and the particle emerges from the analyzer in one of two channels, which are labelled $+$ and $-$ respectively. Likewise, particle 2 impinges in an analyzer with adjustable parameter b, and it too emerges in one of two channels labelled $+$ and $-$. Let x_a be the outcome of analysis of particle 1, hence either $+$ or $-$, and let x_b be the outcome of analysis of particle 2, again either $+$ or $-$. The following probability notation will be used:

$p_\lambda(x_a, x_b \,|\, a, b) =$ the probability of joint outcomes x_a, x_b, provided that λ is the complete state of $1+2$ when they leave the source and a and b are the settings of the respective adjustable parameters.

In terms of $p_\lambda(x_a, x_b \,|\, a, b)$ we can define probability of single outcomes, and also conditional probability in a standard manner:

Quantum entanglement and nonlocality

Figure 1. Particles 1 and 2 propagate from a common source, the complete state of the particles at the moment of departure being λ. The analyzer of particle 1 has an adjustable parameter a, and the analyzer of particle 2 has an adjustable parameter b. Each particle can emerge from its respective analyzer in one of two channels, labelled + and −.

$$p^1_\lambda(x_a \mid a, b) = p_\lambda(x_a, + \mid a, b) + p_\lambda(x_a, - \mid a, b),$$

$$p^2_\lambda(x_b \mid a, b) = p_\lambda(+, x_b \mid a, b) + p_\lambda(-, x_b \mid a, b),$$

$$p^1_\lambda(x_a \mid a, b, x_b) = p_\lambda(x_a, x_b \mid a, b)/p^2_\lambda(x_b \mid a, b),$$

$$p^2_\lambda(x_b \mid a, b, x_a) = p_\lambda(x_a x_b \mid a, b)/p^1_\lambda(x_a \mid a, b).$$

In terms of these probability expressions two distinct independence conditions can be defined, as proposed by J. Jarrett.[7]

1. Parameter Independence: $p^1_\lambda(x_a \mid a, b)$ is independent of b,
   $p^2_\lambda(x_b \mid a, b)$ is independent of a.
2. Outcome Independence: $p^1_\lambda(x_a \mid a, b, x_b) = p^1_\lambda(x_a \mid a, b)$,
   $p^2_\lambda(x_b \mid a, b, x_a) = p^2_\lambda(x_b \mid a, b)$.

The names of these independence conditions are self-explanatory. Furthermore, if the compound event consisting of the choice of parameter a and the occurrence of outcome x_a has space-like separation from the compound event consisting of the choice of parameter b and the occurrence of outcome x_b (as is the case in the experiment of Aspect et al.) then both independence conditions are *prima facie* required by the limitations upon direct causal connectedness imposed by relativity theory. It should be noted that the two independence conditions permit correlations between x_a and x_b, for the complete state λ may imply a deterministic or a probabilistic correlation between the outcomes. But precisely because λ is the complete state of 1 and 2 at the moment of their separation, the specification of the parameter b and the outcome x_b should not, from the standpoint of relativistic locality, have any effect upon the outcome x_a that is not already implicit in λ (and likewise, if the letters a and b are interchanged).

7. J. Jarrett, "On the physical significance of the locality conditions in the Bell arguments," *Noûs* **18**, 569–589 (1984). The terminology and notation used in the text differ from Jarrett's.

Jarrett showed that the conjunction of Parameter Independence and Outcome Independence is equivalent to Bell's locality condition, which says essentially that

$$p_\lambda(x_a, x_b \mid a, b) = p_\lambda^1(x_a \mid a)\, p_\lambda^2(x_b \mid b).$$

Since the experimental result of Aspect et al. contradicts Bell's locality condition (provided the loopholes mentioned above are set aside), it follows that at least one of the two independence conditions must be false. It turns out that the consequences of violation of Parameter Independence are very different from the consequences of violation of Outcome Independence, as the following analysis shows.

1. Suppose that Parameter Independence fails, for example, because $p_\lambda^2(x_b \mid a, b) \neq p_\lambda^2(x_b \mid a', b)$. Then at a moment when particles 1 and 2 are well separated and are about to impinge upon their respective analyzers, an experimenter can make a choice between the parameter values a and a', thereby affecting the probability of the outcome $x_b = +$ for the analysis of particle 2; and if an ensemble of pairs of replicas of the pairs $1 + 2$ is prepared in a sufficiently short interval of time, then the frequency of $+$ outcomes will be affected with virtual certainty by the choice between a and a'. In this way a bit of information is conveyed from the experimenter to an observer of the output of the second analyzer. Most important, *this bit of information can be transmitted faster than light.* Consequently, the failure of Parameter Independence implies the possibility of an unequivocal violation of the special theory of relativity.

2. Suppose that Outcome Independence fails, for example, because $p_\lambda^2(x_b \mid a, b, x_a) \neq p_\lambda^2(x_b \mid a, b)$ for some choice of a, b, from which it follows that $p_\lambda^2(x_b \mid a, b, +) \neq p_\lambda^2(x_b \mid a, b, -)$. Again a message can be sent from an experimenter associated with the analyzer of particle 1 to an observer of the analysis of particle 2, but the procedure must be different from the foregoing. The experimenter must monitor each of the particles 1 of an ensemble of replicas of $1 + 2$, and decide whether to block the propagation of the particle 2 if $x_a = +$, or else to block its propagation if $x_a = -$. The particle 2 must be placed "on hold" (for instance, a photon may be trapped in a light guide) until the monitoring of its partner has been completed. An observer of the output of the second analyzer can infer from the statistics of x_b which decision the experimenter made. In this way a bit of information is conveyed from the experimenter to the observer. Clearly, however, *this information cannot be transmitted faster than light.* Time is required to monitor the outcome x_a, to perform the operation of taking particle 2 "off hold," and to permit the propagation of particle 2 to its analyzer; and the total time required is greater than the time required for a direct transmission of a light signal between the two analyzers.

Since, as noted previously, the experimental results of Aspect et al. agree with quantum mechanics, the Bell locality condition must be violated by the quantum mechanical predictions concerning the test situation. Which, then, of the two conditions, Parameter Independence and Outcome Independence, is violated by quantum mechanics? If the quantum mechanical polarization state of the two photons in their experiment is

$$\Psi = \frac{1}{\sqrt{2}}\left[u_x(1)\otimes u_x(2) + u_y(1)\otimes u_y(2)\right],$$

where $u_x(1)$ represents polarization of photon 1 along the x-axis, $u_x(2)$ represents polarization of photon 2 along the x-axis, and $u_y(1)$ and $u_y(2)$ have analogous meanings. Then it can easily be seen as follows that the quantum mechanical predictions based upon Ψ violate Outcome Independence. Let the analyzers of photons 1 and 2 be idealized sheets of polaroid placed perpendicularly to the paths of the photons, with the transmission axes of each sheet oriented along the x-axis. In the notation introduced above, the parameters a and b of the two analyzers are each chosen to be the angle 0. The outcome $x_a = +$ will be taken to mean that photon 1 passes through the polaroid sheet oriented at an angle a to the x-axis, and $x_a = -$ means that it fails to pass; and the values + and - of x_b have analogous meanings. The quantum mechanical prediction based upon Ψ yields the probability $\frac{1}{2}$ that photon 1 will pass through its polaroid sheet, because the absolute square of the coefficient of the term $u_n(1)\otimes u_x(2)$ in Ψ is $\frac{1}{2}$. If, however, in addition to knowing that the initial quantum state of the two photons is Ψ, one also knows that photon 2 has failed to pass through its sheet of polaroid, prior to a polarization analysis of photon 1, then the term $u_y(1)\otimes u_y(2)$ is picked out from Ψ, thereby ensuring that photon 1 will not pass through its polaroid sheet. If we express these quantum mechanical probabilities in the foregoing notation, taking the complete state λ to be Ψ, we obtain

$$p_\Psi^1(+\,|\,0,0) = \tfrac{1}{2},$$

$$p_\Psi^1(+\,|\,0,0,-) = 0,$$

in contradiction to Outcome Independence.

It can be shown, though a somewhat elaborate argument is needed,[8] that quantum mechanics does not violate Parameter Independence. As pointed out above, a violation of Bell's locality condition (and hence the

---

8. P. H. Eberhard, "Bell's Theorem and the different concepts of locality," Il Nuovo Cimento **46B**, 392–419 (1978); G. C. Ghirardi, A. Rimini, and T. Weber, "A general argument against superluminal transmission through the quantum mechanical measurement process," Lettere al Nuovo Cimento **27**, 293–298 (1980).

possibility of agreement with the results of Aspect et al.) is ensured by the failure of either Parameter Independence or Outcome Independence. But a greater strain in current physical theory would result from failure of the former than of the latter, because only failure of Parameter Independence permits a signal to be transmitted faster than light. We conclude that quantum mechanics is a non-local theory, in the sense of violating Bell's locality condition, but its non-locality "peacefully coexists" with the relativistic prohibition of superluminal signals. Indeed, it may be appropriate to introduce a notation which is familiar in a political context in order to summarize the relation between quantum mechanics (QM) and special relativity theory (SR):

$$\text{QM} \textcircled{\&} \text{SR}.$$

An examination of the structure of the quantum mechanical description Ψ of the pair of photons $1 + 2$ will throw much light upon this peaceful coexistence. First of all, according to Ψ the polarization of photon 1 with respect to the x–y axes is not definite, and the same is true of photon 2. If the state were represented only by $u_x(1) \otimes u_x(2)$, then both photons would have definite polarization along the x-axis; and if the state were represented only by $u_y(1) \otimes u_y(2)$, then both would have definite polarization along the y-axis. But since both terms are present in Ψ, there is a probability $\frac{1}{2}$ that photon 1 will exhibit a polarization along x and a probability $\frac{1}{2}$ that it will exhibit a polarization along y, and similarly for photon 2. It is tempting to regard Ψ as merely a description of the state of the scientist's *knowledge* of the two photons, or alternatively, as a description of an inhomogeneous ensemble of photon pairs, the individual members of which have definite properties that are not described by Ψ. In fact, the temptation is to supplement the quantum mechanical description of the pair of photons by a hidden variables description. But that is precisely what Bell's theorem and the associated experimental results preclude, unless one is willing to accept a hidden variables theory which violates Bell's locality condition. If, however, we concede that Ψ is a complete description of the polarization state of the pair of photons, then we must accept the *indefiniteness* of the polarization of each with respect to the x–y axes as an objective fact, not as a feature of the knowledge of one scientist or of all human beings collectively. We must also acknowledge *objective chance* and *objective probability,* since the outcome of the polarization analysis of each photon is a matter of probability. It is convenient to use a term of Heisenberg to epitomize objective indefiniteness together with the objective determination of probabilities of the various possible outcomes; the polarizations of the photons are *potentialities.*[9] The work

9. W. Heisenberg, *Physics and Philosophy* (New York: Harper and Bros., 1958), 185.

initiated by Bell has the consequence of making virtually inescapable a philosophically radical interpretation of quantum mechanics: that there is a modality of existence of physical systems which is somehow intermediate between bare logical possibility and full actuality, namely, the modality of potentiality.

A further peculiarity of the quantum mechanical view of the physical world is exhibited by Ψ, the existence of an n-particle state (with $n \geq 2$) in which the individual particles are not in definite states. It can be shown that there is no pair of one-photon polarization states w and z such that Ψ is equivalent to the conjoined attribution of w to photon 1 and z to photon 2. Schrödinger called an n-particle state with this peculiarity an "entangled" state. When the photon pair is in the state Ψ, the polarizations of both 1 and 2 with respect to the x–y axes (and actually with respect to all other pairs of orthogonal axes) are potentialities, but the actualization of either one of them automatically ensures the actualization of the other. If, as a result of a polarization analysis (or other appropriate process), photon 1 exhibits a definite polarization along x or along y, then photon 2 would certainly exhibit the same definite polarization if it is subjected to a subsequent polarization analysis; and likewise if 1 and 2 are interchanged. The entanglement of states of an n-particle system thus rests upon the fact that a quantum mechanical state of a system is a network of potentialities, the content of which is not exhausted by a catalogue of the actual properties which are assigned to the system.

The conclusions just reached have a direct bearing upon the correctness of Einstein's physical world view. In a late and careful paper of 1948 Einstein says the following:

the concepts of physics refer to a real external world, i.e., ideas are posited of things that claim a 'real existence' independent of the perceiving subject (bodies, fields, etc.), and these ideas are, on the other hand, brought into as secure a relationship as possible with sense impressions. Moreover, it is characteristic of these physical things that they are conceived as being arranged in a space–time continuum. Further, it appears to be essential for this arrangement of the things introduced in physics that, at a specific time, these things claim an existence independent of one another, insofar as these things 'lie in different parts of space'.[10]

The entanglement of the states of the spatially separated photons 1 and 2 conflicts with Einstein's thesis that "these things claim an existence independent of one another." Of course, Einstein was well aware of the entanglement in the quantum mechanical description of composite physical systems, and in fact his paper with Podolsky and Rosen made essential

10. A. Einstein, "Quantenmechanik und Wirklichkeit," Dialectica **2**, 320–324 (1948), translated by Don Howard.

use of entanglement. But Einstein wished to interpret quantum mechanical entanglement epistemically, as expressing the scientist's knowledge of correlations among the properties of spatially separated system, and entanglement would not characterize the complete states which he envisaged. As we have seen, Bell's theorem and subsequent experimental work seem to preclude this epistemic interpretation, thereby confirming that entanglement is an objective fact about the physical world.

The first of Einstein's theses in the passage quoted, and the one which appears to stand highest in his philosophical hierarchy, is that physical things "claim a 'real existence' independent of the perceiving subject." This thesis *is* consistent with all the conclusions which we drew from an analysis of Bell's theorem, the relevant experiments, and the formalism of quantum mechanics. However, this thesis of physical realism, when separated from the rest of Einstein's theses, leaves open the character of the real existence of physical things. The foregoing analysis led to radical conclusions regarding the character of physical existence: i.e., that there are objective indefiniteness, objective chance, and objective probability, in short, that there is a modality of existence which has been designated as potentiality. It is hard to resist speculating about how Einstein would have reacted to Bell's theorem and the experiments of Aspect et al., but of course we must acknowledge candidly that any guesses about this hypothetical reaction are sheer speculation. A good starting point for speculation is the remarkable passage at the conclusion of the posthumously published Fifth Edition of *The Meaning of Relativity:*

From the quantum phenomena it appears to follow with certainty that a finite system of finite energy can be completely described by a finite set of numbers (quantum numbers). This does not seem to be in accordance with a continuum theory, and may lead to an attempt to find a purely algebraic theory for the description of reality. But nobody knows how to obtain the basis of such a theory.[11]

This passage acknowledges the empirical success of quantum theory and indicates a willingness to incorporate some of the implications of quantum mechanics into his world view. In the spirit of this passage might Einstein have been receptive of the ideas of potentiality and entanglement and might he have been reconciled to the abandonment of the mutually independent existence of spatially separated things? We cannot answer this question, but we can say that one coherent option which remains open to a sympathetic but independent-minded follower of Einstein is to accept his highest thesis, that of physical realism, but to give it a sense which is derived from an analysis of quantum mechanics.

11. A. Einstein, *The Meaning of Relativity,* fifth edition (Princeton, N.J.: Princeton University Press, 1955), 165-6.

One qualification to the last sentence is needed: that the "coherence" of the option is not established beyond all doubt. The problem of the actualization of potentialities – known also as the problem of the reduction of the wave packet and the problem of measurement – is a dark cloud on the horizon. A consequence of the linear dynamical law of quantum mechanics (the Schrödinger equation) is that in the final physical stage of a measurement process, the indexical property of the apparatus (e.g., the direction of a pointer on a dial) may be objectively indefinite – apparently in gross disagreement with laboratory experience. The most dramatic exposition of this difficulty is the famous "cat paradox" of Schrödinger,[12] but the difficulty was well known to other pioneers of quantum mechanics, including von Neumann and Einstein himself. Until this problem is solved one cannot claim that an extended version of physical realism, taking the complete state of a physical system to be a network of potentialities, is a coherent view of the physical world.

3. ON THE PHILOSOPHY OF BOHR

The version of physical realism arrived at in Section 2 recognizes Bohr as the victor in the debate with Einstein, Podolsky, and Rosen over the *physical* question, "Can quantum-mechanical description of physical reality be considered complete?". It does not follow that Bohr would be satisfied with this kind of physical realism, despite its incorporation of ideas inspired by quantum mechanics. Furthermore, Bohr's own reasons for defending the completeness of quantum mechanical descriptions are entirely different from the considerations of Bell's theorem and the experiments which it inspired.

The heart of Bohr's answer to Einstein and his collaborators is the following passage:

Of course there is in a case like that just considered no question of a mechanical disturbance of the system under investigation during the last critical stage of the measuring procedure. But even at this stage there is essentially the question of *an influence on the very conditions which define the possible types of predictions regarding the future behavior of the system.* Since these conditions constitute an inherent element of the description of any phenomenon to which the term 'physical reality' can be properly attached, we see that the argumentation of the mentioned authors does not justify their conclusion that quantum-mechanical description is essentially incomplete.[13]

12. E. Schrödinger, "The present situation in quantum mechanics," in *Quantum Theory and Measurement,* ed. J. A. Wheeler and W. Zurek (Princeton, N.J.: Princeton University Press, 1983), translated from a German article of 1935.
13. N. Bohr, *Atomic Physics and Human Knowledge* (New York: Science Editions, 1961), 60–61.

There is nothing in this passage that suggests a commitment to objective entanglement or to potentiality as a new modality of physical existence. In fact, the thrust of the passage is away from ontological questions and towards questions about knowledge and language – e.g., questions of "possible types of predictions regarding the future behavior of the system" and of "the description of any phenomenon to which the term 'physical reality' can be properly attached." In order to see how Bohr's answer applies to the pair of photons which we have been considering, it is useful to note that the quantum mechanical polarization state Ψ studied in Section 2 can be expressed in different ways, which are mathematically equivalent:

$$\Psi = \frac{1}{\sqrt{2}}\left[u_x(1)\otimes u_x(2) + u_y(1)\otimes u_y(2)\right]$$
$$= \frac{1}{\sqrt{2}}\left[u_{x'}(1)\otimes u_{x'}(2) + u_{y'}(1)\otimes u_{y'}(2)\right],$$

where the $x'-y'$ axes are obtained by rotating the $x-y$ axes by an arbitrary angle about the line of propagation of the two photons. The first of the expressions for Ψ shows that a polarization analysis of photon 1 with respect to the $x-y$ axes would yield polarization information concerning photon 2 with respect to the same axes, without disturbing photon 2; while the second expression for Ψ shows the same with respect to the $x'-y'$ axes. Since polarization analysis of photon 1 with respect to the $x-y$ axes and with respect to $x'-y'$ axes are mutually exclusive, and since the choice determines what kind of prediction can be made concerning photon 2, Bohr maintains that it is not proper to attach the term "physical reality" to polarization of photon 2 with respect to both $x-y$ and $x'-y'$. The limitation upon the attribution of two distinct properties to photon 2 in no way stems from action at a distance, because an analysis performed upon photon 1 does not physically disturb photon 2. There is only a limitation upon possible predictions which an experimenter can make.

In the answer to Einstein et al. just quoted, and repeatedly elsewhere in his essays, Bohr insists that the unequivocal application of terms of theoretical physics to physical situations depends upon the careful specification of experimental arrangements, and he illustrates his dicta with penetrating analyses of interactions between objects and apparatus. Bohr's discussion constitutes an important contribution to one of the central problems of modern philosophy of science, that of the relation between theoretical terms and experience. It also constitutes a contribution to the part of the theory of language called "pragmatics," which – in contrast to syntactics and semantics – specifically studies the circumstances of the language user. Bohr is one of the most remarkable writers on the pragmatics of the languages of the exact sciences, even though this aspect of his thought has not been extensively studied.[14]

Although Bohr's writing is obviously philosophical in the sense of exploring fundamental questions concerning nature and human knowledge, it is far from clear how close he came to formulating a systematic philosophy. Some crucial passages will be cited, however, which at least will indicate the direction in which he was attempting to achieve systematization.

In the treatment of atomic problems, actual calculations are most conveniently carried out with the help of a Schrödinger state function, from which the statistical laws governing observations obtainable under specified conditions can be deduced by definite mathematical operations. It must be recognized, however, that we are here dealing with a purely symbolic procedure, the unambiguous physical interpretation of which in last resort requires a reference to a complete experimental arrangement. Disregard of this point has sometimes led to confusion, and in particular the use of phrases like 'disturbance of the phenomena by observation' and 'creation of physical attributes of objects by measurements' is hardly compatible with common language and practical definition.[15]

It is not relevant that experiments involving an accurate control of the momentum or energy transfer from atomic particles to heavy bodies like diaphragms and shutters would be very difficult to perform, if practical at all. It is only decisive that, in contrast to the proper measuring instruments, these bodies together with the particles would constitute the system to which the quantum-mechanical formalism has to be applied.[16]

While in the mechanical conception of nature, the subject–object distinction was fixed, room is provided for a wider description through the recognition that the consequent use of our concepts requires different placing of such a separation.[17]

Such considerations point to the epistemological implications of the lesson regarding our observational position, which the development of physical science has impressed on us. In return for the renunciation of accustomed demands on explanation, it offers a logical means of comprehending wider fields of experience, necessitating proper attention to the placing of the object–subject separation. Since, in philosophical literature, reference is sometimes made to different levels of objectivity or subjectivity or even of reality, it may be stressed that the notion of an ultimate subject as well as conceptions like realism and idealism find no place in objective description as we have defined it.[18]

---

14. This aspect of Bohr's thought is emphasized by P. Zinkernagel, *Conditions for Description* (London: Routledge & Kegan Paul, 1961), and by A. Petersen, *Quantum Physics and the Philosophical Tradition* (Cambridge MA: MIT Press, 1968).
15. N. Bohr, *Essays 1958-1962 on Atomic Physics and Human Knowledge* (New York: Vintage Books, 1966), 5.
16. N. Bohr (1961; see note 13), 50.
17. *Ibid.*, 91–92.
18. *Ibid.*, 78–79.

Without entering into metaphysical speculations, I may perhaps add that an analysis of the very concept of explanation would, naturally, begin and end with a renunciation as to explaining our own conscious activity.[19]

There have been many attempts to fit these passages and others like them – which are characteristically penetrating, suggestive, and elliptical – into a coherent philosophy.[20] It is beyond the scope of the present paper to assess these efforts of exposition and systematization of Bohr's thought. Some judgments can be made with confidence, however. The second passage is clearly discordant with the interpretation of Bohr as a macro-realist, who attributes objective existence to macroscopic bodies but treats microphysics only as an instrument for predicting observable behavior of macroscopic bodies. This passage shows that Bohr does not reserve a quantum mechanical description for microscopic entities and a classical description for macroscopic ones. Rather, he treats quantum mechanically any physical system which is the object of investigation and uses a classical description for parts of the apparatus of measurement, which are situated on the subject's side of the subject–object separation. The fourth passage is discordant with the interpretation of Bohr as an idealist who regards the contents of consciousness as the fundamental reality, and all physical discourse as merely an instrument or short-hand for summarizing, systematizing, and anticipating these contents.[21]

A. Petersen suggests a very interesting generalization of the two foregoing negative judgments: not only is it incorrect to attribute to Bohr either a macrophysical or an idealistic ontology, but any ontology whatever is alien to his thought. Here are some of Petersen's statements, which are either explicitly or implicitly presented as expositions of Bohr's philosophical ideas:

19. *Ibid.*, 11.
20. Among these are the books mentioned in note 14; and also P. Feyerabend, "Problems of Microphysics," in *Frontiers of Science and Philosophy,* ed. R. Colodny (Pittsburgh PA: University of Pittsburgh Press, 1962); K. M. Meyer-Abich, *Korrespondenz, Individualität und Komplementarität* (Wiesbaden: Steiner, 1965); L. Rosenfeld, "Strife about complementarity," Science Progress **41**, 393–410 (1953); C. A. Hooker, "The nature of quantum mechanical reality: Einstein versus Bohr," in *Paradigms and Paradoxes,* ed. R. Colodny (Pittsburgh PA: University of Pittsburgh Press, 1972); M. Jammer, *The Philosophy of Quantum Mechanics* (New York: Wiley, 1974); J. Honner, "The transcendental philosophy of Niels Bohr," Studies in History and Philosophy of Science **13**, 1–29 (1982); and H. Folse, *The Philosophy of Niels Bohr* (Amsterdam: North-Holland, 1985), reviewed by A. Shimony, Physics Today **38**, no. 10, 108–109 (1985).
21. In Ch. 8, sect. 2 of Folse's book, note 20, there is a good compilation of quotations from Bohr confirming this interpretation.

Bohr's remarks on quantal and thermodynamical irreversibility illustrate his approach to the description problem in physics. Especially, they indicate that he thought this problem to be a purely conceptual one. The question is not what *is* in an ontological sense, but what can be stated unambiguously in physical terms.[22]

In the course of the interpretation discussion [of quantum mechanics], the irrelevance of ontological ideas became increasingly conspicuous, but their elimination has been a slow and difficult process which may still be far from complete.[23]

When it was objected that reality is more fundamental than language and lies beneath language, Bohr answered, 'We are suspended in language in such a way that we cannot say what is up and what is down.'[24]

Petersen's proposals command attention for the intrinsic reason that they provide lucid, unforced, and illuminating interpretations of passages like the first, third, fourth, and fifth above, and for the extrinsic reason that he was Bohr's assistant for seven years. What is not convincing, however, is Petersen's claim that Bohr's replacement of ontology by other concerns is a radically new departure in western philosophy. Bohr's ideas belong quite clearly to an important philosophical tradition: that in which epistemology is studied in deliberate abstention from considerations of ontology, and particularly from suppositions about the ontological status of the knowing subject. Classical philosophers in this tradition are Hume and Kant, but some of the analytic philosophers associated with the Vienna Circle and some of the followers of Wittgenstein can reasonably be said to belong to it. An opposing tradition envisages the meshing of epistemology and ontology in two ways: (1) epistemology aims at (and perhaps partially succeeds in) showing how human beings can obtain knowledge of the world as it is, at least to a good approximation and with a reasonably high degree of reliability; and (2) it aims at exhibiting an ontological niche for the knowing subject. In this tradition are found Aristotle, Leibniz, Newton, Locke, and Whitehead, and perhaps Einstein (the last in a fragmentary way, since he did not attempt to formulate a systematic philosophy that is more comprehensive than he needed for understanding physics). This contrast of two philosophical traditions is crude, as grand intellectual classifications usually are, since they group together diverse philosophies, thereby playing down their nuances and individualities. In spite of the crudity of the classification, however, it throws some light upon Bohr's thought and helps to pose some important criticisms.

Particularly instructive is a comparison of Bohr with Kant, since both in their respective ways were critical of claims to knowledge of things in

22. A. Petersen (1968; see note 14), 157.
23. *Ibid.*, 63.
24. *Ibid.*, 188.

themselves. It should be emphasized that their critiques are directed not only against a "naive realism," according to which perception yields direct knowledge of the existence of things in themselves and their properties, but also of "critical realism," according to which such knowledge can be obtained indirectly and inferentially. Kant's doctrine that space and time are forms of intuition imposed upon appearances by the faculty of sensibility, and that causality, substance, and other fundamental concepts are categories imposed upon experience by the faculty of understanding, undermines the legitimacy of all the inferences upon which critical realism relies in order to achieve indirect knowledge of the existence and properties of the things in themselves. The Kantian doctrine excludes a spatio-temporal theater in which both things in themselves and phenomena are located, and it is incompatible with a causal treatment of the relation between these two types of entities – e.g., it precludes regarding phenomena as mental events that are aroused in the knowing subject (itself a thing in itself) by its interaction with other things in themselves. No claim is being made here that the Kantian doctrine of space, time, and the categories holds up under critical scrutiny; indeed, when the synthetic *a priori* judgments which Kant adduces in geometry and in pure natural science are undermined, his argument for the ideality of space and time and for the imposition of the categories by the understanding loses its force. Nevertheless, the power and coherence of Kant's epistemology may be acknowledged by the following conditional statement: if his doctrine of the origin of space, time, and the categories were correct, then the impossibility in principle of human theoretical knowledge of the properties of the things in themselves would follow. The solidity of this conditional statement makes Kant the exemplary exponent of the tradition in which epistemology is developed in deliberate abstention from ontology.

Since Bohr rejects Kant's affirmations of synthetic *a priori* knowledge, one may properly ask how Bohr can prove in principle the illegitimacy of the essential reasoning of critical realism. Bohr nowhere explicitly confronts this question as it has just been posed. There are, however, many statements, particularly about complementarity, which can be construed as answering his own reformulation of the question. For example, in a late and carefully written article Bohr says,

In quantum physics, however, evidence about atomic objects obtained by different experimental arrangements exhibits a novel kind of complementary relationship. Indeed, it must be recognized that such evidence which appears contradictory when combination into a single picture is attempted, exhausts all conceivable knowledge about the object. Far from restricting our efforts to put questions to nature in the form of experiments, the notion of *complementarity* simply characterizes the answers we can receive by such inquiry, whenever the interaction between the measuring instruments and the objects forms an integral part of the phenomena.[25]

This passage contains no philosophical locutions like "properties of things in themselves," but there is a surrogate for them in Bohr's phrase "combination into a single picture," and the argument which he presents for the impossibility of such a picture can be construed as a reason for rejecting in principle the feasibility of the critical realism.

It must be emphasized, however, that Bohr's argument is limited in scope. It is directed against a "single picture" of the kind that is drawn in classical mechanics, in which all the properties of a physical object are simultaneously assigned definite values. Nothing in his argument precludes a "single picture" of the kind envisaged in the realistic interpretation of quantum mechanics presented at the end of Section 2, as a reasonable extension of Einstein's physical realism. In this picture some properties of a physical object are actual, but others (and indeed most) have the status of potentialities. Bohr never seems to consider this possibility explicitly, and his arguments do not suffice explicitly or implicitly to rule it out. Remarkably, however, Heisenberg – who at one time was close to Bohr – recognizes the possibility of this kind of realistic interpretation of quantum mechanics and somewhat tentatively endorses it:

Kant had pointed out that we cannot conclude anything from the perception about the 'thing-in-itself.' This statement has . . . its formal analogy in the fact that in spite of the use of the classical concepts in all the experiments a nonclassical behavior of the atomic objects is possible. The 'thing-in-itself' is for the atomic physicist, if he uses this concept at all, finally a mathematical structure; but this structure is – contrary to Kant – indirectly deduced from experience.[26]

At the conclusion of Section 2 a fundamental difficulty of the realistic interpretation of quantum mechanics was mentioned: that is, the difficulty of explaining the actualization of potentialities within the framework of quantum mechanics. Much effort is currently being devoted to this problem. Among the proposals under examination are nonlinear variants of the dynamical law of quantum mechanics, stochastic variants of quantum mechanical dynamics, and non-quantum mechanical behavior of the space–time field itself.[27] No decisive progress has been achieved along any of these lines, but it is premature to conclude that these efforts are doomed to fail. One can say conditionally, however, that if sustained efforts to

25. N. Bohr (1966; see note 15), 4.
26. W. Heisenberg (1958; see note 9), 91.
27. A survey of these proposals is given in A. Shimony, "Events and Processes in the Quantum World," In *Quantum Concepts of Space and Time*, ed. C. Isham and R. Penrose (Oxford: Oxford University Press, 1986).

provide a realistic account of the actualization of quantum mechanical potentialities should prove to be unsuccessful, then there would be *a posteriori* reasons to conclude that the program itself is misconceived and to return to Bohr. He would presumably object to the locution "actualization of potentialities," just as he objected, in the first passage above, to locutions like "creation of physical attributes of objects by measurement." It is conceivable that Bohr's ideas will be vindicated in this roundabout way. But a precondition to a full vindication of Bohr's philosophy is to fill the lacunae of his exposition and to transform his suggestions into a systematic and coherent world view.

COMMENT

In this essay and elsewhere I have cited the EPR argument as an articulation of Einstein's philosophy and of his view of the completeness of quantum mechanics from 1935 onward, in spite of documentary evidence that he was dissatisfied with the wording of the argument [Don Howard, "Einstein on Locality and Separability," *Studies in History and Philosophy of Science* 16 (1985), pp. 171ff, and Arthur Fine, *The Shaky Game: Einstein, Realism and the Quantum Theory* (Chicago: University of Chicago Press, 1986), pp. 36ff]. I agree with Howard and Fine that there are subtleties in other passages by Einstein that cannot be found in the EPR argument, but these never seem to contradict the basic ideas of EPR. Furthermore, once an ambiguity is removed, EPR is an intrinsically powerful argument, independently of Einstein's ultimate ratification of it. The ambiguity appears in the premiss, "If, without in any way disturbing a system, we can predict with certainty . . . the value of a physical quantity, then there exists an element of physical reality corresponding to this physical quantity"; "can predict" may be understood in the strong sense of having data at hand sufficient for the prediction, or in the weak sense of being able to make a measurement that would provide data sufficient for the prediction. The EPR argument goes through only if "can predict" is understood in the weak sense, but Einstein's physical realism and opposition to anthropocentrism strongly suggest that the weak sense was his intended meaning. Bohr can be understood as maintaining that EPR's premiss is valid only if the strong sense of "can predict" is used, and then, of course, one never is in a position to ascribe the existence of elements of reality simultaneously to two quantities that cannot be jointly measured. Einstein could argue (properly, I believe) that this reply entails a commitment to anthropocentrism.

PART C
Complex systems

14
The methodology of synthesis: parts and wholes in low-energy physics

I. ASPECTS OF THE PROBLEM

One of the most pervasive features of the natural world is the existence of reasonably stable systems composed of well-defined parts which are to a large extent unchanged by entering into composition or leaving it. The problem of parts and wholes is to understand with the greatest possible generality the relation between the components and the composite system.

The parts–wholes problem has an ontological aspect, which concerns the properties of the components and the composite system without explicit consideration of how knowledge of them is obtained. Among the ontological questions are the following: Is there an ultimate set of entities which cannot be subdivided and which are therefore "atomic" in the etymological sense? If the properties of the components are fully specified, together with the laws governing their interactions, are the properties of the composite system then fully determined? In particular, are those properties of composite systems which are radically different from those of the components, and which might properly be characterized as "emergent," also definable in terms of the latter? Do composite systems belong, always or for the most part, to "natural kinds"? Is the existence of natural kinds explicable in terms of the laws governing the components? Are both the possible taxonomy and the actual taxonomy of natural kinds thus explicable? Is there a hierarchy of "levels of description" – i.e., microscopic, macroscopic, and possibly intermediate – such that laws can be formulated concerning a coarser level without explicit reference to the properties at a finer level of description?

The parts–wholes problem also has an epistemological aspect. Suppose that the most precise and best-confirmed laws turn out to govern relatively simple systems – as is indeed mostly the case in physics – but that the systems of interest are enormously complicated combinations of

This work originally appeared in R. Kargon and P. Achinstein (eds.), *Kelvin's Baltimore Lectures and Modern Theoretical Physics,* Cambridge, MA: MIT Press, 1987. Reprinted by permission of the publisher.

simple components. Then there will be insuperable experimental difficulties in gathering knowledge about all the initial conditions of the parts, and insuperable mathematical difficulties in deducing from the basic laws the properties of the composite system. To what extent can the composite system be said to be understood in terms of the laws governing its parts? And if there is independent phenomenological knowledge of laws on a coarse level of description, how do we know that these are in principle derivable from the laws on a finer level?

Aside from the ontological–epistemological distinction, there is a subdivision of the parts–wholes problem according to domains of investigation. There is the domain of inorganic systems (the physical sciences), that of organic systems (biology), that of systems endowed with minds (psychology), and that of groupings of human beings (social sciences). There is no *a priori* reason to believe that the questions listed above should be answered in the same way in all these domains, and it may not even be appropriate to pose the same questions in all of them. Of course, even the separateness of these domains is an aspect of the parts–wholes question. If the properties of an organism can in principle be defined in terms of those of the constituent molecules, and biological laws can in principle be derived from those of physics, then the domain boundary between the inorganic and the organic loses its fundamental significance and is maintained only for reasons of professional specialization. Likewise for the other domain boundaries.

The parts–wholes problem is so vast and ramified that no apology is needed for restricting attention to a small part of it. I shall consider almost nothing outside of physics, and within physics almost exclusively those branches in which the energies involved are typically of the order of an electron volt per particle or less: atomic physics, molecular physics, solid-state physics, and in general the physics of bulk matter. Low-energy physics is roughly Democritean. Electrons and stable nuclei have properties which can be studied in isolation, and to a very good approximation they keep these properties unchanged when they enter atoms, molecules, fluids, and solids. Hence electrons and nuclei behave as "building blocks," even though high-energy investigations reveal a rich internal structure of the latter. Democritus's characterization of his atoms as infinitely hard, indivisible, and immortal certainly does not apply without drastic reservations to these building blocks. Nevertheless, they preserve their integrity in low-energy interactions to such an extent that Democritus's vision of explaining the full variety of the natural world in terms of the combinations of a very few kinds of unchanging bodies is realized to a remarkable extent. By contrast, high-energy physics – according to which particles are created and annihilated and the nucleon components (quarks) are removed from confinement either with great difficulty or not at all – is radically non-Democritean.

This essay consists largely of summaries of the treatment of parts and wholes in five areas of low-energy physics, concerning atoms, molecules, fluids, infinite Coulomb systems, and spin systems. Roughly Democritean answers are given in each area to most of the ontological questions concerning parts and wholes (if the divisibility of nuclei is set aside). I wish to call the attention of philosophers of science to the great variety and subtlety, and the often surprising nature, of the derivations of properties of composite systems from those of the components. Ontologically, the theory of the wholes is reduced to the theory of the parts, but the pejorative overtones of "reductionism" – a suggestion of flattening and loss of distinctive features – is certainly inappropriate. Epistemologically, the reduction of the theory of the wholes to that of the parts has to be understood with serious qualifications. In the derivations that will be mentioned, the first principles of low-energy physics are usually supplemented with secondary principles, which can reasonably be regarded as consistent with the former but rarely rigorously inferrable from them. The methodology of the physical understanding of complex systems turns out to have intricacies that are seldom discussed and are difficult to formulate clearly.

This essay does not pretend to be authoritative, for the simple reason that I am not an expert in any of the branches of physics from which my examples are drawn. I nevertheless hope to convey my enthusiasm for inquiry in the territory that is reconnoitered.

2. FIRST PRINCIPLES

The first principles of low-energy physics fall into three groups: those of nonrelativistic quantum mechanics, those concerning the dominant forces among the building blocks, and the principles of statistical mechanics.

A

Quantum mechanics is a framework theory that at present is commonly believed to apply to every physical system from elementary particles to the whole of the cosmos. My skepticism about the universal validity of quantum mechanics is largely irrelevant to the present discussion, since there is little doubt that quantum mechanics holds with great accuracy within the domain of low-energy physics. As a framework theory, quantum mechanics has to be supplemented by detailed information about the constitution of any particular system to which it is applied before inferences can be made about the behavior of the system and before experimental results can be predicted. Nevertheless, the framework which quantum mechanics provides is rich.

(i) It prescribes the general characteristics of the space of states of any system, most commonly summed up by saying that there is a one-to-one correspondence between the pure states and the rays of a Hilbert space. Implicit in this characterization of the space of states is the superposition principle.

(ii) It prescribes the general structure of the class of observables (as they are commonly called, though a less anthropocentric name such as "dynamical variables" would be preferable).

(iii) It prescribes the rule for calculating the probability distribution of possible values of an observable, contingent upon its actualization (by measurement or possibly by other means), when the state is known.

(iv) Nonrelativistic quantum mechanics offers a general law for the temporal evolution of the pure state of an isolated system, provided that a sufficient characterization of the system is given (namely, its Hamiltonian operator).

(v) Quantum mechanics itself has a remarkable principle of composition. If A and B are two systems, associated respectively with the Hilbert spaces H_A and H_B, then the Hilbert space associated with the composite system $A + B$ (provided that their dynamical interaction does not change them internally) is the tensor-product space $H_A \otimes H_B$. Only in special cases is it possible to write a vector ϕ belonging to $H_A \otimes H_B$ in the simple-product form $u \otimes v$, where u belongs to H_A and v to H_B, and only in these special cases is it correct to say that the pure state represented by ϕ is equivalent to the attribution of pure states to each of the components A and B. In general, A and B have an "entangled" state (to use Schrödinger's expression), so that there is no complete characterization of A without reference to B and vice versa. It is evident that quantum-mechanical entanglement requires a partial retrenchment from the Democritean conception of the parts–wholes relationship, but it leaves open the possibility of realizing other aspects of this conception. The quantum-mechanical ground state of a many-particle system is determined by its Hamiltonian operator, which consists of kinetic-energy contributions for each particle and potential-energy contributions for each pair of particles. Such a Hamiltonian is in the spirit of Democritus – *mutatis mutandis* – since it contains no contributions from the whole that cannot be traced explicitly to the parts. If the quantum-mechanical ground state then turns out to account for the physical properties of the system at a temperature of absolute zero (assumed for the present in order to avoid complicators due to thermal excitations), then it is reasonable to say that a Democritean description of the many-particle system has been realized, even though the ground state is entangled.

(vi) Finally, there are symmetrization principles; i.e., if there are n identical bosons (integer-spin particles) in the system, then any vector in the Hilbert space representing a physically allowable state must be invariant under the interchange of two of these bosons, and if there are n identical fermions (particles with half-integer spin), then any vector representing a physically allowable state must change sign under exchange of two of these fermions. The Pauli principle, which prohibits two fermions to occupy the same single-particle state, is a direct consequence of this antisymmetrization. Both the symmetrization rule and the antisymmetrization rule are modifications of the tensor-product principle, since each restricts the space of states of a composite system to an appropriate subspace of the full tensor-product space.

B

One of the great simplifications of the restriction to low-energy physics is that the dominant forces among the constituents are electromagnetic. Furthermore, these forces are treated essentially classically, by potentials, rather than by postulating a quantized electromagnetic field which interacts with matter fields. Nevertheless, much complexity remains, for electromagnetic forces are exhibited in electron–nucleus interaction, electron–electron repulsion, nucleus–nucleus repulsion, orbit–orbit interaction, spin–spin interaction, electron spin–electron orbit coupling, electron orbit–nuclear spin coupling, and interaction with external electromagnetic fields. The potential terms in the total Hamiltonian can be of any of these types. In addition, of course, there are kinetic-energy terms. When a specified set of nuclei and a specified number of electrons are given, usually together with certain constraints upon the configuration, then the electromagnetic-force laws and the general principles of quantum mechanics yield a definite Hamiltonian. A large set of physical questions (What are the stationary states of the system? What is the ground state? What is the energy of the ground state? At what energy above the ground state does dissociation occur? What are the degeneracies of the various allowable energies? What are the geometrical properties of the system in the ground state? and so on) then become answerable in principle.

C

Statistical mechanics is a science of systematically extracting a relatively small amount of reliable physical knowledge of statistical matters from an ocean of ignorance. It treats mainly two classes of systems. Systems of the first class consist of very many parts, typically falling into a small number of types (e.g., a homogeneous or a heterogeneous gas, or an alloy). Because of the very large number of parts, it is practically unfeasible to gather exact information about the initial conditions of the system. And even if, *per impossible,* a complete knowledge of the initial conditions were given, it would be humanly impossible to solve the equations of motion (classical or quantum mechanical) in order to infer exactly the state at a later time. The other class of systems considered by statistical mechanics consists of open systems, interacting with an environment the exact constitution of which is not known. (If the constitution were known in detail, then the system plus the environment might constitute a single system of many parts that would fall in the first class just mentioned.) The reliable information that one wishes to exact from the immense background of ignorance consists of probability distributions of especially interesting quantities concerning the system. When the system consists of a large number of parts, then typically the probability distributions are very sharply peaked, so that interesting quantities concerning the system as a whole, such as energy, state of condensation, and magnetization, can be predicted with virtual certainty. Statistical mechanics thus provides a powerful instrument for making inferences about wholes from the properties of parts, though it is an instrument that usually requires supplementation by secondary principles. The enterprise of statistical mechanics is greatly expedited if the environment with which the system exchanges energy is negligibly affected by the exchange, so that it can be considered a reservoir. When the system is in equilibrium with the reservoir – which is a concept that can be defined either phenomenologically or probabilistically – then the probability distribution over the space of states can reasonably be shown to be the canonical distribution,[1] which can be written as follows for classical systems:

$$\rho(x) = \frac{1}{Z(\beta)} e^{-\beta E(x)},$$

where x represents the state in an appropriate space of states. $E(x)$ is the energy of the system when it is in the state represented by x, β is the inverse of the product of Boltzmann's constant k and the absolute temperature T, and $Z(\beta)$ is a normalizing factor defined in such a way that the integral of $\rho(x)$ over the space of states is 1. Explicitly,

$$Z(\beta) = \int e^{-\beta E(x)} \, dx.$$

If $Z(\beta)$ is considered as a function of β, it is called the partition function; some indication of its importance and utility will be conveyed below. There is also a quantum-mechanical version of the canonical distribution, which will not be needed explicitly in this essay even though it is implicit in much of the discussion of low-energy physics.

1. See, for example, Khinchin 1949.

3. ATOMS

The structure of atoms is one of the best-understood kinds of composition of wholes from parts in all of physics. Some important features of this composition can be understood quantitatively from first principles, but other features are described only by semi-empirical rules, for which there is some explanation but by no means a rigorous general derivation.[2]

Well established is the possibility of describing the ground state of any atom by a configuration, which tells how many electrons are in single-electron states of given principal quantum number n and given orbital angular momentum quantum number l. The ascribability of definite configurations to each species of atom is the essence of the shell model of the atom. For example, for aluminum the configuration is $1s^2 2s^2 2p^6 3s^2 3p$, where the letters s, p, and d conventionally stand for the values 0, 1, and 2, respectively for the quantum number l; the number preceding a letter prescribes the value of n; and the superscript after the letter gives the number of electrons with the specified values of n and l. The wave function ψ_{al} (a vector in the Hilbert space of square-integrable functions), representing the ground state of aluminum, is a superposition of a number of vectors, each of which is a product of single-particle wave functions, of which two have $n = 1$, $l = 0$; two have $n = 2$, $l = 0$; six have $n = 2$, $l = 1$; two have $n = 3$, $l = 0$; and one has $n = 3$, $l = 1$. In each term in the superposition, spin states for each of the 13 electrons are given. Furthermore, the superposition is so contrived that ψ_{al} is antisymmetric under exchange of any two electrons, as required by principle (vi) above (Section 2A). This is already an enormous amount of information, and it is derived almost rigorously from first principles, together with the specification of the charges of the electrons and nuclei and the spin $\frac{1}{2}$ of the electron. The value $\frac{1}{2}$ for the spin not only implies that the electron is a fermion but also fixes the dimensionality of the spin space associated with each electron to be 2. Consequently, when a fixed one-particle spatial wave function is given – characterized by n and l together with one other quantum number, the magnetic quantum number m – there is a further option of making the spin either up or down along a specified axis, thereby choosing between two orthogonal directions in the two-dimensional spin space. Hence, a pair of electrons can be characterized by n, l, and m without violating the exclusion principle. Another ingredient in the derivation of the configuration is the approximate spherical symmetry of the effective potential which each electron feels, due to the small size of the nucleus in comparison with the atomic radius and due also to the effective "smearing" of the charge distributions of all the other

2. See, for example, Bethe and Jackiw 1968.

electrons. This treatment of the electronic charge distribution is an instance of a mean-field approximation, which will recur below. Because the effective potential is nearly spherically symmetric, the angular momentum of the electron is a conserved quantity and hence l is a good quantum number. Further exploitation of spherical symmetry yields rigorous information about angular momentum: for example, that for fixed n a given value l of the angular momentum quantum number is compatible with exactly $2l + 1$ values of the component of the angular momentum along any specified axis, and hence $2l + 1$ values of the magnetic quantum number m (i.e., $m = -l, \ldots, l$). That fact was implicit in the configuration stated above for aluminum – e.g., there are six electrons in the one-electron state 2p, because the p is an abbreviation for $l = 1$, which permits three possible values of m $(1, 0, -1)$, for each of which there are two possible spin orientations. That is why the closed shell 2p has six electrons. To be sure, the multiplicity of 3p is only 1 in aluminum, but that is because the 3p shell is incomplete; in all the atoms from argon onward there are six 3p electrons.

Although the configuration provides much information about the ground state, there is more to be determined, and almost everything else is much harder to extract from first principles. One wishes to know, for example, the radial dependence of the single-particle wave functions, which in turn determines such important properties as the average size of the atom. To find the radial dependence, one must know the effective potential felt by a single electron and use it to solve the differential equation in the radial variable r which is obtained from the time-independent Schrödinger equation. But the potential depends upon the wave functions of all the electrons. How does one emerge from a maze in which the potential depends upon the wave functions and the wave functions depend upon the potentials – even if one sets aside all the mathematical difficulties of solving the differential equations? A way out is provided by the self-consistent field method of Hartree and Fock. A sequence of successive approximations is applied in practice, and it is reasonable to expect rapid convergence to a situation in which wave functions and potentials fit each other self-consistently.[3]

Except in the simplest atoms (through beryllium), the configuration does not determine how the orbital angular momenta l_i and the spins s_i of the individual electrons are combined to yield a total orbital angular momentum quantum number L for the whole set of electrons, a total spin S, and also a total angular momentum J. In principle the possible combinations could be checked to see which yields the lowest-lying energy,

---

3. A good treatment is found in Merzbacher 1970, pp. 535–539.

but that requires detailed knowledge of the wave functions, which – as just noted – is a formidable obstacle. There are, however, Hund's rules based on spectroscopic evidence, which are quite reliable. The first two of Hund's three rules are the following:

1. The LS with the largest S compatible with the configuration has the lowest energy.
2. In the case where the largest S is associated with several different allowable values of L, the largest L has the lowest energy.

There is no generally valid proof for these rules, though they have been confirmed for a number of atoms by detailed calculations. But much plausibility can be given to them, especially to the first, by general arguments from first principles. The larger S is, the more parallel are the spins. When the spins are parallel, they make a symmetrical contribution to the complete wave function, and hence the antisymmetrization principle for electrons implies that the spatial part of the wave function is antisymmetric. But that has the general effect of keeping the electrons, on the average, farther from one another than would be the case with symmetrical spatial wave functions, and therefore the potential energy of electrostatic repulsion among the electrons is diminished, thus lowering the energy.[4]

The mathematical complications of determining the ground-state wave function of an atomic species should not make us lose sight of a fundamental Aristotelian fact: that the species is a natural kind. All atoms of a given species (i.e., a given number of neutrons and protons in the nucleus, and as many electrons as protons for electric neutrality) have the same ground state when there are no external perturbations, and they have the same array of excited states. The *identity* of all atoms of a given species ensures that they all emit and absorb electromagnetic radiation at the same frequencies and have the same size, shape, and internal motion. Weisskopf (1979, p. 87) emphasizes two other features of atoms: stability ("The atoms keep their specific properties in spite of heavy collisions and other perturbations to which they are subjected") and regeneration ("If an atom is distorted and its electron orbits are forced to change by high pressure or by close neighboring atoms, it regains its exact original shape and orbits when the cause of distortion is removed"). Atoms thus exhibit form in Aristotle's sense, and even have the tendency to maintain this form, which phenomenologically is like his final cause. But the Aristotelian form is achieved by Democritean means – by interactions among the electrons and the nucleus, which leave these building blocks intact. Of course, one additional element is required for the achievement of form:

4. A good discussion is given in Baym 1969, p. 458.

the principles of quantum mechanics. Classical physics did not have the resources to explain natural kinds of composite systems, even when natural kinds of the indivisible building blocks were postulated, whereas Aristotelian science accounted for natural kinds uneconomically, by postulating an irreducible principle for each. In contrast to both, quantum mechanics ensures that the formal properties of very many kinds are implied by the properties of a small number of kinds of components.

4. MOLECULES

On the whole, molecular structure is more difficult to understand than atomic structure, primarily because molecules lack the spherical symmetry of the single-nucleus atom. Nevertheless, there is one principle of molecular composition that was extracted from laboratory experience and is simple enough to be taught in elementary chemistry: the principle of valence.[5] The valence of an element is defined phenomenologically as the number of atoms of hydrogen that one atom of the element can combine with or take the place of in forming compounds. From H_2O, HCl, NH_3, and CH_4 we infer that O, Cl, N, and C, respectively, have valences 2, 1, 3, and 4. Very soon, however, one finds that valence is not an intrinsic characteristic of an element, since in CO carbon behaves as if it has valence 2, and iron would have to be assigned valence 2 because of FeO and valence 3 because of Fe_2O_3. Clearly, valence is a crude and complicated principle of composition – and there are wonderful monsters like xenon-fluoride to increase the complication.

The hope of physical chemists is to put the valence concept on a firm basis, with all appropriate qualifications, by using the principles of composition of low-energy physics, especially quantum mechanics and the electromagnetic forces. Even before the new quantum mechanics, G. N. Lewis proposed to interpret chemical bonding in terms of the sharing of a pair of electrons by each of two atoms in a molecule, and the valence of an atom was interpreted as the number of electrons available for sharing. W. Heitler and F. London later gave a quantum-mechanical treatment of this idea in which the valence electrons are those that are not paired in the atom's configuration, where pairing means having the same spatial wave function (or "orbital") but opposite spins. In the bonding of two atoms in a molecule, a valence electron from one of the atoms is free to combine with a valence electron of the other in such a way that their combined spin state is the singlet state

$$|\uparrow\rangle_1|\downarrow\rangle_2 - |\downarrow\rangle_1|\uparrow\rangle_2.$$

5. See, for example, Baym 1969, pp. 486–498; Roby 1973.

This is a vector in the two-electron spin space which obviously changes sign under the exchange of electrons 1 and 2. Consequently, the overall antisymmetrization of electrons 1 and 2 is achieved by a symmetric spatial wave function for the two electrons. The effect of the symmetry of the spatial wave function can best be seen if one considers a coordinate r along an axis passing through the nuclei of the two atoms, utilizing the fact that the nuclei are more massive than the electrons and hence more precisely localizable without violating the uncertainty principle. An antisymmetric spatial wave function will have an amplitude close to zero halfway between the nuclei, and a symmetric spatial wave function will usually have a large amplitude and even be peaked near the midpoint. Therefore the symmetric spatial wave function makes the two electrons, on the average, closer to each other than does a comparable antisymmetric one, and hence the contribution of the repulsive electrostatic potential of the two electrons (a positive energy) is larger in the symmetric case than in the antisymmetric; but this positive contribution is more than compensated by the fact that, on the average, the electrons spend more time close to the nuclei in the symmetric case, thereby increasing the attractive contribution. Hence the singlet spin state is favored for the achievement of the lowest possible energy. This is an excellent argument, except for some hand-waving concerning the relative magnitude of the competing repulsive and attractive contributions. In the case of the H_2 molecule, Heitler and London were able to perform a quite accurate quantitative calculation from first principles to confirm the hand-waving argument. The extrapolation to more complex molecules is always somewhat risky, however.

Without any consideration of more detailed calculations it is evident that some modification of the foregoing account of valence is needed. The ground-state configuration of carbon is $1s^2 2s^2 2p^2$, which provides two unpaired electrons in the 2p shell (unpaired because there are three mutually orthogonal spatial wave functions available in this shell – $m = 1$, $m = 0$, and $m = 1$). Usually, however, carbon exhibits a valence of 4, though it was noted that in CO it exhibits valence 2. Pauling (1931) offered a resolution of this and similar difficulties by arguing that in molecular bonding one must consider not only the atomic ground-state configuration but also the lower-lying excited configurations. The reason is that the energy expended to raise some of the electrons to excited levels is more than compensated for by the availability of additional unpaired electrons. This suggestion is a fruitful new secondary principle of composition in molecular physics, known as "the hybridization of molecular orbitals." It is not merely an *ad hoc* device to save the Heitler–London theory of spin pairing, since good calculations of molecular geometry can be made by means of hybridization theory. Hybridization has to be used with tact, and sometimes must be refined by additional secondary principles. For example, Pauling and Keaveny (1973, p. 93) answer as follows an argument that hybridization theory breaks down for iron-group elements:

It has been recognised that the orbital occupied by a bonding electron may be either expanded or contracted considerably with respect to the orbital in the isolated atom. This expansion or contraction is accompanied by a contraction or expansion of other orbitals occupied by unshared electrons. Expansion or contraction of one orbital decreases or increases its shielding of nuclear charge for another orbital.

Machine calculations and empirical data seem to support this refinement of hybridization theory.

Machine calculations of wave functions are not imperturbable, objective judgments independent of all appeal to secondary principles, because the choice of initial guesses of wave functions in an iteration procedure is crucial for the rate at which convergence occurs. If a principle such as hybridization of molecular orbitals is a good approximation to the truth, then the initial wave function that it suggests will lead to rapid convergence. If not, the convergence will be too slow to be useful.

5. COULOMB SYSTEMS IN THE THERMODYNAMIC LIMIT

A vast range of phenomena can be described quite well by the formalism of thermodynamics. Essential to this formalism is the possibility of characterizing a system at a macroscopic level in terms of a small number of extensive variables: V, U, and $N_1, \ldots, N_k$, where V is the volume, U is the total energy, and N_i is the mole number of the ith chemical constituent. By *extensive* is meant that when the system is spatially subdivided into subsystems (not too small) the value of the extensive variable of the system as a whole is the sum of the values of that variable in all the subsystems. Furthermore, it is assumed that there is an entropy S, which is a linear homogeneous function of the extensive variables and which therefore is extensive too. The maximization of S relative to constraints determines the thermodynamic state of the system.[6]

Extensiveness as a constructive principle is so often taken for granted in ordinary applications (for example, the energy necessary to heat a kilogram of milk one degree is 1,000 times the energy needed to heat a gram of it one degree, largely independent of the shape of the container) that it takes some reflection to see that it is thoroughly nontrivial. Extensiveness depends, for example, on the negligibility of surface forces, which of

6. See, for example, Tisza 1966, pp. 102–193.

course is not always the case. In order to avoid the complication of the surface, the standard procedure is to idealize by going to the thermodynamic limit, in which there is infinite volume and infinitely many particles, but to take the limit in such a way that

$$\lim(N/V) = \text{density}$$

exists. It remains to be established, however, that U and S are extensive variables in the thermodynamic limit, so that $U = Vu$, where u is the internal energy per unit volume, and $S = Vs$, where s is the entropy per unit volume. There is no *a priori* assurance that the constructive principle of thermodynamics is consistent with the constructive principles of atomic and molecular physics, in which the properties of the composite are determined by the properties of component particles and their interactions. One might properly worry, for example, that if the fundamental forces among particles are long-range, then putting together more and more particles will not determine an internal energy proportional to volume, even if the density is fixed.

These questions turn out to be remarkably difficult. Using classical mechanics, one can prove the existence of the thermodynamic limit, provided that the intermolecular forces are repulsive at short distances and attractive at long distances but that the attractiveness falls off very rapidly as the distance increases.[7] Coulomb forces, which are by far the most important of electromagnetic forces at low velocities, do not have these characteristics, and the physically important question is whether the thermodynamic limit holds for a neutral Coulomb system (equal densities of positive and negative charges). The answer is no if classical mechanics rather than quantum mechanics is taken as the framework theory. But, most remarkable, the answer is yes if quantum mechanics is used. The proof is extraordinarily difficult, but hinges upon two simple facts: that the exclusion principle keeps fermions from coming too close to one another, thereby providing an effective repulsive core if the negatively charged particles are electrons (or, more generally, if either the positively charged particles or the negatively charged particles are fermions of one kind); and that electrostatic shielding in a neutral Coulomb system counteracts the long-range character of the Coulomb force.[8] The very existence and stability of bulk matter with ordinary macroscopic properties depends crucially upon an intimate quantum-mechanical principle that underlies the first of these two facts. In 1519–1522 Magellan proved that the Pacific Ocean existed by sailing across it, but 450 years had to elapse before Dyson, Lenard, Thirring, Lebowitz, and Lieb proved by quantum

7. See, for example, Thompson 1972, pp. 67–71.
8. A survey of the argument is given in Lieb 1976.

mechanics that something like the Pacific Ocean *could* exist. Thus science progresses.

6. NORMAL FLUIDS

In a very large, spatially extended system, it is trivial to define macroscopic physical variables by summing or averaging over a very large number of microscopic variables. The local matter density $\sigma(\mathbf{x})$ of a fluid can be defined classically by summing the masses of the particles in a small region $R_{\mathbf{x}}$ centered about x, which is large enough to contain on the average many particles but is small in comparison with the macroscopic dimensions of the system, and then dividing by the volume of $R_{\mathbf{x}}$. An analogous definition can be given quantum-mechanically, but it makes use of expectation values of positions of particles. Likewise, a local mean velocity $\mathbf{v}(\mathbf{x})$ of the fluid can be defined quantum-mechanically as well as classically. And similarly for other macroscopic quantities such as pressure and temperature, which also may be local quantities, varying with $\mathbf{x}$. Enough has been said to make it clear that *local* can be *macroscopic*. What is nontrivial, and clearly not simply a matter of definition, is whether there are any laws of nature that can be formulated solely in terms of a set of macroscopic quantities. If there are such laws, then the microscopic level is not the only level of description at which laws are exhibited. Actually, we are quite confident that there are laws at a macroscopic level of description, because of the great empirical success of such disciplines as fluid dynamics. The difficult problem is to explain why this is so.

An outstanding example is the Navier–Stokes equation, which is the dynamical law for a viscous fluid in nonturbulent motion:

$$\sigma(\mathbf{x})[\dot{\mathbf{v}}(\mathbf{x}) + (\mathbf{v}(\mathbf{x})\,\mathrm{grad})\,\mathbf{v}(\mathbf{x})]$$
$$= -\mathrm{grad}\,p(\mathbf{x}) - \eta_1\,\mathrm{curl}\,\mathrm{curl}\,\mathbf{v}(\mathbf{x}) + \eta_2\,\mathrm{grad}\,\mathrm{div}\,\mathbf{v}(\mathbf{x}).$$

The matter density $\sigma(\mathbf{x})$ is the mass per unit volume in the neighborhood of $\mathbf{x}$; $\mathbf{v}(\mathbf{x})$ is the local velocity; $p(\mathbf{x})$ is the local pressure; and η_1 and η_2 are friction constants. There have been many derivations of this equation, but I find that of Fröhlich (1973) impressive for its generality and lucidity. The derivation begins with the exact time-dependent Schrödinger equation for a system consisting of N particles of specified mass, but restates this equation in terms of the von Neumann density operator Ω instead of the wave function, as is appropriate where there is imperfect knowledge of the true quantum state. In other words, Ω is the quantum-mechanical analogue of the distribution function of classical statistical mechanics. Ω contains information about the correlated behavior of all the N particles, but for practical purposes most of these correlations are of no interest. It therefore is useful to define reduced density operators, Ω_1, Ω_2, Ω_3, etc., concerning (respectively) single particles, pairs of particles, triples of particles, etc. From the original dynamical equation for Ω, a hierarchy of equations for Ω_1, Ω_2, Ω_3, etc. follows easily. The complication of this hierarchy lies in the fact that the equation in Ω_1 contains a term dependent on Ω_2, the equation for Ω_2 contains a term dependent on Ω_3, and so on. The macroscopic quantities in the Navier–Stokes equation depend only upon Ω_1 and Ω_2, and with appropriate additional assumptions the equation itself can be derived. First, it is assumed that in equilibrium there is rotational and translational invariance, so that the pair-correlation function $P(\mathbf{x}, \mathbf{y})$, which is essentially the probability of jointly finding particles located at $\mathbf{x}$ and $\mathbf{y}$, becomes at equilibrium a function $P_e(|\mathbf{x} - \mathbf{y}|)$ which depends only on the distance between the two points. Second, it is assumed that the interaction between particles is short-range, so that large changes of σ, $\mathbf{v}$, and p do not occur over distances in which the interaction is non-negligible. And third, attention is restricted to situations in which the deviation of $P(\mathbf{x}, \mathbf{y})$ from the equilibrium pair-correlation function $P_e(|\mathbf{x} - \mathbf{y}|)$ is a function of $\mathbf{x}$ only via the distance $|\mathbf{x} - \mathbf{y}|$ and via linear dependence upon the macroscopic fields $\sigma(\mathbf{x})$ and $\mathbf{v}(\mathbf{x})$. Since these assumptions are quite mild, Fröhlich's derivation has great generality. In principle, values of η_1 and η_2 can actually be calculated from Ω_2, though to do so the hierarchy of equations for Ω_i must be terminated at $i = 2$ by an approximation, and the resulting dynamical equation for Ω_2 must be solved. For practical purposes it usually is sufficient to know that the form of the Navier–Stokes equation is legitimate and to depend upon measurement to supply the friction constants.

7. PHASE TRANSITIONS IN SPIN SYSTEMS

Most of the discussion of composition so far has drawn only upon the first two classes of principles summarized in section 2, those of quantum mechanics and of electromagnetism. These are the principles that one would expect to dominate at low temperature, when thermal fluctuations are negligible. At elevated temperatures, however, the thermal fluctuations become more and more important as a disordering influence. The derivative principles of composition that one expects to find at finite temperatures are likely to be the result of competition between the ordering tendencies implicit in the rules of quantum mechanics and the disordering tendencies due to thermal excitations. Splendid instances of such competition are provided by phase transitions, such as from a solid to a fluid phase, from a superfluid to a normal fluid, or from a ferromagnetic to a nonferromagnetic arrangement of spins in a crystal.

Below 1,043°K (the Curie temperature), a single iron crystal exhibits net magnetization in the absence of an external magnetic field; above this temperature it does not do so. This phenomenon is interpreted as the preferential alignment of the unpaired electronic spins (each of which behaves as a little magnetic dipole) of the atoms below the Curie temperature, whereas no preferred direction survives above it. Qualitatively, this interpretation is in accordance with the foregoing remark that elevation of temperature has a disordering effect. It is nevertheless remarkable that there is a sharp transition from the ferromagnetic phase (in which magnetization occurs in the absence of an external field) to the nonferromagnetic phase, and that discontinuities occur in the specific heat and in other quantities.

A more detailed explanation requires a statistical-mechanical calculation.[9] First a reasonable expression is written for the energy of a specified configuration of spins:

$$E[\mu_1, ..., \mu_N] = \sum_{1 \le i < j \le N} \phi(|\mathbf{r}_i - \mathbf{r}_j|)\mu_i\mu_j - H \sum_{i=1}^{N} \mu_i,$$

where each μ_i can be either $+1$ or -1, according as the ith spin is up or down with respect to a given direction (ordinarily that of the external magnetic field, if it is nonvanishing), H is the intensity of the field, and ϕ is a function that depends upon the distance between the locations of the ith and jth spins. In principle the calculation of the partition function

$$Z(\beta, H) = \sum_{\text{all configurations}} e^{-\beta E[\mu_1, ..., \mu_N]}$$

can be carried out when the function ϕ is given, and from Z the magnetization, the specific heat, and other macroscopic quantities of interest can be calculated by differentiations and other mathematical operations. For almost all choices of ϕ the calculation of Z is extremely difficult. A great simplification is the mean-field assumption, according to which each spin effectively interacts with the mean field produced by all the other spins together, with the result that

$$\sum_{1 \le i < j \le N} \phi(|\mathbf{r}_i - \mathbf{r}_j|)\mu_i\mu_j \cong \frac{\text{const}}{N} \sum_{1 \le i < j \le N} \mu_i\mu_j.$$

This calculation can be performed quite easily, and it exhibits a phase transition: the spontaneous magnetization per spin, as a function of the absolute temperature T, has the following form:

$$m_0 = [3(1 - T/T_C)]^{1/2} \quad \text{for } T \text{ close to but less than } T_C,$$
$$m_0 = 0 \quad \text{for } T > T_C.$$

9. See, for example, Thompson 1972, chapters 4–6.

This derivation of a phase transition is an incomplete triumph. In the first place, the behavior of the magnetization, the specific heat, and other quantities near T_C do not agree with experiment. Furthermore, the mean-field assumption is entirely implausible physically, since a crystal can be enormously large in comparison with a single atom, and it makes no sense that the far-off spins should contribute as much to the total field felt by a given spin as the ones closer by.

A more plausible model is the Ising model, in which a given spin is assumed to interact only with its nearest neighbors:

$$E[\mu_1, \ldots, \mu_N] = -J \sum_{\substack{i=1 \\ j = \text{nearest neighbors}}}^{N} \mu_i \mu_j - H \sum_{i=1}^{N} \mu_i.$$

The one-dimensional version of this model was analyzed by E. Ising in 1925 and shown not to exhibit a phase transition. (Ising erroneously argued that the same was true in two dimensions.) In 1944 L. Onsager performed a rigorous calculation for the two-dimensional Ising model and found that

$$m_0 = [1 - (\sinh 2\nu)^{-4}]^{1/8} \quad \text{for } T < T_C,$$
$$m_0 = 0 \quad \text{for } T > T_C,$$

where $\nu = J/kT$ and $\sinh 2\nu_C = 1$. An intuitive explanation can be given for the failure of a phase transition in one dimension and its occurrence in two. In one dimension, the occurrence of a single spin flip due to a thermal perturbation interrupts the long-range order; in two dimensions, each spin has four neighbors, and if one is deviant because of a thermal perturbation, the other three are likely to maintain the long-range order, except in the improbable event of two or more becoming deviant. (The political implications of this argument are obvious.) A beautiful variant of this reasoning explains how the periodic structure of a crystal is maintained over lengths many orders of magnitude greater than the lattice spacing, so that deviations from exact periodicity do not become cumulative with distance (Peierls 1979, pp. 85–91).

The spectacular work of the last two decades by K. Wilson, B. Widom, M. Fisher, and L. Kadanoff on phase transitions and critical phenomena[10] has exhibited the power of several further principles of composition. Their main results concern the values of critical exponents and relations among them, these exponents being numbers that characterize the singular behavior of thermodynamic quantities, such as susceptibilities and specific heats, in the vicinity of singular points in thermodynamic space (e.g.,

10. See, for example, Ma 1976.

critical points and lines of phase transition). These critical exponents turn out to depend crucially upon a small number of parameters, notably the dimension d of the system and the "spin dimensionality" D or its analogue.[11] What is striking is that the values of the critical exponents are independent of many microscopic features that one might intuitively think to be important, such as the geometry of the lattice array and the strengths of the interactions. The explanation for the insensitivity of the critical exponents to these factors is that near the singular points fluctuations are very large. (In some cases the fluctuations can even in a sense be visible; near the critical point of a fluid there is "critical opalescence," in which giant fluctuations cause unusually large scattering of light.) But if the extent of the average fluctuation is much larger than the average spacing between atoms, then the details of the interactions become irrelevant or are swamped out. This is the essence of the hypothesis of universality for each class characterized by a few parameters. It is at first surprising, but then upon reflection quite reasonable, that general statements can be made about whole classes of macroscopic systems in virtue of giant fluctuations, which are disordering factors. The neglect of fluctuations is the chief respect in which the mean-field treatment of phase transitions fails. The rigorous exploitation of the hypothesis of universality requires the machinery of the renormalization group (a group of transformations in each of which a change of scale plays a crucial role); in a few cases, quantitative calculations of critical exponents have actually been achieved by this machinery.

Although the universality results just mentioned concern only phase transitions and critical phenomena, they are probably indications of a principle of much greater generality. This is the principle enunciated by P. Anderson (1984, p. 85) and called by him "the principle of continuation." Anderson writes:

In a very deep and complete sense we can refer back whole classes of problems involving interacting fermions to far simpler and fewer problems involving only noninteracting particles. While there can be great or small quantitative differences between a real metal, for instance, and a gas of noninteracting electrons, the essentials of their qualitative behavior – specific heat, T dependence of susceptibility, T dependence of various transport coefficients, etc. – are the same.

Later, Anderson states the conjecture that the renormalization group "affords at least one way of putting mathematical teeth into the basic concept

11. As already pointed out, the qualitative behavior of the Ising model depends upon whether d is 1 (in which case there is no phase transition) or 2 or 3 (in which case there is a transition). D is 1 if the spin can be either up or down in a fixed direction; it is 2 if the spin can be a vector confined to a plane; it is 3 if it is a vector that can point in any direction in space. There are generalizations of this meaning of spin dimensionality when one is dealing with other types of transitions between order and disorder.

of continuation . . . , and perhaps it is even *the* most fundamental way of doing so" (p. 167).

8. ONTOLOGICAL COMMENTS

On the basis of the foregoing summaries a number of propositions can be asserted with some confidence concerning the ontological aspects of the parts–wholes relationship in physics. All the ontological questions listed in section 1 will be addressed, though the first only briefly. Furthermore, a few additional matters not anticipated in the list of questions will be touched upon.

A. It is an open question in elementary-particle physics, which has been bypassed in this essay, whether there is an ultimate level of noncomposite particles. Electrons and other leptons may have no internal structure, since they exhibit pointlike behavior in scattering experiments;[12] the same may be true of quarks, which are the components of protons and other heavy particles. Whatever may be the answer to this question, the Democritean equating of noncompositeness with immortality is surely abandoned. Electron–positron pairs can be created and annihilated, even though each member of a pair is created or annihilated as a whole rather than by fabrication and dismantling.

B. The Democritean picture is transformed by the fact that, from the point of view of quantum field theory (the best framework of fundamental physics that we have), a particle is a quantum or excitation of a field, and a field in some sense is a holistic entity, given over all of space–time. Although this statement seems inconsistent with a Democritean answer to the parts–wholes problem, this *prima facie* judgment must be carefully qualified by locality considerations. The propagation of a field is presumably governed by relativity theory, which forbids direct causal connection between two points with spacelike separation. As a result, there is complete freedom to specify a classical field over a spacelike surface, constrained only by smoothness and good behavior at infinity. Except for these restrictions, a classical relativistic field on the whole spacelike surface is determined by the field values on an exhaustive set of disjoint subregions of the spacelike surface, and hence a residuum of the Democritean point of view concerning parts and wholes is found in classical field theory. Analogous (but suitably modified) statements can be made about a quantized relativistic field.

C. From the standpoint of fundamental physics, the low-energy domain to which we restricted attention in sections 3–7 is derivative. The

12. See, for example, Bransden et al. 1973, p. 210. For the view that the electron may be composite, see Greenberg 1985.

elementarity of electrons may be an open question, but there is no doubt that all nuclei other than that of ^1H are composite, even if one does not probe the quark structure of the protons and neutrons. Nevertheless, there are several crucial physical facts which ensure that the level of low-energy physics has a certain autonomy, allowing it to be investigated with little attention to deeper and finer levels, and has a quite firmly Democritean character. One fact is the scale of energies in nature. Typically, the energy required to excite a nucleus is of the order of 100,000 electron volts, whereas the excitation energy of an atom is of the order of one electron volt and that of a molecule even less; excitations in solids can be much less. Consequently, in energy transactions typical of low-energy physics, the nuclei are not disrupted or even excited (though there are unstable or nearly unstable nuclei for which this statement must be amended). A second fact is that the lightest of the particles with nonvanishing rest mass, the electron, has (according to the relativistic equation $E = mc^2$) an energy equivalent of about 500,000 electron volts, and therefore the creation of an electron–positron pair has an energy threshold far beyond the domain of low-energy physics. A third fact is that even though there is not such hindrance to the production of low-energy photons, since photons have zero rest mass, their existence can largely be neglected in the low-energy domain, where the semi-classical radiation theory makes few predictions that differ from those of quantum electrodynamics.[13]

D. The properties of atoms, molecules, normal fluids, infinite Coulomb systems, and spin systems are derivable in principle from the properties of their components and the physical laws governing the interactions of the components. It is not strictly correct, however, to say that the laws underlying these derivations are none other than the principles of low-energy physics summarized in section 2. One reason for caution is that corrections from relativistic quantum mechanics and gravitational theory are required in order to obtain in full detail the properties of the systems surveyed; for instance, relativistic effects are already found in the hydrogen atom, and they become increasingly important with increasing atomic number. Furthermore, the rigorous derivations of phase transitions in spin systems did not assume electromagnetic interactions among the elementary spins, but rather mathematically concocted interactions that hold only between pairs of spins of a restricted class (e.g., nearest neighbors or next-nearest neighbors). The results of Onsager and his successors are not to be depreciated for this reason. What that work shows is that long-range order can be accounted for in terms of interactions with strictly short range, which is a more severe restriction than the falling

13. See, for example, Sargent et al. 1974, p. 97.

off with distance of electromagnetic forces. Consequently, the motivation for postulating some kind of hitherto unobserved long-range interaction for the purpose of accounting for long-range order is removed. (Similarly, the success of the Bardeen–Cooper–Schrieffer theory of superconductivity showed that electromagnetic forces in the framework of quantum mechanics could account for that kind of long-range order, without the postulation of a new fundamental long-range force.)

E. Many of the properties of composite systems which are implied by the laws governing the components are emergent, in the sense that they are qualitatively radically different from the properties of the components. The long-range order of a ferromagnet is an example of an emergent property, as are the rigidity of a crystal and the viscosity of a fluid. The term *emergent* is sometimes used to mean "underivable in principle from the properties of the components," but this is not the meaning that I adopt, nor would it be semantically economical for anyone maintaining a roughly Democritean point of view to do so.

F. The ontological reduction of the properties of the composite system to those of the components, as asserted in paragraph D, does not constitute a renunciation of the entanglement of states of composite systems, which was one of the peculiarities of quantum mechanics emphasized in section 2. Quite the contrary, in many instances the ontological reduction is possible only because of entanglement – e.g., in the role of hybridization in molecular bonding.

G. Composite systems fall into natural kinds, as the discussion of atoms pointed out explicitly and that of molecules tacitly. The same can be said for nuclei and the elementary particles composed of quarks, though these systems lie outside the domain of this essay. The stable configurations permitted by quantum mechanics and the force laws determine in principle the complete taxonomy of possible kinds of stable systems composed from the given components. In the case of relatively simple composite systems – the composite elementary particles, the nuclei, the atoms, and the smaller molecules – the actual taxonomy is also determined, provided that the environmental conditions are suitably specified. (In the case of the heavier nuclei, the requisite environmental conditions are satisfied only in sufficiently hot stars.) The reason is that one can reasonably assume enough trial encounters to have occurred among the components that the space of possible configurations has been well explored in the time interval under consideration. In the case of complex systems, such as macromolecules, there is certainly not enough time within the relevant intervals (e.g., the existence of the earth) to explore the space of possible configurations, and hence the actual taxonomy is in large part the result of the contingencies of evolutionary history. It should be added that within the framework that has been sketched, imperfections in natural kinds (such as flaws in crystals and errors in DNA) are as comprehensible as the near perfections. The Democritean treatment of the central Aristotelian doctrine of natural kinds must be reckoned as one of the great triumphs of modern physics.

H. One of the remarkable features of the physical world is the existence in many situations of several well-defined levels of description for the same physical system. Three levels were considered in the analysis of normal fluids and spin systems, but can be found in many other types of systems. The deepest level – that of relativistic field theory – was only briefly mentioned above, and its existence is acknowledged but not normally exploited by working condensed-matter physicists. The second level is the nonrelativistic quantum-mechanical many-body treatment of interacting electrons and nuclei. The third and coarsest level is that of macroscopic description, in which the fundamental variables for spin systems are temperature, applied magnetic-field strength, and magnetization, and those for fluids are local density, local velocity, and local pressure (all these local quantities being defined by averages over regions large compared to an atomic volume, and all of them possibly varying with position). The remarkable feature of a level of description is not that there are quantities at that level which are defined by summing or averaging over the quantities at a deeper level of description, but that there are laws which govern the specified level without supplementary information from a deeper level. The Navier–Stokes equation for fluid flow and the equation of state $M = M(H, T)$ for a spin system are formulated entirely in terms of appropriate macroscopic variables. To be sure, there are parameters entering these equations which are not supplied by the macroscopic theory and which must either be measured empirically or derived by resorting to a deeper level, but there are no physical variables in these equations except those of the macroscopic level.[14]

I. As pointed out in the summary of Fröhlich's derivation of the Navier–Stokes equation, the derivation of a macroscopic equation from the laws governing a deeper level is not absolutely general; it holds only in a range of circumstances which fortunately is very wide. Sometimes this failure of generality is cited as an objection against the thesis of reducibility of macroscopic physics to physics at a deeper level of description. The

14. The Ω_i in section 6 could be considered to constitute a hierarchy of levels of description intermediate between the second and third levels listed here, except that the dynamical equation for Ω_i refers to Ω_{i+1}. With suitable assumptions, the hierarchy of equations can be truncated at the kth without serious error (this was in fact done for $k = 2$ in section 6), and then there is an autonomous intermediate level with its own dynamical law. Grad 1967 contains a fine discussion of levels of description.

most common version of this objection is that classical thermodynamics is a nonstatistical theory, whereas statistical fluctuations require changes in the thermodynamics which Boltzmann and Gibbs purport to derive from statistical mechanics. My answer to this objection is that excessive rigidity concerning the concept of reducibility can mask the truly wonderful relations between levels of description. From an Olympian point of view, unclouded by the difficulties of mathematical inference, the laws at the deeper level imply, for each set of system constitutions, initial conditions, and boundary conditions, a definite set of values or possibly a definite set of probability distributions of the macroscopic quantities; the extent to which the putative macroscopic laws are satisfied is then made explicit. If the agreement is good (by a reasonable standard), in a set of circumstances that is wide (by a reasonable standard), then a striking relation holds which captures the intuitive and practical sense of "reduction." What constitutes "reasonable standards" obviously cannot be specified at the level of generality of the present discussion, but the treatment of fluctuations in statistical mechanics shows that sensible things can often be said about this question.

9. EPISTEMOLOGICAL COMMENTS

The epistemological problem that has most preoccupied philosophers of science is the legitimation of scientific principles on the basis of empirical evidence. Some variant of the hypothetico-deductive method is recognized widely to be appropriate for this purpose, though, of course, there are many different opinions about the details of the method and the cognitive claims that can be justifiably made for it. The deductive steps in the hypothetico-deductive method are usually given less attention than the other steps; deduction is supposed "in principle" to be well understood, even though in practice it encounters enormous technical difficulties. In other words, there is a tendency to take an Olympian point of view for granted in discussing the deductive steps of the hypothetico-deductive method.

In the enterprise of reducing macrophysical theories to microphysical theories and, in general, explaining the properties of composite systems in terms of those of the components, the principles governing the components are taken as established, and therefore the problem of deduction emerges from the background and assumes a central role. The special epistemological problems concerning the relation between parts and wholes in physics arise just because humans are debarred by their intellectual limitations from taking an Olympian point of view. If there is not a satisfactory surrogate for Olympian deduction, then human beings are not entitled to assert that macrophysics is reduced to microphysics and that wholes are understood in terms of parts, no matter how matters appear to the deities. The five illustrations discussed in sections 3–7 contain a number of procedures, none of which by itself provides a satisfactory surrogate for Olympian deduction but which work together remarkably well to this end.

One procedure is systematic approximation to the exact solutions of difficult mathematical equations – for example, the Hartree–Fock iteration method mentioned in section 3. Approximations might be thought to be purely mathematical, carried out without reference to the physical problems that posed the difficult equations. In practice, however, the sublimation of the mathematics from the physics is incomplete, and physical considerations direct and supplement purely mathematical ones in several ways. First of all, the starting point of an approximation method is very important, and observational data together with intuitive understanding can suggest a good beginning. Conversely, a qualitatively wrong starting point – e.g., one with a different symmetry from that of the phenomenon of interest – will ensure that the approximation method will never converge to the true solution; a barrier of singularity separates a state with the wrong symmetry from those with the correct symmetry (Anderson 1984, pp. 27–30). Even if the starting point is not radically wrong in this sense, a clumsy choice will entail convergence so slow that the outstanding features of the true solution will not be recovered by a computation of reasonable duration. Another way in which physical considerations guide and direct mathematics is by indicating that an approximation method has been carried far enough in a situation where no rigorous proof exists of convergence or where no rigorous bounds can be put upon the error of truncation. Example: In the ground state of a system with finitely deep potential wells, the energy is a minimum; if the energy computed with an approximate solution is very little above the experimentally measured energy, then it is reasonable to conclude that the approximation is good and has captured the major features of the true solution.

A second procedure is the performance of simplifying steps in the course of physical analysis of a problem. This procedure is quite different from the construction of a model, in which the simplifications are made initially in characterizing the constituents of the system and the forces among them. The procedure to which I am now referring is exemplified at several points in section 6 – e.g., it was assumed that changes of σ, $\mathbf{v}$, and p are small over distances of the order of the interaction range between atoms, and that the dependence of the pair-correlation function $P(\mathbf{x} - \mathbf{y})$ on powers of $\mathbf{v}$ higher than the first is negligible. These simplifications are made quite naturally in the course of the analysis, but would be hard to insert into a rigorously treated model. Usually these simplifications are intuitively appealing, and when they seem particularly significant they are often dignified with the name of *Ansatz*.[15] Simplifications are made plausible not only by physical intuition but also by the fact that there is an empirically well-established *terminus ad quem,* a phenomenological equation that is known to work well for normal liquids that are not driven to the point of turbulence. This empirical evidence provides inductive support for the supposition that the simplifications are valid broadly and not merely in highly idealized special cases.

In the two procedures considered so far, informal inductions are made in order to remedy the shortcoming of human deductive powers. Disagreement with empirical evidence would then constitute *prima facie* evidence that the truncation was premature or the simplification unjustified. Conversely, agreement with empirical evidence provides inductive support for the legitimacy of the truncation or the simplification, though the informality and the lack of rigor must be acknowledged; it is certainly possible that good agreement is achieved, not because of the smallness of the terms neglected relative to those retained, but because of fortuitous cancellations among the neglected terms. Confirmation is always riskier than disconfirmation. As in inductive inference generally, however, when good agreement is found in a variety of experimental situations, and when the physical picture becomes more and more coherent, then confidence in the confirmation justifiably increases.

A third procedure is the construction of rigorously soluble models possessing some of the crucial qualitative features of real systems – for example, the short-range character of the microscopic magnetic moments that compose an actual ferromagnet. As noted in sections 7 and 8, the great significance of these results is the rigorous establishment of emergence; that is, the exhibition of macroscopic properties radically different from those of the constituents. Conceptually, though not technically, it is a larger step to establish the possibility of emergence at all than to show that real composite systems behave qualitatively in the same way as the models.

A fourth procedure is to implement what Anderson calls "the principle of continuation," thereby providing the missing linkage between models and real systems. For this purpose the renormalization group is very powerful, but an exposition of that is beyond the scope of this essay and its author. Evidently, however, Anderson's suggestion that the renormalization

---

15. A famous example is L. Boltzmann's *Stosszahlansatz* in the kinetic theory of gases, which says essentially that the velocity distribution of a molecule that has just undergone a collision is the same as that of a randomly selected molecule. As a matter of fact, this *Ansatz* is automatically satisfied by an appropriate model (Grad 1967, p. 56).

group is not merely a technique for studying a special body of phenomena but an instrument of great generality in many-body physics deserves very careful study.

The systematic deployment of these procedures for the purpose of understanding the properties of composite systems in terms of their components deserves the epithet "the methodology of synthesis," which has previously been applied to statistical physics (Toda et al. 1983, p. v). The epistemological significance of the physics of composite systems is summed up in this phrase.

One final epistemological comment is appropriate. Throughout this essay, physical first principles have been assumed to be given. Of course, they are not given, but have to be established by ingenious experimentation and profound analysis of the experimental results. In view of the great mathematical difficulties of determining what experimental prediction a theory makes concerning a many-body system, it is obviously desirable to carry out the critical experiments for assessing first principles upon the simplest possible systems.[16] But a reasonably successful execution of a physics of composite systems within a given framework of first principles, carried out in accordance with the "methodology of synthesis" sketched above, should be regarded as very impressive supporting and reconfirming evidence for those principles. Contrariwise, the failure to execute successfully a physics of composite systems within a proposed framework in spite of much effort (for which we now have a good standard of strenuousness) should be regarded as serious disconfirming evidence. The investigation of the properties of composite systems would usually be classified as "normal science," but it should be obvious from the foregoing discussion that "normal" cannot be equated with "mundane" or "routine."

Acknowledgment. I wish to thank Peter Achinstein, Howard Stein, William Klein, and Charles Willis for their stimulating comments and questions.

REFERENCES

Anderson, P. W. 1984. *Basic Notions of Condensed Matter Physics.* Menlo Park, Calif.: Benjamin/Cummings.

Baym, G. 1969. *Lectures on Quantum Mechanics.* New York: Benjamin.

16. It may be objected that a many-body system supplied the data that inspired the beginning of quantum theory: the radiation in a black cavity. This objection is not decisive, however, because black-body radiation is in fact extremely simple, consisting of non-interacting excitations.

Bethe, H. A., and R. W. Jackiw. 1968. *Intermediate Quantum Mechanics,* second edition. Reading, Mass.: Benjamin/Cummings.

Bransden, B. H., D. Evans, and J. V. Major. 1973. *The Fundamental Particles.* London: Van Nostrand Reinhold.

Fröhlich, H. 1973. "The Connection between Macro- and Microphysics." *Rivista del Nuovo Cimento* 3: 490–534.

Grad, H. 1967. "Levels of Description in Statistical Mechanics and Thermodynamics." In *Delaware Seminar in the Foundations of Physics,* vol. 1, ed. M. Bunge. New York: Springer-Verlag.

Greenberg, O. W. 1985. "A New Level of Structure." *Physics Today* 38, no. 9: 22–30.

Khinchin, A. 1969. *Mathematical Foundations of Statistical Mechanics.* New York: Dover.

Lieb, E. 1976. "The Stability of Matter." *Review of Modern Physics* 48: 553–569.

Ma, Shang-keng. 1976. *Modern Theory of Critical Phenomena.* Reading, Mass.: Benjamin.

Merzbacher, E. 1970. *Quantum Mechanics,* second edition. New York: Wiley.

Pauling, L. 1931. "The Nature of the Chemical Bond. Application of Results Obtained from the Quantum Mechanics and from a Theory of Paramagnetic Susceptibility to the Structure of Molecules." *Journal of the American Chemical Society* 53: 1367–1400.

Pauling, L., and I. Keaveny. 1973. "Hybrid Bond Orbitals." In *Wave Mechanics,* ed. W. Price et al. New York: Wiley.

Peierls, R. 1979. *Surprises in Theoretical Physics.* Princeton University Press.

Roby, K. R. "Mathematical Foundations of a Quantum Theory of Valence Concept." In *Wave Mechanics,* ed. W. Price et al. New York: Wiley.

Sargent, M., M. O. Scully, and W. E. Lamb. 1974. *Laser Physics.* Reading, Mass.: Addison-Wesley.

Thompson, C. 1972. *Mathematical Statistical Mechanics.* New York: Macmillan.

Tisza, L. 1966. *Generalized Thermodynamics.* Cambridge, Mass.: MIT Press.

Toda, M., R. Kubo, and N. Saito. 1983. *Statistical Physics* I. Berlin: Springer-Verlag.

Weisskopf, V. F. 1979. *Knowledge and Wonder,* second edition. Cambridge, Mass.: MIT Press.

15
Some proposals concerning
parts and wholes

In the world as we know it there is an immense variety of systems having identifiable parts. If one focuses only upon inorganic systems one can recognize a descending hierarchy of cosmos, metagalaxy, galaxy, star (possibly with attendant planets), macroscopic bodies, molecules, atoms, elementary particles; or, more accurately, one can recognize a number of alternative physical hierarchies. If one focuses upon organisms, one finds multi-cellular plants or animals, organs and suborgans, cells, and organelles; and there are molecules which are identifiable parts of cells. The taxonomy of systems can be extended further by considering entities which are less concrete than the foregoing. For example, space–time has regions of space–time as parts, and points as members.

In view of this variety, it is probably unwise to adopt a unitary position either of holism or of individualism concerning the ontological status of parts and wholes in systems generically. Furthermore, one should not characterize the philosophical position of a historical figure simplistically, without taking into account his treatment of different types of systems. Newton, for example, is often taken to be an arch-analyst, since he analyzes the behavior on any composite system in terms of the particles which compose it and the forces upon these particles (due either to each other or to external agents). Newton's ontological position, however, is holistic in at least two ways. First, his atoms are indivisible, even though they have geometrically discriminable parts. And second, Newton treats absolute space holistically, since he recognizes that all points are intrinsically identical and therefore their principle of individuation can only be location in space as a whole. This holistic conception of space is essential to Newton's defense against one of Leibniz's objections against absolute space, namely, that if the points are intrinsically identical then by the principle of the identity of indiscernibles there cannot be a multiplicity of them.[1]

Presented at a Workshop on Parts and Wholes at the University of Lund, Sweden, in June 1983. Reprinted by permission of the Swedish Council for Planning and Coordination of Research.
1. See H. Stein, "Newtonian Space–Time," *Texas Quarterly* 10: 174 (1967).

Epistemological and methodological problems concerning parts and wholes should also be treated pluralistically, with attention to the great differences among types of composite systems. I shall elaborate my plea for pluralism by looking at four different types of composite systems: (1) composite microphysical systems in pure quantum states; (2) macrophysical systems undergoing phase transitions; (3) organic cells considered as cybernetic systems; (4) typical human societies.

1

If quantum mechanics has unlimited validity, then in principle any physical system can be characterized by a pure quantum state. In practice, it is rarely fruitful to ascribe a pure quantum state to a macroscopic body, but it is often very fruitful to do so for composite microscopic systems, such as a deuteron of an atom. One of the fundamental principles of quantum mechanics is the superposition principle, which roughly says that if u_1 and u_2 are pure states of system I, then there is a physically possible system which "overlaps" both u_1 and u_2 and which is designated as $u_1 + u_2$. The intuitive meaning of overlapping can be conveyed by considering a situation in which an observable property A has the definite value a_1 in state u_1 and the definite value a_2 in state u_2. If a_1 and a_2 are different, then A has no definite value in $u_1 + u_2$; A would only be a potentiality in that state, realizable only in two possible ways, as a_1 or a_2, with appropriate probabilities.

Now consider another system II, two possible states of which are v_1 and v_2, in which observable property B has respectively the distinct values b_1 and b_2. Furthermore, consider the possible states of the composite system I + II. Among the possible states of I + II are the thoroughly non-holistic state $u_1 v_1$ (the state in which I is in u_1, II is in v_1) and the equally non-holistic state $u_2 v_2$. Now the superposition principle says that a possible state of I + II is $W = u_1 v_1 + u_2 v_2$, and in fact there are situations in which this state can be prepared in a laboratory. It can easily be shown that W is not factorizable – that it cannot be expressed in the form yz for *any* state y of I and *any* state z of II. In the locution of Schrödinger, systems I and II are "entangled". Several comments are appropriate.

a. Enganglement is intimately bound up with the role of potentiality in quantum mechanics. The potentialities for I and II are linked when I + II is in the state W, because the actualization of A to have the value a_j (where $j = 1$ or 2) entails the actualization of B to have the value b_j.

b. Entanglement is not a matter of dynamical interaction. Dynamical interaction occurs in classical physics and is retained in quantum mechanics, but entanglement is a strictly quantum mechanical concept. If two classical systems interact dynamically, each has a definite pure state at every instant of time, but the state of one of them at $t + dt$ depends not only on its own state at t but also upon the state of the other at t. If I + II is an entangled system at t, then it is incorrect to ascribe *any* pure state to I at time t and *any* pure state to II.

c. It does not seem possible to interpret entanglement merely epistemically – i.e. that our knowledge of I is inseparable from our knowledge of II. It is an ontological matter. Of course, an argument and evidence are needed to support this bald assertion.

2

Phase transitions – e.g., from a vapor to a liquid, from a liquid to a crystal, from a non-ferromagnetic to a ferromagnetic crystal – are among the most dramatic of physical phenomena. Since a phase transition affects a macroscopic body collectively, and since it apparently occurs in a discontinuous manner (no ferromagnetism above the Curie temperature, spontaneous ferromagnetism below it), one may plausibly suppose that these phenomena provide instances of holism within physics. A very striking example to the contrary was provided by Onsager in 1949. The two-dimensional Ising model consists of a plane array of elementary spins, each of which is either up or down. Each spin interacts only with its nearest neighbors, and the contribution to the total energy of the system by a pair of nearest neighbors is less when they have the same alignment than when they have opposite alignment. At absolute zero all the spins would align in the same direction, since that is energetically favored; but at elevated temperatures the random distribution of thermal energy will cause some of the nearest neighbors to be oppositely aligned. The important question is whether the model exhibits a long range ordering of the alignment of spins below a critical temperature, and a discontinuous loss of long range order above the temperature. Onsager rigorously demonstrated that the answer to this question is positive, thus providing a model of the ferromagnetic phase transition. Several comments are appropriate.

a. Onsager's demonstration was a triumph of the analytic point of view. A collective phenomenon was derived not merely in terms of mechanically describable interactions among components, but in terms of interactions extending only to nearest neighbors.

b. Onsager used the essentially probabilistic methods of statistical mechanics, but because the number of components in the model was large (effectively infinite), statements could be made with virtual certainty about the macroscopic behavior of the system.

c. Actual physical systems are much harder to analyze mathematically than the Ising model, and therefore it is not correct to claim that the phase transitions in actual substances have been explained analytically. There have been many partial results, however, and the analytic explanation of phase transitions is (in Lakatos's locution) a "progressive research program".

d. Some phase transitions cannot be understood except in terms of quantum mechanical entanglement – notably, the transitions in superconductors and superfluid helium, of course at very low temperatures. It is remarkable, however, how much of an understanding of phase transitions can be achieved without invoking the holism of quantum mechanical entanglement.

3

Until fairly recently a fairly strong negative case could be made for the inexplicability of biological reproduction, growth, repair, and goalseeking in physico-chemical terms, though there were arguments – quite sterile in retrospect – on whether reductionists or holists should bear the burden of proof. The dramatic discoveries of molecular biology in the last few decades have changed the character of the debate entirely. Reduction of biology to chemistry and physics is not only conceivable, but is a magnificently progressive research program. A great popular exposition of the case for reduction is presented in Monod's *Chance and Necessity* – an indispensable reference in a workshop on parts and wholes. Again, a few comments are appropriate.

a. Monod explains the elaborate cybernetic functioning of the cell in terms of sequences of two- and three-body interactions, typically among substrates and enzymes. In this sense his treatment is anti-holistic. However, when he invokes the stereo-specific recognition of a substrate by an enzyme and when he relies upon the stability of molecular states, he implicitly assumes quantum mechanical holism *at the molecular level*. D'Espagnat refers to Monod's failure to note the role of quantum mechanical entanglement in his overall scheme as "une lacune d'importance".

b. Fröhlich[2] has suggested that there may be cooperative phenomena *at the cellular level* which can be understood only in terms of quantum mechanical holism, even though the temperature of the cell is far above that of superconductors. Whether this conjecture is true or not, it points to open questions about the exact contribution of quantum mechanical and non–quantum mechanical operations in the total cybernetic functioning of the cell. These questions may be particularly important in neurology.

c. The sequencing of very many two- and three-body chemical reactions for the purpose of maintaining the structure and operation of the cell is a fantastic feat of chemical engineering. According to the neo-Darwinian

2. H. Fröhlich, *Journal of Collective Phenomena* 1: 101 (1973).

view which Monod espouses, this feat is the result of trial and error in the evolutionary laboratory, understandable in principle without invoking any orthogenetic evolutionary mechanism. Although I strongly agree with Monod on this point, I feel that our understanding is very sketchy and needs to be enriched by plausible reconstructions of prebiotic evolution. (Some assessment by experts of the hypothetical reconstructions offered by Eigen and by Prigogine and his school would be welcome.)

4

Collective descriptions of the composition and functioning of human societies are undoubtedly valuable in the social sciences. It would be misguided to abstain from using concepts like social organization, culture, and macro-economic structure because of ontological scruples about holism. One does not need to postulate an ontology of "the superorganic" in order to recognize the utility of collective concepts and the possible validity of principles formulated in terms of these concepts. The examples cited in sections 2 and 3 show that collective behavior in macrophysical systems and in biological cells can often be explained in great detail in terms of the properties and the interactions of the parts, and the explanatory power of schemes admitting only few-body interactions has been noted. An ontology in which individual human beings have a fundamental mode of existence, while societies, institutions, cultures, etc. have only a derivative existence, should suffice for the social sciences. The reasonableness of this claim is reinforced by reflecting upon the actual and potential richness of the psychological states of individual human beings: an entire culture, with its language, literature, rituals, etc. can be internalized within one human psyche. That human beings are biologically social animals does not imply that the society has a more fundamental ontological status than individual human beings, or even an independent status. DNA is a "social molecule", functioning as a template for the construction of RNA, which in turn guides the construction of the proteins needed in the life of the cell; but the social nature of DNA does not endow the cell with a holistic ontological status. At least, there is no need to do so for the purpose of understanding causal sequences in the cell. The more precisely causal sequences are understood in the social sciences, I believe, the more clearly will the ontological primacy of individual human beings be evident. Finally, I shall note without comment the speculation that a human group, perhaps all mankind, may enter a quantum mechanically entangled state ("the Maharishi effect").

Proposals (1), (2), (3), and (4) concern four very different types of composite system. I have deliberately refrained, however, from making any proposals about the relation between the whole of a sentient organism and its parts. To a behaviorist or a strict materialist, this relation need not be any different in principle from the relation between the cell and its molecular constituents, discussed in section 3, although there are obviously unsolved problems about the details or cybernetic organization of multi-celled organisms, particularly those endowed with nervous systems. It is just because I find the behaviorist and materialist accounts of sentience utterly unconvincing that the parts–whole problem concerning sentient organisms seems extraordinarily baffling. For many years Whitehead's philosophy of organism and the related monadology of Leibniz seemed promising to me, but perhaps their obscurities outweigh their promise. I have also thought about the conjecture (which seems to be one of Bohm's proposals) that quantum mechanical entanglement provides a kind of constructive principle capable of application even in psychophysics, a domain far from the locus of the triumphant empirical applications of quantum mechanics. But an inner voice cautions against such *Schwärmerei.* In short, we do not know.

COMMENT

An efficient way to refine the initial set of proposals which I made to the Workshop on Parts and Wholes is to comment on some contributions which seem to me largely but not completely correct.

Hans-Rainer Duncker argued eloquently for a holistic conception of the structure of organisms, and he gave a beautiful extended example of holism in the interrelation of oviparity, lung characteristics, and metabolic efficiency in birds. The conclusion which he draws is that "the basic functional levels of organisms have been analyzed by the experimental, so-called causal-analytic science; the higher functional levels are only accessible to comparative disciplines like morphology and ethology." It seems to me possible to agree to this conclusion if it is construed methodologically, and to dissent if it is construed ontologically. I could not decide, on the basis of Duncker's written and oral presentations, whether he intends an ontological interpretation, but the points I have to make are relevant whether they are to be taken as criticism or as clarification.

If we restrict our attention to organisms which do not manifest mentality, then the holistic construction and performance can be considered to be a triumph of engineering. All the elements operate in accordance with physical laws, and the meshing of these elements – by chemical and electrical cybernetics – is equally governed by these laws. Michael Polanyi[1]

---

1. Michael Polanyi, *Knowing and Being* (U. of Chicago Press, Chicago, 1969), especially the essay "Life's Irreducible Structure."

has argued that the irreducibility in principle of biology to physics stems from the fact that engineering is also irreducible, for the blueprint or design is not a matter of physics (a clever use of the mechanists' claim that organisms are machines). Polanyi is correct that the blueprint is not required by the laws of physics but is only a contingency compatible with these laws. He does not succeed, however, in making a case that a supervening law, consistent with but transcending physics, is needed to explain the genesis of the blueprint. He plays down the fecundity of chance to produce viable integrated systems of chemical cybernetics, given the incredibly large number of combinations that were tried out during the history of the biosphere.[2] (Pasteur won the great battle against the advocates of spontaneous generation, but their fundamental error was the omission of one crucial ingredient from their flasks – *time*.) What is convincing in Polanyi's account is that scientists are unable in practice to discover and to grasp the *modus operandi* of an organism and of a machine unless the blueprint is understood teleologically and as a Gestalt. For this reason Duncker's statement, quoted above, is correct if interpreted methodologically.

One can also accept Duncker's insistence upon a theory of a "historically developed organism" without in any way retracting that ontologically biology (in the absence of mentality) is fully understandable in physical terms. Evolution is unavoidably historical, because each possible course of development of the biosphere from geological conditions at some base time (e.g., four billion years ago), including the course that actually occurred, had almost infinitesimal probability, in strongest contrast with the simple systems treated in the thermodynamics of irreversible processes, in which the actual course (if characterized macroscopically) has much greater probability than all the other courses combined. No biologist could predict from first principles, together with the geological initial conditions, that the species of chickens would emerge, since this species occupies an infinitesimal part of the relevant phase space. But as Laszlo Tisza has remarked, the biologist should have no inferiority complex *vis-à-vis* the physicist, for although the latter can predict quantitatively the trajectory of a planet, the former can reliably predict that the hen's egg will hatch only into the chickens' tiny sector of an immense phase space. His remark must be qualified in one respect: the *explanation* for this reliability is to be found in the physical principles which govern the pre-programmed sequence of steps of reproduction and embryological development.

2. Jaques Monod, *Chance and Necessity* (Random House, New York, 1972); M. Eigen and P. Schuster, *The Hypercycle: A Principle of Natural Self-Organization* (Springer, Berlin–Heidelberg–New York, 1979).

All that I have said so far assumed the exclusion of mentality. The program of physicalist explanation of mental phenomena seems to me hopeless, and since mentality does seem to be efficacious in the performance of higher animals, it follows that a purely physicalist account of this performance could not be adequate. If one is a dedicated evolutionist, this statement about the functioning of animals endowed with mentality is hard to reconcile with the confident assertion made earlier that the structure and behavior of lower organisms are comprehensible physically. Whitehead's philosophy of organism offers a possible solution to the problem. According to him, the fundamental concrete entities of the world are occasions of experience, which must be characterized in terms borrowed, though with great extrapolation, from human psychology. Physics provides a veridical but coarse-grained description of concrete reality, abstracting from its fundamental protomental character: "The notion of physical energy, which is at the base of physics, must then be conceived as an abstraction from the complex energy, emotional and purposeful, inherent in the subjective form of the final synthesis in which each occasion completes itself."[3] When the fundamental occasions have a monotonous, repetitive character, the physical description serves as a good approximation to the complete truth. Since physics suffices to explain the formation of stable complex structures, including macromolecules and mutually catalytic systems of these, those systems which biologists recognize as organisms could be composed of the Whiteheadian occasions, but in such a way that the protomental nature of these occasions is effectively in abeyance. In this way a biological level of description would be superimposed upon a physical level of description, which in turn is superimposed upon a fundamental protomental level of description, but little would be lost by disregarding the fundamental level (just as little or nothing is lost in the treatment of chemical phenomena by neglecting the internal structure of protons and neutrons). But this is not the whole Whiteheadian story. The possibility remains, when circumstances are right, that the fundamental protomentality of the constituent occasions of an animal become "coördinated" instead of being washed out by incoherence and randomization.

Thus in a man the living body is permeated by living societies of low-grade occasions so far as mentality is concerned. But the whole is coördinated so as to support a personal living society of high-grade occasions. This personal society is the man defined as a person.[4]

3. A. N. Whitehead, *Adventures of Ideas* (Macmillan, New York, 1933), chapter 11, p. 239.
4. *Ibid.*, chapter 13, p. 267.

In this way Whitehead offers a theory which accounts coherently for both the manifestly animate and the apparently non-animate sectors of the great chain of being, as the theory of evolution requires. Whitehead's account commands respect in spite of great difficulties implicit in it.[5] At the minimum, it provides a calibration for measuring the imaginativeness of proposals concerning the mind–body problem.

Both Joachim Israel and Mihailo Marcovič are correct on a descriptive level when they emphasize the holistic character of human societies, and the latter made some helpful suggestions about the multiple meanings of "holism". Again, however, it is essential to distinguish questions of methodology from those of ontology. If the deduction of features of the biosphere from physical first principles is hopeless because of the role of chance in evolution, the deduction of the structure of actual human societies and of the content of their cultures from the principles of individual psychology is also hopeless. Consequently the social sciences do and should have methodological autonomy. But there is an obverse side to this argument. If the enormous complexity and variety of organism (again setting aside mentality) can be understood in terms of the physical interactions of their components, why cannot the structure and variety of human societies be understood in terms of the interactions of human individuals? Individual human beings are much more complex "atoms" than are the literal atoms of physics, and furthermore they are "atoms" in which social instincts are genetically implanted. Hence there are immense resources for understanding social facts in terms of the characteristics of individual human beings and the interactions among individuals, and the need to postulate an ontologically irreducible social element has never been established.

I shall offer one illustration to strengthen my thesis. No social fact is more pervasive than a language. One is born into a linguistic environment almost just as into a physical environment. Furthermore, linguistic phenomena provide opportunities for reliable prediction which are rather rare in the social sciences. We can predict with confidence that if no nuclear war or comparable catastrophe occurs within the next fifty years, then the majority of the inhabitants of the region presently demarcated as China will speak Chinese. Of course no linguist or psychologist could predict from psychological first principles the emergence of a language having the rules of Chinese. We have a situation analogous to the predictions concerning chickens, discussed previously. The aspect of the analogy which seems to me most fundamental is that the reliability of the respective

5. A. Shimony, "Quantum Physics and the Philosophy of Whitehead," in *Boston Studies in the Philosophy of Science,* eds. R. S. Cohen and M. Wartofsky (Humanities, New York, 1965).

predictions – biological in one case, social in the other, but in both cases descriptively holistic – rests upon mechanisms governing individual components. The generic linguistic capacity of small children, and their propensity to learn by some kind of induction the rules of the language to which they are exposed, are the facts of *individual psychology* which underlie the *social fact* of linguistic continuity.

Patrick Suppes is right in citing quantum mechanics to exemplify "the delicate dance from parts to wholes and back again," but not in the details of his citation. The question of the reduction of chemistry to quantum physics is a matter of principle, not of computational feasibility. It is an obfuscation to conflate these two issues. By contrast, a serious challenge to the reduction in principle would be posed if there do not exist in nature certain superpositions of molecular states which one would expect to be realizable according to the quantum mechanical superposition principle. An interesting argument to this effect has been given in the first chapter of a recent book by Hans Primas.[6] There is great tension within quantum mechanics concerning holism. The quantum correlations of nucleons in a nucleus, of photons in EPR-type experiments, and of electrons in atoms and superconductors require a holistic account of the quantum state of composite systems. On the other hand, the occurrence of definite events at the conclusion of measurements indicates a limitation to holism. It appears difficult to explain this limitation unless the linear dynamics of quantum mechanics (the time-dependent Schrödinger equation) is modified. At one time hidden variables theories constituted a promising avenue to resolve this tension, but the disconfirmation of the family of local hidden variables theories has largely (though not completely) blocked this avenue. New ideas are needed, and they will have to be deep and subtle concerning the whole–parts relationship.[7]

6. Hans Primas, *Chemistry, Quantum Mechanics and Reductionism* (Springer, Berlin–Heidelberg–New York, 1981).
7. A. Shimony, "Metaphysical Problems in the Foundations of Quantum Mechanics," *International Philosophical Quarterly* 18: 3 (1978).

16
The non-existence of a principle of natural selection

The theory of natural selection is a rich systematization of biological knowledge without a first principle. When formulations of a proposed principle of natural selection are examined carefully, each is seen to be exhaustively analyzable into a proposition about sources of fitness and a proposition about consequences of fitness. But whenever the fitness of an organic variety is well defined in a given biological situation, its sources are local contingencies together with the background of laws from disciplines other than the theory of natural selection; and the consequences of fitness for the long range fate of organic varieties are essentially applications of probability theory. Hence there is no role and no need for a principle of the theory of natural selection, and any generalities that may hold in that theory are derivative rather than fundamental.

I. THESIS

The main thesis of this paper is that the biological theory of natural selection does not contain a general principle of natural selection and has no need for such a principle.

This thesis in no way denies that natural selection is one of the primary mechanisms shaping the constitution of the biosphere; and in no way is it an endorsement of any of the well-known alternatives to or modifications of the theory of natural selection, such as creationism, Lamarckianism, or orthogenesis. Nor does my thesis entail that no general principles are to be found in the theory of evolution as a whole, of which the theory of natural selection is a part. On the contrary, the two other commonly acknowledged parts of the theory of evolution, namely the theories of heredity and variation, do have general principles. The exact content of the principles of heredity and variation has changed as evolutionary biology developed from Darwin and Wallace to the present, but the existence of some principles governing those phenomena has been an invariant in the history of the subject. In particular, their existence is independent of the question of the reducibility of biology to physics, which will not be taken

These two papers originally appeared in *Biology and Philosophy* 4 (1989), pp. 255-73 and 280-86. © 1989 Kluwer Academic Publishers. Reprinted by permission of Kluwer Academic Publishers.

up in this paper. What I am claiming is that the theory of natural selection has no principle coordinate with those of heredity and variation, and this fact also is independent of the reducibility of biology to physics.

There have been many and varied attempts to formulate a general principle of natural selection, typically formulated in terms of the concepts of fitness, adaptation, or differential reproduction rate. All of these attempts have been misguided, in my opinion, not because they omit some profound or subtle conceptual component, but because the principle which they are trying to grasp simply does not exist. I believe, furthermore, that my thesis is implicit in the insistence of the neo-Darwinians upon the utterly opportunistic character of the evolutionary process and in their rejection of any general guiding plan in evolution, e.g., Simpson (1963, pp. 190–212), Monod (1971, pp. 23–44 and 118–120), Mayr (1982, pp. 516–517 and 588–590). I construe the neo-Darwinians as trying to say meta-theoretically that the evolution of the biosphere, subsequent to the establishment of the genetic code, is governed by the principles of heredity and variation and the laws of physics, and is constrained by biological and environmental boundary and initial conditions, but *not constrained otherwise:* within these constraints *let happen what happens.* The qualification "meta-theoretically" is needed because the non-existence of a certain kind of biological principle is not the kind of statement that can be made in the object language of biology, but it can be made in a suitable meta-language.

The non-existence of a general principle of natural selection does not preclude the existence of a theory of natural selection – if the word "theory" is used in its etymological sense of "viewing" or in the sense of "a scheme or system of ideas and statements." The theory of natural selection is the systematization of knowledge about the process of natural selection, and it is a very rich theory. It contains penetrating *dicta* about fitness, adaptation, and differential reproduction rates; wonderfully illuminating analyses of the adaptedness of innumerable biological varieties; and a body of mathematically formulated propositions which seem to me (admittedly no expert) both impressive and promising. I maintain, however, that the theory of natural selection is not built upon a general first principle, in the way that the theory of electromagnetism is built upon Maxwell's equations. Such generalities as it does possess are derivative rather than primary. They hold at the level of special circumstances and conditional formulations; for at these levels the injunction "let happen what happens" can and often does eventuate in new constraints. I have elsewhere (Shimony, 1989) called the theory of natural selection a "null theory," because it has no first principle. This locution, however, is perhaps misleading and provocative, for it suggests that the entire theory is trivial, which is completely alien from my intention. Rather, I regard the status of the theory of natural selection – a rich theory without its own general first principle – to be one of the most remarkable phenomena of human knowledge. There are branches of physics which have a similar status, specifically the kinetic theory of gases and statistical mechanics, as C. S. Peirce (1877) recognized soon after the publication of *The Origin of Species*. The relations between these branches of physics and the theory of natural selection deserve more attention than they have received so far, but I shall not try to examine them extensively in the present paper.

2. CLASSIFICATION OF NATURAL FACTS

In stating the main thesis of this paper I used the word "principle" and therefore took for granted that its meaning is roughly understood. Some comments on this term and several related ones will be valuable, however, as preparation for an argument that no principle of natural selection exists.

There is a loose hierarchical structure in the natural sciences as presently constituted, which may be partly due to sociological and historical factors, but which surely is in large part the reflection of a hierarchy (or network of hierarchies) in nature. There is usually no practical difficulty in distinguishing singular facts from general facts in a clearly delimited branch of science. It is a general fact that in the absence of air resistance and other perturbing forces the motion of a macroscopic body slightly above the earth's surface is governed by the differential equation

(1) $$d^2\mathbf{r}/dt^2 = \mathbf{g},$$

where $\mathbf{r}$ is the position of the particle and $\mathbf{g}$ is the local gravitational acceleration. Within the science of ballistic motion this general fact is classified as a "law," though the singling out of laws from among the class of general facts is a rather conventional matter. (For example, it is a general fact that the specific resistivity of all samples of pure copper at $0°$ Celsius is 1.55 microhms cm, but this general fact is seldom if ever dignified by the title of "law.") The ballistic differential equation (1) determines only a family of physically permissible trajectories of a body, and the selection of an actual trajectory is accomplished only when some singular facts are given, typically the initial conditions $\mathbf{r}(t_0)$ and $\dot{\mathbf{r}}(t_0)$ at some specified time t_0. When one branch of science is subsumed under a more general branch, a general law of the former may be seen from the standpoint of the latter to involve a singular fact. For example, the constant $\mathbf{g}$ in the ballistic differential equation (1) is due to a singular fact, namely, the mass distribution of the earth; and from this distribution, together with Newton's laws of mechanics and universal gravitation, $\mathbf{g}$ can be computed.

Frequently the singular facts which are contrasted with the general facts of a theory are called "contingencies," in the sense that both their truth and their falsity are consistent with the generalities. Of course, a fact which is contingent in this sense may be logically implied by a general fact in conjunction with other singular facts; for instance, $r(t_0)$ and $\dot{r}(t_0)$ can be determined from equation (1) together with initial conditions $r(t_1)$ and $\dot{r}(t_1)$, where t_1 is a time different from t_0.

The subsumption of a law of a specialized science under those of a more general science is usually accomplished by conjoining to the latter not a set of singular facts, but a somewhat general contingency, the truth and falsity of which are both consistent with the laws of the more general science. For example, in the elementary theory of molecular reactions there is a conservation law: the number of atoms of a given species entering a reaction must equal the number of atoms of that species emerging from the reaction. This law is a consequence of the laws of nuclear physics together with the condition that the nuclei of all atoms involved are stable and the energy exchanges are too low to induce nuclear reactions. This is a condition which holds in one set of circumstances and fails to hold in another, while the laws of nuclear physics are valid in both.

So far I have not commented on the word "principle," which occurs in the main thesis of this paper. The word does not seem to have a canon-ical meaning in physics. One speaks of "the principle of least action," "the principle of conservation of energy," and "the principle of Galilean relativity," but one does not commonly find the word "principle" substi-tuted for "law" or "equation" in locutions like "Newton's second law," "Maxwell's equations," and "the Schrödinger equation." I have the im-pression that "principle" is reserved for laws of great generality, which can be readily paraphrased in prose without a proliferation of mathemat-ical symbols. In this paper I shall follow a simple usage that goes back to Aristotle (or his translators into Latin): that a principle of a theory is a general fact not derived from any other propositions within the theory, and hence fundamental. This proposed usage would permit one to call the second law of thermodynamics a "principle" within thermodynamics, but not within statistical mechanics. Likewise evolutionary biology has principles of variation, even though specialists in mutagenesis can derive them (though not with full rigor) from the laws of molecular physics.

Complications in the classification of natural facts result from the role of chance in natural phenomena. There are laws (some having the status of principles) of statistical mechanics, quantum mechanics, and other statistical theories, which are formulated in terms of probability or which govern probability distributions. That a physical system in equilibrium with a thermal bath is properly described by the canonical distribution is a general law of statistical mechanics, applying to any system which satisfies the specified conditions, but in any application the canonical distribution only determines the probabilities of the various possible states of the system. If a theory postulates a probabilistic law concerning a certain type of process, then a singular proposition about the outcome of such a process is contingent in a different sense from that used earlier: the truth value of the proposition is not only undetermined by the law, but also undetermined by the conjunction of the law with a complete set of initial and boundary conditions. I shall abstain from discussing the deep question of determinism, i.e., whether nature is governed by a set of laws sufficient to determine the truth value of every singular proposition about an event E if the truth values of all singular propositions concerning events in the past of E are specified.

A final complication concerning the relation of singular facts and generalities arises from the magic of the law of large numbers in probability theory, which has been popularized by the locutions "order out of chaos" (Prigogine and Stengers, 1984) and "order out of disorder" (Careri, 1984). It is a general fact about salt crystals that their space lattice is face-centered cubic. It is also true that the formation of a crystal is a chance process involving enormously many elementary interactions among sodium and chlorine atoms and the atoms of the environment. In broadest outline, the occurrence of the standard NaCl crystal as an equilibrium configuration of its constituents is a consequence of the laws governing the interactions of pairs of atoms together with the validity of the canonical distribution. Of course, the canonical distribution assigns non-zero probabilities to many-atom states which differ radically from the normal lattice structure of NaCl crystals, but the total probability of all these abnormal states is utterly negligible. Lattice imperfections, which do commonly occur, are the visible manifestation of the probabilistic genesis of the crystal order, but these imperfections are consistent with the normal overall lattice structure. With these reservations it is correct to claim that a general fact concerning salt crystals emerges from the statistics of an enormous set of contingent facts, and indeed the entire discipline of condensed matter physics is largely devoted to such derivations of generalities from contingencies.

I agree with the theorists of prebiotic evolution that the most spectacular instance which nature offers of the emergence of order from disorder is the evolution of life.[1] This assertion comes close to a commitment to the reducibility of biology to physics, but I would not assert reducibility

1. Agreement with Prigogine on this general thesis (which is common to theorists of prebiotic evolution) does not entail acceptance of his claim that life is an instance of "dissipative structures." For a critique of Prigogine on this point see P. W. Anderson (1984, pp. 262–271).

without careful qualifications (some of them concerning technical diffi-culties posed by quantum mechanics, which is the central framework the-ory of current physics, and some of them concerning the potential and actual roles of mentality in the biosphere). Fortunately, the argument of this paper is independent of the reducibility or non-reducibility of biology to physics, and any remarks I make on this problem will be intended to broaden the perspective rather than to provide essential steps in the demonstration of my thesis. If evolutionary biology is construed to be restricted to the study of evolution of new forms of life from earlier forms (thus excluding the consideration of prebiotic evolution) then the funda-mental generalities concerning the biosphere, notably the general laws of heredity and variation, qualify as principles of evolutionary biology, even if from a broader standpoint they are instances of the emergence of order out of disorder. My thesis is that evolution *within the biosphere* is not governed by a principle of natural selection which is coordinate with the principles of heredity and variation.

3. ANALYSIS OF A DARWINIAN FORMULATION

It would be tedious and inefficient to examine individually a large num-ber of proposed formulations of a principle of natural selection. And even an exhaustive assessment of the formulations in the literature would not *ipso facto* suffice to demonstrate my thesis, for it could always be maintained that there is a principle of natural selection yet to be found which is subtler than the formulations given heretofore. My strategy will be first to exhibit how a carefully phrased formulation by Darwin fails to qualify as a principle; and then I shall try to show that the failure of Darwin's formulation has a generic character that is not likely to be remedied by variations and refinements.

In the Summary of Chapter 4 of *The Origin of Species,* entitled "Nat-ural Selection; or the Survival of the Fittest," Darwin writes:

if variations useful to any organic being ever do occur, assuredly individuals thus characterized will have the best chance of being preserved in the struggle for life; and from the strong principle of inheritance, these will tend to produce offspring similarly characterized. This principle of preservation, or the survival of the fit-test, I have called Natural Selection. (Darwin, 1943, p. 98)

This passage refers both to heredity and variation, as it should, since without these two features of life there would be no domain in which nat-ural selection could be operative; but it clearly does not conflate the prin-ciple of natural selection with the principles of heredity and variation. Darwin's formulation has the further virtue of carefully refraining from making a positive statement about the outcome of a useful variation, but instead twice uses probabilistic locutions, i.e., "will have the best chance," and "will tend to produce." This caution anticipates the explicit or implicit employment of probability that is common in recent formulations, e.g., Brandon (1978) and Mills and Beatty (1979). It completely absolves Darwin from the well known objection that the principle of natural selection is tautologous on the ground that the criterion for greatest fitness is leaving the most descendants.

The crux of my argument is that there is a natural subdivision of Darwin's formulation into two exhaustive propositions, each of which clearly fails to have the status of a principle of natural selection. (i) The first proposition is that a probability of being preserved is ascribable to each variety of an organism, presumably in some biological situation. (Darwin uses the word "chance" to mean "probability," as was common in Victorian writing, but I have substituted the word "probability" in order to conform to current technical usage.) (ii) The second proposition is that the probabilities of being preserved are somehow efficacious in shaping the long-range fate of the varieties (Darwin's locution being "will tend to produce offspring similarly characterized").

Proposition (i) is suggestive but obscure, and formidable difficulties stand in the way of giving it a satisfactory explication. The ascription of probabilities to varieties is evidently relative to the biological situation: the features of the generic organism from which the varieties ramify, the geological features of the immediate environment of the varieties, the meteorological features over fairly long intervals of time, the biota of the immediate environment, and influences that may impinge from remote parts of the world. One of the greatest difficulties in explicating proposition (i) is to give a criterion for the proper level of detail in specifying the biological situation. One option, which at least has the virtue of not being arbitrary, is to say that a maximal specification is required – i.e., the microstate of the entire set of entities which may impinge upon the organism of interest. If nature is deterministic, then this specification of the situation would in principle imply definite truth values of all the propositions concerning the future, and the concept of probability would apply only in degenerate form (probability unity for truth, probability zero for falsity, and no intermediate degrees of probability). If nature is indeterministic, then a maximal specification would permit intermediate degrees of probability for future propositions, but it is hard to see how anything general and interesting could be said about these probabilities, since a maximally specified situation by definition would be unique. A second option is to use a "coarse-grained" specification of a biological situation, of the sort that is offered in typical descriptions by population biologists. There is no *a priori* guarantee, however, that the probabilities of preservation of the competing varieties are the same in different coarse-grained specifications of the same situation, and indeed the contrary is very likely to be the case. But if there is a multiplicity of probabilities, which if any is efficacious in shaping the eventual statistics of the competing varieties? One can only hope that nature is cooperative in overcoming this difficulty, by contriving an approximate invariance of the probabilities when the level of detail in specifying the situation is changed within broad limits.[2] Gambling devices under normal conditions of operation display this type of invariance. For example, the mass distribution (and perhaps even a set of the first few moments of this distribution) is a sufficient specification of a normally tossed die, so that additional details about the surface of the gambling table or the mode of tossing or the currents of the air do not change the probabilities of the possible outcomes of a toss. Much progress has been made in proving mathematically that a die exhibits the type of invariance under consideration, e.g., Hopf (1934) and von Plato (1982); but even without a mathematical demonstration evidence for it is provided by the fact that frequencies of outcomes are stable when the conditions of tossing are varied. Biological situations are so complex that a mathematical demonstration of the invariance of probabilities under change of specification cannot be expected, but one might reasonably hope that population statistics in an ensemble of similar situations will provide empirical evidence for approximate invariance.

For the purpose of establishing my thesis solutions to the great difficulties mentioned in the foregoing paragraph are not needed. All the solutions which I have listed and can envisage support the thesis. If the option is taken of defining the probabilities of the biological outcomes of interest only relative to a maximal specification of the situation, then one has adopted the microscopic point of view of Laplace's demon (or a modern indeterministic version thereof). But the theory of natural selection is a macroscopic theory, which is irrelevant to the point of view of a Laplacean demon. If the second option is taken and a coarse-grained specification of the biological situation is accepted, then there is a fork. (a) There is a multiplicity of probabilities, relative to different coarse-grained specifications, and in this case the probabilities which are efficacious for the future history of the varieties are indefinite. If so, then proposition (i) which was disengaged from Darwin's formulation simply fails to be true: *unique* probabilities of being preserved cannot be ascribed to the varieties in the population of interest. (b) There is approximate invariance of the probabilities of being preserved when the level of

2. I believe that the propensity interpretation of probability is legitimate in certain contexts (such as that of classical mechanics) only if this invariance is exhibited. See Poincaré (1913, p. 163–173).

detail of the specification is changed within broad limits. In this case, the probabilities in question are sufficiently well defined that one can pass on to proposition (ii) of Darwin's formulation. But this approximate invariance of the probabilities is a *contingency*. It holds in certain situations and fails to hold in others, and the singular facts about the situation determine whether it holds or not. There may in fact be fairly broad and well defined classes of situations in which the invariance holds approximately (as in the example of a die), but whether a biological situation belongs to one of these classes is again a contingency. (c) Something intermediate between (a) and (b) may hold, for example that there are definite *comparative* probabilities as the level of detail of specification is changed, even though there are no invariant quantitative probabilities. In my opinion, the propensity interpretation of probability has underplayed the interest and methodological importance of comparative probabilities, but this will be discussed elsewhere. If the ambition is curtailed to treat evolution as a stochastic process in the sense understood by mathematicians, and one is content to understand the lineaments of evolution qualitatively, then comparative probabilities may suffice to provide great illumination. But even in a weakened version of the invariance of probabilities under the change of the level of detail this invariance would still be a contingency, holding or failing to hold as a result of the singular facts of an individual situation.

To summarize, there is no principle of natural selection implicit in proposition (i) of Darwin's formulation. Whether the probabilities of being preserved can be ascribed to the varieties of an organism, and with what definiteness and what qualifications, are local and contingent matters. When the probabilities are well defined, no general principle is needed to ensure that this is so; and when they are not well defined the situation is so chaotic that nothing other than chance itself, i.e., the absence of principle, is at work. Is it conceivable that over and above the local contingencies there is some general principle that shapes or influences or somehow affects the determination of the relevant probabilities? Yes, it is conceivable, provided that one accepts a metaphysics in which the future is efficacious in a way that is not already implicit in the present. Efficacity of the future is asserted, for example, by the advocates of orthogenesis and the Omega principle (see Mayr, 1982, pp. 528–531). Darwin himself and the authors of the neo-Darwinian synthesis consistently rejected a metaphysics of the efficacy of the future. Thus committed, they are obliged to recognize that probabilities of preservation are determined – to the extent that they are determined at all – by present circumstances, and hence by contingencies. Therefore, if there is a principle of natural selection, its locus cannot be in propostion (i) of Darwin's formulation.

We now turn to proposition (ii), that the probabilities of being pre-served are somehow efficacious in shaping the long-range fate of the varieties. Could proposition (ii) be the locus of a principle of natural selection? This question can be subsumed under the more general question: what is the efficacy in the long run of any ascription of probabilities? The answer to this question depends upon the explication of the concept on probability, which requires investigations tangential to the subject of this paper. My treatment will therefore be very condensed, though I think sufficient for the purpose at hand.

If the frequency interpretation is accepted, then the answer is trivial. This is an ontic interpretation, in that it identifies probability with a fact in the world, namely a long-range relative frequency of a specified property in an infinite ordered sequence of individual cases, e.g., Reichenbach (1938). The linkage of probability to long-range behavior is established by definition. The weakness of the frequency interpretation, both in analyses of stochastic processes in nature and in evaluations of hypotheses upon evidence, is the problematic relation between the frequency in a finite segment of the infinite sequence and the limiting frequency in the sequence as a whole. It is not surprising that two of the most eminent advocates of the frequency interpretation, C. S. Peirce (1932) and Karl Popper (1957), abandoned the frequency interpretation in favor of a different ontic interpretation, that of propensity.

The propensity interpretation ascribes an ontological status to the tendencies or propensities of the various possible outcomes of a singular chance event, such as the toss of a coin or the decay of a nucleus. A long sequence of chance events is itself a singular event, even though it is composite; and the propensities of the possible compositions of the sequence are determined by the propensities of the component events, with due attention, of course, to conditionalization. When the component events are independent, or if the dependencies are properly restricted, there are laws of large numbers which assign probability close to unity to a narrowly circumscribed set of typical compositions, and probability close to zero to the union of all the atypical compositions. The laws of large numbers supply the linkage between the elementary probabilities and the long-range behavior of a sequence of events. The linkage is apodictic, following simply from the axioms of probability theory, which are obeyed by the propensity interpretation as well as by other interpretations of probability. But the consequent in the apodictic linkage is itself probabilistic: it is an assertion of probability concerning the composition of the sequence. The ontological status of a propensity is not lost just because it has a numerical value which is close to unity or close to zero. There are serious epistemological questions concerning the rationality of decisions that are to be made on the basis of evaluations of propensities, but I do not wish to be drawn into these. I only want to make a negative assertion: that there is nothing outside of probability theory that induces the frequencies of sequences of events to "ripen" properly in accordance with the probabilities of the component events.

Finally, there are epistemic interpretations of probability, notably logical probabilism and personalism. I follow Carnap (1945) in broad outline in recognizing that both an ontic and an epistemic interpretation of probability are legitimate, but differ from him in preferring propensity to frequency within the class of ontic interpretations and preferring a kind of personalism (called "tempered personalism", Shimony, 1970, Sect. 3) to logical probabilism among the epistemic interpretations. For present purposes the only point that has to be made is that there is agreement among all epistemic interpretations: the laws of large numbers lie strictly within the compass of probability theory. There is no non-probabilistic principle, supplementary to the axioms of probability and the assignment of the relevant conditional probabilities, that shapes the statistics of long sequences of events.

To summarize: there is no principle of natural selection implicit in proposition (ii) of Darwin's formulation. Whenever proposition (i) holds and definite probabilities are ascribed to the events of biological interest, proposition (ii) says correctly that these probabilities are linked to long-range frequencies. But the same kind of linkage occurs in a gambling casino. No principle extrinsic to probability theory is needed to infer from the slight edge in favor of the house in each play of a roulette wheel that there is a very high probability of a net profit to the house in a year of play, and none is needed in the context of biology. It is easy to lose sight of this fundamental similarity because the elementary events are much more complex in evolutionary biology than in gambling games, because the ascription of definite probabilities in biological situations is often dubious, and because probabilistic independence of events is the exception rather than the norm in evolutionary biology, whereas the contrary is the case in honestly run casinos. None of these complications annul the fact that in evolutionary biology the linkage between the probabilities of being preserved (when these are well defined) and the long-range survival rates of the competing varieties is provided by probability theory alone, without the guidance of any extrinsic principle.

My argument is completed by reiterating that propositions (i) and (ii) exhaust Darwin's formulation, and hence if there is no genuine principle of natural selection implicit in either proposition, then there is none in the formulation as a whole.

4. MORE GENERAL ARGUMENT FOR THE THESIS

I wish now to show why the foregoing analysis of Darwin's formulation is generic. For this purpose it is valuable to use Sober's observation that evolutionary theory must consider both the *sources of fitness* and the *consequences of fitness*. Like most modern evolutionary theorists, he defines "fitness" in terms of the probabilities of surviving and reproducing. In particular, in the case of sexual reproduction he expresses the reproductive fitness w as

$$w = \sum_{i=0}^{\infty} \frac{ip_i}{2},$$

where p_i is the probability that the organism will produce i offspring (Sober, 1984, p. 45). He then says that the aim of source laws concerning fitness is "to show how properties of organism and environment can render certain traits selectively advantageous" (Sober, 1984, pp. 58–59). The aim of consequence laws is "to show what can be inferred from facts about fitness" (Sober, 1984, p. 47). Sober's distinction is obviously exhibited in Darwin's formulation, for proposition (i) in his formulation is an assertion about the sources of fitness, when fitness is explicated probabilistically, and proposition (ii) is an assertion about the consequences of fitness. In every other formulation that I have encountered some proposition can be extracted about sources of fitness and some about consequences. It is a virtue of Sober's treatment that the distinction is articulated explicitly, and once it is articulated it seems to be a matter of conceptual analysis, unpacking the very concept of natural selection. The result of this conceptual analysis is to permit a generalization of the foregoing argument that Darwin's formulation does not constitute a genuine principle of natural selection.

First consider sources, that is, features of organism and environment whereby the fitness of each variety is determined. It cannot be taken for granted that fitness is definite, because the probabilities entering into the formula for fitness may not be well defined (as discussed in Sect. 3). But if they are well defined, they depend upon a multitude of factors: the life strategy of the organism, the efficiency of integration of variant features with the common features of the organism, the opportunities and threats of the environment, etc. The sources of fitness are singular facts, though of course the general laws of physics, chemistry, and biology constitute a fixed framework within which the singular facts are efficacious. For example, whether one pattern of hairs on a fly's legs endows it with a higher probability of survival than another pattern depends upon the physical laws of viscous fluids, since the function of the hairs seems to be to expedite liberation from such fluids (Muller, 1950); and *mutatis mutandis* regarding the laws of aerodynamics, optics, chemical kinetics, etc. Although I have emphasized the role of singular facts concerning the sources of fitness, there may be generalities which deserve the name of "source laws." A common and not very profound kind of source law is exhibited by a class of situations with singular facts of a common character. Among animals which rely upon speed for predation or for escape, the fastest variety – *ceteris paribus* – has a higher probability of survival. Although this statement is general, and hence can be classed as a law, its generality is limited, and therefore it does not provide a counter-example to my claim that biological probabilities are determined by contingencies. The only example that I can suggest without limitation in scope within the biosphere (although it is probabilistic in character, and hence limited in another respect) is that a variety of which nothing is specified except that it deviates from the modal type of a population has a smaller probability of survival than the modal type. The plausibility of this source law stems from the basic biological principle that a functioning organism is a highly integrated cybernetic system of immense complexity, together with the rough probabilistic judgment that a random modification of a complex integrated system is more likely to be disruptive than advantageous. In sum, all the kinds of source laws that have been surveyed or that I have seen suggested have a derivative status, and none serves as a general principle of natural selection.

My conclusion concerning the sources of fitness is this: after one has taken into account all the singular facts of a biological situation, and all the relevant scientific laws that have been borrowed from disciplines other than the theory of natural selection, and all the derivative generalities that can reasonably be classified as source laws, there remains no shaping or governing or constraining of the values of fitness that could be ascribed to a general principle of natural selection. It is hard to see what general principle could shape or govern or constrain the ascription of probabilities to varieties of an organism unless a teleological element is postulated in the evolutionary process, and that would transgress the spirit of the theory of natural selection.

The study of the consequences of the fitness of competing varieties, which includes but is not exhausted by the study of consequence laws, is very intricate. Fitness is explicated in terms of probabilities of survival and reproduction, and as discussed previously these probabilities are very difficult to compute and may not even be well defined. Furthermore, there are complex correlations of biological events, since the biosphere behaves collectively. For these reasons, evolutionary biologists commonly resort to models, which are based upon plausible assumptions about the processes of interest and are checked by comparisons with population statistics. A detailed study of the methodology of practicing evolutionary biologists is beyond the scope of this paper. The point that I wish to make here is only that when they study the consequences (as contrasted with the sources) of fitness, they are applying probability theory, though often in an approximate and conjectural way.

An exemplary consequence law is Fisher's (1958, p. 37) "fundamental theorem of natural selection": "the rate of increase of fitness of any organism at any time is equal to its genetic variance at that time." The word "organism" here actually refers to a heterogeneous population, to each variety of which is ascribed a definite fitness independent of the frequency of the variety in the population. Fisher (1958, p. 38) notes that "the theorem is exact only for idealized populations, in which fortuitous fluctuations in genetic composition have been excluded." The proof given this idealization is very simple, and roughly consists of noting that if there is heterogeneity, the fitter varieties will increase on the average faster than the less fit varieties, giving a net increase of fitness. Fisher (1958, p. 40) also investigates the consequences of removing the idealization, and he shows that deviation of the rate of increase W from the value asserted in the theorem will be small unless $1/W$ is large compared to the number of individuals breeding per generation. Fisher's theorem is a straightforward exercise in probability theory, and he makes the following important statement about its epistemological status:

The statement of the principle of Natural Selection in the form of a theorem determining the rate of progress of a species in fitness to survive (this term being used for a well-defined statistical attribute of the population), together with the relation between this rate of progress and its standard error, puts us in a position to judge of the validity of the objection which has been made, that the principle of Natural Selection depends on a succession of favourable chances. The objection is more in the nature of an innuendo than of a criticism, for it depends for its force upon the ambiguity of the word chance, in its popular uses. The income derived from a Casino by its proprietor may, in one sense, be said to depend upon a succession of favourable chances, although the phrase contains a suggestion of improbability more appropriate to the hopes of the patrons of his establishment. It is easy without any very profound logical analysis to perceive the difference between a succession of favorable deviations from the laws of chance, and on the other hand, the continuous and cumulative action of these laws. It is on the latter that the principle of Natural Selection relies. (Fisher, 1958, p. 40)

Fisher's use of the locution "principle of Natural Selection" in this passage does not constitute a disagreement with my thesis, because there is a difference in the intended meaning of "principle." He is obviously using the word to mean "a general law or rule," whereas my meaning (as stated in Sect. 2) is "a general fact not derived from any other propositions within a theory, and hence fundamental." If the latter sense is adopted, then Fisher can be read as claiming (with pride) that his fundamental theorem is not a principle precisely because it is a theorem.

There is, however, a limitation in the scope of Fisher's theorem. His proof assumes that the fitness values of the varieties are frequency independent, and counter-examples to his conclusion have been exhibited in cases when this assumption fails to hold (Sober, 1984, pp. 178–186). But this limitation does not undercut my thesis. To determine whether and how the fitness values depend upon frequencies is part of the investigation of the sources of fitness, and it has already been argued that sources are not the locus of a principle of natural selection. And once fitnesses are determined the investigation of their consequences is strictly a matter of probability theory, no matter how complex the functional dependencies of fitness may be.

There are other striking general results in the theory of natural selection besides Fisher's "fundamental theorem," for example, "in an uncertain environment species will evolve broad niches and tend toward polymorphism" (Levins, 1966). Typically the reasoning for these results is informal, and the analyses of the sources and of the consequences of fitness are intermingled. I conjecture that informality and intermingling are largely responsible for the widespread conviction that the theory of natural selection depends upon a principle of natural selection, which is subtle and difficult to articulate. I further believe that each of the important general results of the theory of natural selection can, if sufficient care is taken, be subdivided into considerations of sources and of consequences of fitness, and when this is done the feeling that a principle of natural selection has been tacitly employed in the reasoning will simply evaporate.

5. COMPARISONS WITH OTHER THESES

My thesis has been confused by some readers and auditors with theses of other writers on natural selection, and there is a danger that other mistaken conflations will be made. In order to avoid such errors of interpretation it will be useful to make some explicit comparisons between my thesis and some others which resemble it in certain respects.

(a) According to Waddington (1957, p. 64), "The general principle of natural selection . . . merely amounts to the statement that the individuals which leave most offspring are those which leave most offspring." Since I follow Brandon (1978), Mills and Beatty (1979) and others (including Darwin himself, as seen in the quotation from him in Sect. 2) in defining fitness in terms of probabilities of survival and reproduction, rather than in terms of actual numbers of offspring, I do not accept Waddington's analysis.

(b) Popper has taken several positions concerning a principle of natural selection, one of which is the following:

I regard Darwinism as metaphysical, and as a research program. It is metaphysical because it is not testable . . . . And yet the theory is invaluable . . . . Although it is metaphysical, it sheds much light upon very concrete and very physical researches. It allows us to study adaptation to a new environment in a rational way: it suggests the existence of a mechanism of adaptation, and it allows us even to study in detail the mechanism at work. (Popper, 1976, pp. 178–186)

Since one of the distinctions which Popper made between himself and the Vienna Circle was a refusal to regard metaphysical statements as nonsensical, it is reasonable to interpret this passage as saying that there is a principle of natural selection with content, but it is not testable (presumably because scenarios for adaptedness can be invented for any feature of a viable organism). My position, by contrast, is that there is no principle of natural selection with content (and incidentally I believe that at least some metaphysical statements can be confirmed or disconfirmed by the use of the hypothetico-deductive method). I do agree with Popper's metatheoretical assertion that the theory of natural selection as a whole has great heuristic value in directing the search for mechanisms of adaptation.

(c) The "neutralist" hypothesis of Kimura (1968) and King and Jukes (1969) asserts that at least at the molecular level many or most of the differences among varieties are selectively neutral, i.e., the probabilities of leaving offspring differ little among the varieties. No commitment is thereby made concerning the factors governing the determination of these probabilities; whereas I make the commitment that the only factors are contingencies and laws borrowed from theories other than that of natural selection, with no principle of natural selection playing a role. Furthermore, I say that it is a contingency whether definite probabilities exist at all in a biological situation, and of course there is a great difference between having equal probabilities and having ill-defined probabilities.

(d) A number of evolutionary biologists have warned against the expectation that there are general laws governing evolutionary phenomena. For example, Stebbins writes:

Evolutionary biology is so complex that attempts in the near future to build syntheses around the framework of rigid, all-inclusive generalizations or laws will continue to be self-defeating and will lead to disputes and confrontations that generate more heat than light. (Stebbins, 1982, p. 14)

He prefers to speak of a "modal theme" rather than a law, where the theme is "designed to fit precisely the most common situations, but less common relationships can be expressed as variations of the central theme" (Stebbins 1982, p. 2). These statements have something in common with my assertions above that the fitnesses of organic varieties usually depend upon singular facts, if they are defined at all, and my denial that any general principle of natural selection shapes or governs or constrains the probabilities entering into the formulae for fitness. But my polemic was not against generalizations *per se,* but only against the proposal of a generalization which would serve as a principle in the theory of natural selection. If generalizations (or statistical statements which come close to them) can be obtained as *results* from contingencies and from laws borrowed from physics, chemistry, biochemistry, etc., these can be celebrated as instances of the emergence of order out of disorder. If, for a moment, we relax our concentration upon evolution within the biosphere and acknowledge prebiotic evolution as part of the domain of natural selection, then indeed we seem to have exceptionless generalizations which are the results of evolution, notably the universality of the genetic code.

I conclude with some remarks about ideas of Fisher and Peirce, which seem to be very close to my thesis.

Not only Fisher's "fundamental theorem of Natural Selection," but all the other *consequences* of fitness presented in his book *The Genetic Theory of Natural Selection* are essentially applications of probability theory. I know no passage in the book which comes close to saying that the *sources* of fitness are not governed by any law which can be properly regarded as non-derivative and hence as a principle of natural selection. But his dedication to the particulate theory of heredity, which he essays to conjoin to the theory of natural selection (Preface and Chapter 1), makes it reasonable to attribute to him the opinion that the sources of fitness are to be found in the laws of heredity and variation, together presumably with singular facts. I would therefore not be surprised by the occurrence among his innumerable papers of passages which are very close to my thesis.

The passage from Peirce which was referred to but not quoted in Sect. 2 is the following:

Mr. Darwin proposed to apply the statistical method to biology. The same thing has been done in a widely different branch of science, the theory of gas. Though unable to say what the movements of any particular molecule of gas would be on a certain hypothesis regarding the constitution of the class of bodies, Clausius and Maxwell were yet able, by the application of the doctrine of probabilities, to predict that in the long run such and such a proportion of the molecules would, under given circumstances, acquire such and such velocities, that there would take place, every second, such and such a number of collisions, etc.; and from these propositions they were able to deduce certain properties of gases, especially

n regard to their heat relations. In like manner, Darwin, while unable to say what
he operations of variation and natural selection in every individual case will be,
demonstrates that in the long run they will adapt animals to their circumstances.
Peirce, 1877)

Peirce seems to subsume the theory of natural selection under the theory
of probability. The passage is too brief to conclude with confidence from
t alone that he rejected, as I do, the idea of a principle of natural selec-
ion extrinsic both to probability theory and to the scientific laws govern-
ng the physical and biological theatre in which selection occurs. However, in a famous cosmological paper Peirce writes eloquently of chance
and claims that its operations are prior to law:

Without going into other important questions of philosophical architectonic, we
can readily foresee what sort of a metaphysics would appropriately be constructed
from these conceptions . . . . It would suppose that in the beginning – infinitely
emote – there was a chaos of unpersonalized feeling, which being without connection or regularity would properly be without existence. This feeling, sporting
here and there in pure arbitrariness, would have started the germ of a generalizing
endency. Its other sportings would be evanescent, but this would have a growing
virtue. Thus, the tendency to habit would be started; and from this, with the
other principles of evolution, all the regularities of the universe would be evolved.
Peirce, 1932, parag. 33)

This passage shows reverence for chance and a recognition of its fertility.
It is a pioneering statement about the possibility of the stochastic emergence of order from disorder. To the extent that Peirce's thought about
he biological doctrine of natural selection remained consistent with his
cosmology, he could not have admitted a general law which is operative
at the fundamental level in the stochastic process of evolution, for he
egards generalities as derivative facts. In Peirce's immense and often
disordered body of writing there may very well be passages which express
a contrary opinion (and indeed the elliptical phrase, "the other principles
of evolution," in the passage just quoted is unsettling). But I believe that
ny thesis of the non-existence of a principle of natural selection fits the
main current of his thought. It is honorable to be an epigone of Peirce.

REFERENCES

Anderson, P. W.: 1984, *Basic Notions of Condensed Matter Physics,* Benjamin/
Cummings, Menlo Park, CA.
Brandon, R.: 1978, 'Adaptation and Evolutionary Theory,' *Studies in the History
and Philosophy of Science* **9**, 181–206. Reprinted in Sober (1984a).
Careri, G.: 1984, *Order and Disorder in Matter,* Benjamin/Cummings, Menlo
Park, CA.

Carnap, R.: 'The Two Concepts of Probability,' *Philosophy and Phenomenologi-cal Research* **5**, 513–532.

Darwin, C.: 1943, *The Origin of Species,* Modern Library, New York.

Fisher, R. A.: 1958, *The Genetical Theory of Natural Selection,* Dover, New York.

Hopf, E.: 1934, 'On Causality, Statistics and Probability,' *Journal of Mathematics and Physics* **17**, 51–102.

Kimura, M.: 1968, 'Evolutionary Rate at the Molecular Level,' *Nature* **217**, 624–626.

King, J. L. and T. H. Jukes: 1969, "Non-Darwinian Evolution,' *Science* **164**, 788–798.

Levins, R.: 1966, 'The Strategy of Model Building in Evolutionary Biology, *American Scientist* **54**, 421–431. Reprinted in Sober (1984a).

Mayr, E.: 1982, *The Growth of Biological Thought,* Harvard University Press, Cambridge, MA.

Mills, S. and J. Beatty: 1979, 'The Propensity Interpretation of Fitness,' *Philosophy of Science* **46**, 263–286. Reprinted in Sober (1984a).

Monod, J.: 1971, *Chance and Necessity,* Vintage Books, New York.

Muller, H. J.: 1950, 'Evidence of the Precision of Genetic Adaptation,' *The Harvey Lectures,* 1974–8, Charles C. Thomas, Springfield, IL, 165–229.

Peirce, C. S.: 1877, 'The Fixation of Belief,' *Popular Science Monthly* **12**, 1–15. Reprinted in C. S. Peirce: 1934, *Collected Papers* V, C. Hartshorne and P. Weiss (eds.), Harvard University Press, Cambridge, MA.

Peirce, C. S.: 1932, *Collected Papers* II, C. Hartshorne and P. Weiss (eds.), Harvard University Press, Cambridge, MA.

Plato, J. von: 1982, 'Probability and Determinism,' *Philosophy of Science* **49**, 51–66.

Poincaré, H.: 1913, *The Foundations of Science,* The Science Press, Lancaster, PA.

Popper, K.: 1957, 'The Propensity Interpretation of the Calculus of Probability, and the Quantum Theory,' in *Observation and Interpretation,* S. Körner (ed.), Butterworths Scientific Publications, London.

Prigogine, I. and I. Stengers: 1984, *Order out of Chaos: Man's New Dialogue with Nature,* Bantam Books, New York.

Reichenbach, H.: 1938, *Experience and Prediction,* University of Chicago Press, Chicago.

Shimony, A.: 1970, 'Scientific Inference,' in *The Nature and Function of Scientific Theories,* R. G. Colodny (ed.), Pittsburgh University Press, Pittsburgh.

Shimony, A.: 1989, 'The Theory of Natural Selection as a Null Theory,' in *Statistics in Science,* D. Costantini and R. Cooke (eds.), Kluwer, Dordrecht, Holland.

Simpson, G. G.: 1963, *This View of Life: The World of an Evolutionist,* Harcourt, Brace & World, New York.

Sober, E.: 1984, *The Nature of Selection,* MIT Press, Cambridge, MA.

Sober, E. (ed.): 1984a, *Conceptual Issues in Evolutionary Biology,* MIT Press, Cambridge.

Stebbins, G. L.: 1982, 'Modal Themes: A New Framework in Evolutionary Synthesis,' in *Perspectives in Evolution,* R. Milkman (ed.), Sinaur, Sunderland, MD.

Waddington, C. H.: 1957, *The Strategy of the Genes,* Allen and Unwin, London.

Reply to Sober

The paper (Shimony, 1989) which Elliott Sober (1989) has commented upon grew out of my efforts to understand the structure of the theory of evolution and to explain it to students. The absence of anything approaching a broadly accepted axiomatization of the theory of evolution makes it difficult to articulate this structure. But one should not complain about this fact. It may very well be that the theory of evolution resists axiomatization or would be little served by it. At a minimum, a good *informal* articulation of the structure of the theory is a precondition for an adequate formalization.

Let me try, then – before attempting to answer Sober's criticisms – to lay down some *desiderata* for an informal articulation of any theory.

(a) The *domain* of phenomena to which the theory applies should be indicated. (There are different ways in which this step may be performed. A theory applying to a very broad domain may be presupposed, and the scope of the theory of interest may be indicated by specialization; or the domain may be described phenomenologically, in a language which is fairly characterized by practitioners as an "observation language"; or some combination of these procedures may be used.)

(b) The *fundamental concepts* of the theory should be presented explicitly. (These will be somehow connected to experience, but in this note I can reasonably ask to be excused from entering into the intricate problem of how this linkage is accomplished. In any case, what makes a concept fundamental is that it is not explicitly defined in terms of other concepts within the theory; definability in terms of concepts in a more basic or more general theory is not precluded, but that is another matter, which can be set aside in the formulation of the theory under consideration.)

(c) A set of propositions, expressed in terms of the fundamental concepts mentioned in (b) and possibly also of phenomenological concepts and those of other scientific theories, should be designated as *underived* within the theory, and these are its *principles*. (Steps (b) and (c) should not be rigidified. We have plentiful examples from geometry and physics of equivalent formulations of the same theory which differ in the choice of principles and fundamental concepts. In other words, even though the systematic formulation of a theory requires the specification of underived propositions, there need be no commitment to the *intrinsic* fundamental character of one set and the *intrinsic* derivative character of all the others.)

(d) The *framework* within which the theory is presented should be indicated. (The framework includes all the scientific theories which are taken for granted and drawn upon in the formulation of the theory under consideration. Thus, even when biology is formulated as an autonomous discipline, setting aside the question of whether it can in principle be reduced to chemistry and physics, biologists can still make free use of these two sciences, as well as of geology and to some extent astronomy. The framework also includes logic, mathematics, statistics, and scientific methodology, which are disciplines greatly differing in character from the natural sciences – whatever their character may be. Finally, in my opinion there is a metaphysical component in the framework, which includes scientifically relevant propositions concerning such matters as causation, chance, necessity, possibility, universality, and contingency. In Section 2 of my paper I emphasized the distinction between a general law and a contingency, which is commonplace in most of the natural sciences, but I may add that this distinction is one of the important instances of a metaphysical component in the natural sciences.)

(e) The *epistemological status* of the underived propositions of the theory should be made explicit. (Since I am an empiricist, I believe that the justification of all the underived propositions of a well-founded scientific theory is ultimately provided by experience; but how directly, with what degree of support, with what qualifications, etc. must be examined in each situation separately. It should be evident that making the specification of the epistemological status of principles a *desideratum* in the formulation of a theory is by no means tantamount to foundationalism; there is no commitment that the principles are more directly or more certainly justified than the derivative propositions of the theory.)

(f) The texture of discourse of the theory should be sufficiently tight that clear reasoning is possible from the principles to derivative propositions and applications. (Again rigidity is not implied. The reasoning in question need not be mathematical in character, and it need not be rigorously deductive. Nevertheless, the theory should make it possible to proceed in a controlled manner, so that one knows what to anticipate on the basis of the theory and one has some idea where to place the blame if expectations are frustrated by experience.)

After this preliminary listing of the *desiderata* for an informal exposition of a theory I shall turn to Sober's criticisms.

(1) Sober correctly says that my argument depends upon my meaning of "principle". I was explicit and wrote, "a principle of a theory is a general fact not derived from any other propositions within the theory." (I

occasionally used the locution "first principle" in exactly the same sense. It was probably confusing to do so, and I ask Sober and my readers henceforth to delete the adjective "first" in the few places where I used the two-word phrase.) Sober seems to be unhappy with my usage, which has the consequence that "every nontrivial mathematical result in the theory of natural selection fails to be a first principle for that very reason." He also writes, "These results require proof because they are not obvious. But this is not enough to show that they are not first principles. Everything depends on the kind of proof they receive". I conjecture that Sober wants to use the word "principle" to mean something like "important general proposition", and if so, then I fully agree with him that Fisher's Fundamental Theorem and a vast array of other propositions of evolutionary biology are principles in that sense. It may be, however, that Sober wishes to convey something else by his unhappiness about my definition: that the evolutionary character of biology makes the identification of principles either labile or unfruitful. The last sentence of the passage which I have just quoted can be construed this way, as well as the statement, "My own reaction is to distinguish the singular historical proposition that terrestrial life obeys Mendelian principles from the general laws that describe what a system will be like if it is Mendelian." If I am reading him correctly, I certainly am sympathetic. But I believe that I have already indicated my sympathy both in my paper and in *desiderata* (b), (c), and (d) above. Mendelian genetics can be formulated as an autonomous theory, and if so, then Mendelian laws of inheritance are underived propositions in the theory – hence principles in my sense. On the other hand, one can formulate a more general biological theory in which sexual reproduction governed by Mendelian laws is seen as a singular evolutionary development. Although in such a theory it would not be correct to say that the Mendelian laws are "derived," nevertheless arguments can be given that they are understandable evolutionary results, even of quite high probability given the circumstances. I believe that Sober and I are in agreement not only that both of these options exist, but that a clear understanding of biology requires the recognition of the possibility of freely moving between formulations with different starting points. What I am claiming, however, is that the non-existence of a principle of natural selection is an invariant, independent of the details of the formulation of evolutionary biology. I could not prove the truth of this claim definitively, but I became convinced of it partly because of the elusiveness of the sought-for principle, and partly because it is neither a source law nor a consequence law, and hence there is no locus for it.

(2) Sober characterizes my argument for the non-existence of a principle of natural selection as a destructive dilemma: "Any general law in the theory must be either a source law or a consequence law. No source

law can be a first principle, ... No consequence law can be a first principle . . . ." (his first paragraph). To undermine my argument he demonstrates at length that a source law can be a principle – in my sense of the word, I believe. (He also argues that a consequence law can be a principle, but this argument is very brief and does not seem to be central to his case.) His strategy does not succeed, however, because my argument does not rest upon the assertion that no source law can be a principle. It is conceivable to me that a law of considerable generality might be discovered concerning fitnesses, and this law might be taken as a postulate in some formulation of evolutionary biology. A good candidate might be a suitably refined version of the following: of several variants of a kind of organism, that variant which is most efficient in utilizing the energy of its nutrients will have the highest probability – *ceteris paribus* – of reaching maturity. By its content this proposition is a source law; and it is a principle in one formulation of evolutionary biology just because it is taken as underived in that formulation. But my argument is not damaged, because this source law is not a principle supplied *by* the theory of natural selection; instead, it is a principle supplied *to* the theory of natural selection by its framework, which surely includes other branches of biology. When I denied the existence of a principle of natural selection I did not deny the existence of a theory of natural selection. It is a rich theory, which proceeds by garnering information from its framework and from the examination of specific biological situations – hence information about source laws and about sources in their singularity – and using that information to draw conclusions about the probable fates (consequences) of organic varieties. Darwin's formulation of the principle of natural selection, which I analyzed in Section 3 of my paper, is essentially a scheme of the reasoning employed in the theory of evolution. I noted two parts of the scheme: (i) says that fitnesses are somehow fixed, without saying how this is done, and (ii) says that there are consequences of the fitnesses. I deny, however, that the scheme constitutes a principle, because it is neither a premiss nor a rule of inference for the reasoning which it schematizes. If the fitnesses are given – whether singularly or by source laws – then consequences are determined by using the apparatus of the framework mentioned in *desideratum* (e), above all by probability theory.

(3) Sober's footnote 2 asks whether I am not making the unreasonable demand "that a first principle should affect probabilities even after all singular facts are taken into account." I tried to answer a question of this kind in advance in Section 3 of my paper, where I discussed the level of detail appropriate in specifying a biological situation. I tried to make clear the unreasonableness of demanding the microstate of the situation – "all the singular facts", in Sober's locution. In other words, I agree with Sober's rejection of a "Laplacean description" (1984, 120ff).

(4) Quite apart from Sober's doubts about my usage of "principle", he is skeptical about my attribution of principles to the theories of heredity and variation and none to the theory of natural selection. He writes, "I do not see that the laws of heredity and variation are any more independent of contingency than the laws of natural selection. Both sets posit probabilities and confront the interpretational problems that Shimony mentions." However, in listing similarities between the theories of heredity and variation on the one hand, and the theory of natural selection on the other, Sober neglects their crucial differences. Whether the theories of heredity and variation are formulated as autonomous theories or embedded into a more comprehensive theory of biological evolution, they both impose constraints upon the temporal development of the biosphere. Without the principle that offspring mainly resemble parents, or some suitable refinement of this principle (by Mendel or Morgan or contemporary molecular biologists), there would be no way to make inferences about the phenotype of the offspring from a description of the phenotypes of the parents. The principles of the theories of heredity and variation are the underived statements of the constraints which these theories impose; and it is legitimate to use the word "principle", even though the constraints in question are restricted to terrestrial life and even though they are probabilistic. The theory of natural selection, by contrast, *imposes no constraints of its own.* It is rather the systematic study of temporal development of the biosphere after all constraints from elsewhere are acknowledged: the constraints imposed by the general laws of physics and chemistry, the constraints imposed by the general laws of other branches of biology, and the constraints imposed by the relevant singular facts of the biological situation. In order to provide a substantial refutation to my claim that there is no principle of natural selection a critic would have to show just what *unborrowed constraint* upon the temporal evolution of the biosphere that theory imposes. If this challenge to critics cannot be met, then one should accept the truly remarkable fact that there is a systematic theory of natural selection without a principle of natural selection. How can this be? Well, in a few phrases one can answer: the magic of probability, the fertility of chance, the emergence of order out of disorder, etc. But there is no substitute for a detailed study of the theory to turn these phrases from promissory notes into a wealth of knowledge.

(5) Sober makes a long negative comment on my neo-Darwinian characterization of the evolutionary process as utterly opportunistic, and he concludes with the remark, "the idea that adaptations are not programmed to emerge does not imply that the theory of natural selection has no first principles." I certainly do not quarrel with this denial of an implication. But in what way does it point to the *content* of the sought-for principle of natural selection, its *locus,* or its *utility* for systematizing our knowledge

about the evolutionary process? In the end, Sober refrains from taking a stand about the existence or proper formulation of the principle of natural selection, and thus he leaves me no target. I have some proclivity to be a Don Quixote in my philosophical enterprises, but I find it difficult to tilt against a merely possible windmill.

In summary I wish to say that my argument for the non-existence of a principle of natural selection is part of my attempt to understand as clearly as possible the structure of the theory of evolution as a whole. What is particularly difficult is to have orderly thought about a process which is intrinsically disorderly. One condition for orderly thought is to keep clearly in mind the distinction between underived and derived propositions. Sober tries to play down this distinction, in part perhaps because the practical investigations of evolutionary biologists do not depend upon it. But philosophical understanding is more demanding than practice. There is a famous remark of Darwin that "Herschel says my book is the 'law of higgledy-piggledy'" (Darwin and Seward, 1903; 1, 191). Presumably Herschel intended his remark as a criticism, but had he meant it both as a compliment and a characterization of the process of natural selection, he would have shown great insight. In spite of this strange character of the process of natural selection, however, a philosopher should not allow his analysis of the process to be higgledy-piggledy.

REFERENCES

Darwin, F. and A. C. Seward: 1903, *More Letters of Charles Darwin,* John Murray, London. (I thank Michael Ruse for this reference.)
Shimony, A.: 1989, 'The Non-existence of a Principle of Natural Selection', *Biology and Philosophy* **4**, 255–273.
Sober, E.: 1984, *The Nature of Selection,* MIT Press, Cambridge, MA.
Sober, E.: 1989, 'Is the Theory of Natural Selection Unprincipled? A Reply to Shimony', *Biology and Philosophy* **4**, 275–279.

PART D
Time

17
Toward a revision of the protophysics of time

I. INTRODUCTION

The first purpose of this paper is to raise questions about the epistemo-logical status of some of the principles of protophysics, particularly of the protophysics of time (Section 2). Reasons will be given for doubting that their status can be adequately clarified unless there is a retrenchment of the protophysical program. The second purpose is to suggest a revision of the protophysical program (Sections 3 and 4): instead of making the methodology of space and time measurements strictly prior to physics, this methodology should be developed in tandem with physics. A discussion of Newton's treatment of absolute time will give the proposed revision some historical and conceptual specificity. The third purpose is to indicate briefly the implications of the proposed revision for relativistic kinematics and for epistemology in general (Section 5).

2. THE EPISTEMOLOGICAL STATUS OF PROTOPHYSICAL PRINCIPLES

Some quotations from P. Janich's *Protophysics of Time* will be useful starting points for analysis.

[I]t was my intention to . . . formulate as a first part of the reconstruction of phys-ical terminology those established postulates (*Festsetzungen*) which in linguistic and non-linguistic respects are at the basis of the art of measurement as practiced by physicists. As the guiding principle of these reconstruction efforts I meant to apply the principle of methodical order according to which – on the basis of extrascientific speech and action, i.e., independently of the results of modern physics – all linguistic and non-linguistic steps required for an operative definition were to be enumerated in such a way that in doing so neither gaps nor circles would occur.[1]

[P]rotophysics asserts nothing about nature, but rather about production pro-cesses as suitable means for particular ends; it also makes no statements about

1. P. Janich (1985), p. xxi.

nature in the sense that it would assert that certain states of affairs could not be real. Protophysical norms exclude no empirical states at all.[2]

[P]rotophysics can claim for itself that (1) through the complete operative definition of its basic vocabulary, through freedom from contradiction and through uniqueness, the step-wise followability of the production instructions is secured; that (2) through recourse to the historical achievements of a non-scientific artisan's trade, the followability of the proposed procedural instructions is an historical fact; that (3) protophysical norms systems can be considered as adequate reconstructions of the scientific art of measuring; and that (4) up to now no single rival theory is known which, with consideration of pragmatic or action-theoretical aspects, enables non-circularly a historical and systematic understanding of the scientific art of measuring as it is in fact practiced.[3]

A number of protophysical themes are presented in these passages, compactly but authoritatively: the themes of methodical order, of the crucial role of the nonscientific artisan, of norms, of the reconstruction of the art of scientific measuring, and of the abstention from reliance upon the discoveries of empirical science. What has not been quoted, but what P. Lorenzen, P. Janich, and their colleagues claim to be implicit in the program just summarized, is that Euclidean geometry and classical kinematics are entrenched as protophysical principles in a way that is not shaken by the discoveries of relativistic electrodynamics, mechanics, and gravitational theory.[4]

Before inquiring about the epistemological status of the principles of protophysics, I must note an argument of Janich against H. Pfarr about a preliminary point. Janich states,

Pfarr makes the mistake of not distinguishing between the *linguisitic presuppositions of statements* and the *nonlinguisitic conditions of actions,* here of protophysical established (*normierten*) actions. . . . the 'first' propositions that are 'presupposed' as definitions are tied to nonlinguistic, technical production of devices. These production processes for their part are prescribed by instructions and do not have linguistic presuppositions with a truth value.[5]

I note, however, that protophysicists do somehow make a bridge between instructions and declarative sentences, and eventually they espouse a set of principles in declarative form. Janich speaks, for example, of "kinematic theorems"[6] and "propositions about velocity changes in a clock-free kinematics."[7] In discussing the epistemological status of the principles

2. *Ibid.*, p. 83.
3. *Ibid.*, pp. 83–84.
4. P. Lorenzen (1976), p. 69; Lorenzen (1987), p. 237.
5. P. Janich (1984), pp. 194–95.
6. P. Janich (1985), p. 112.
7. *Ibid.*, p. 124.

of protophysics, I am referring to a body of declarative sentences, whatever their relation to instructions in imperative form may be.

The exact content of the principles of protophysics is a subtle matter. What is crucial is the relation between the norms of spatial and temporal measurement and the realizability of these norms in scientific work. The protophysicists are evidently trying to walk a very fine line. On the one hand they insist upon the ideal character of protophysical norms: "'Geometry'. . . is the theory of (ideal) points, straight lines, planes and solids, as they become determined by ideal norms for the process of shaping things (with their sides, edges, and corners)";[8] or "The intersubjectivity of measurement results . . . is based upon the non-empirical undisturbedness of measuring instruments, i.e., upon an undisturbedness not depending on measuring results but rather one normatively secured by man."[9] On the other hand, the protophysicists do not abstain completely from the question of the realizability of their norms. Janich says, "In fact, protophysics could not make good its claim to be a rational reconstruction or foundation of physics if it did rest on unfollowable norms. Therefore we must clarify whether the protophysical norms are known to be unsatisfiable."[10]

What, then, is being asserted about the realizability of the norms of spatial and temporal measurement? Are they realizable *in principle,* and if so, exactly what does this mean? Furthermore, if the answer is positive (after a suitable clarification of the phrase "in principle"), what is the epistemological justification for the positive answer?

I shall now list three possible proposals that could be made to answer these questions, partly for the purpose of textual exegesis and partly in order to sharpen the epistemological issues. The first two proposals are quite clearly not intended by the protophysicists, but it is useful to list them in preparation for the third, which may be close to their intentions.

Proposal (i). The principles of protophysics are analytic, or true *per definitionem,* because of the norms that are laid down concerning length-measuring devices, time-measuring devices, and so forth. The choice of certain norms rather than others, and hence of certain definitions rather than others, is strongly suggested by the procedures of prescientific technology, but the epistemological status of the protophysical principles does not depend upon the motivation for adopting the norms, but only upon the commitments themselves. Although this proposal has the virtue of

8. P. Lorenzen (1976), p. 61: "'Geometrie' . . . ist die Theorie von (idealen) Punkten, Geraden, Ebenen und Körpern, wie die durch ideale Normen für Bearbeitungsverfahren von Dingen (mit ihren Seiten, Kanten und Ecken) bestimmt werden."
9. P. Janich (1985), p. 81.
10. *Ibid.*, p. 83.

clarity, it cannot be what the protophysicists have in mind. This proposal makes the choice of geometry, chronometry, and kinematics a matter of convention, limited only by the requirement of consistency, and it has the effect that there are many alternative geometries, chronometries, and kinematics, none having an essentially different epistemological status from the others. The primacy and the entrenchment which the protophysicists claim for their principles obviously cannot be justified by a conventionalist proposal.

Proposal (ii). The declarative sentences that constitute the principles of protophysics are inductions from life-world experience. The norms that are laid down for length-measuring and time-measuring devices are indeed satisfied in prescientific technology with the precision required for practical purposes, and the value of laying down and following the norms in that context is indubitable. In the life-world figures which are constructed by the same precept are geometrically isomorphic, provided that we make corrections for the disturbances and imperfections of performance. We can and do change scale in drafting and in shop work with great practical efficacy, thereby lending support to Lorenzen's constructive "Formprinzip,"[11] which is the basis of Euclidean geometry. And we can successfully move clocks without disturbing them, thereby to good approximation achieving agreement between clocks in relative motion. Clearly, however, this proposal is contrary to the intention of the protophysicists, in view of their claim that the justification of their norms is nonempirical. Furthermore, if questions of textual exegesis are set aside, the proposal is unsatisfactory for two reasons. First, the extent of the evidential base of the induction is obscure, just because of the proviso that corrections must be made for disturbances which occur in life-world technology. Second, any induction is uncertain, and the uncertainty is aggravated in the inductions in question by the fact that they are extrapolations. The norms derived from the life-world are to be applied arbitrarily far beyond it, to figures of astronomical dimensions and to clocks transported by arbitrarily high velocities.

Proposal (iii). The entrenchment of the protophysical norms is justified by a type of reasoning which some inductive logicians call a "vindicatory argument": that nothing can be lost, and possibly something can be gained in our search for experimentally based scientific knowledge by adherence to these norms.[12] Many passages in the writings of the

---

11. P. Lorenzen (1982), p. 102.
12. The prototype of this argument in inductive logic is Reichenbach's justification of the "straight rule" for estimating the limit of an ideally infinite sequence of ratios of occurrences of a certain type of event in finite segments of an infinite ordered population. Nothing is lost, he says, and something may be gained in assuming that the limit exists, for if it does not exist then no finite segment can reveal the general statistical

protophysicists suggest an adherence to some form of vindicatory argument. For example, the third passage cited in this section asserts that no rival theory "enables non-circularly a historical and systematic understanding of the art of measuring," thus suggesting that something could be lost, and nothing possibly could be gained, by abandoning the protophysical norms. The point at which Janich seems to me to come closest to an explicit avowal of a vindicatory argument is in his struggle with a problem that causes him great difficulty, defining operationally a clock which is undisturbed by motion at high velocities:

should the 'behavior' of clocks be defined operationally in a methodical way, or should it be left open for empirical inquiry? To the best of my present knowledge I cannot answer the question. Admittedly the interior measurement of velocity often is not possible theoretically, for instance in astronomy. The answer depends on the claims raised for logical properties of physical theories. At any rate, neither does physics have conclusive arguments in favor of its own, that is the empiristic solution. The well known answers in physical textbooks suffer from a lack of operational meaning of the relative velocity of two observers (and other operational definitions too).[13]

I take him to be saying that he cannot prove the norms of the protophysics of time to be satisfiable in principle, and yet that there is no substitute for them, since a physics without these norms will not have its own fundamental terms under control.

The genre of vindicatory arguments seems to me to be potentially a valuable resource for epistemology, but one which is very difficult to actualize definitively. It is not easy to prove with anything like rigor that nothing will be lost and something will possibly be gained in a certain enterprise by following a certain prescription. One needs to know whether alternatives to the presently formulated protophysical norms have been so exhaustively explored as to exhibit that none of them will permit the construction of physics in a way that sufficiently agrees with the principles of methodical order. The proposal to be sketched in Section 4 is intended as an existence proof of a satisfactory alternative, and in the course of the sketch that vague word "sufficiently" will be somewhat clarified. Quite apart from this alternative proposal there is a body of evidence that makes

composition of the population, and if it does exist then the ratio in a sufficiently large segment will come within any preassigned small distance of the limit (1938, chapter 5). Although this version of vindicatory argument does not seem to me successful, I do believe that there is a role for a suitably refined argument of this genre in inductive logic and in epistemology generally. There is also some conceptual connection between the family of vindicatory arguments and the "regulative" use of the ideas of reason (unlike the "constitutive" use of the categories of the understanding) according to Kant (Appendix to the Transcendental Dialectic).

13. P. Janich, "Time measurement and kinematics," p. 18.

me skeptical about the possibility of a definitive vindicatory argument for current protophysical principles: namely, the richness of the results of relativity theory, which were historically obtained by *not* following the norms of the protophysicists, even though the protophysicists work very hard to disengage these results from a methodology which they regard as fallacious and to incorporate them into their own framework.[14]

3. A DIFFICULTY IN JANICH'S PROTOPHYSICS OF TIME

The constructive work of Janich is too intricate to summarize here.[15] For the purposes of this paper, however, it will suffice to refer to several of his crucial definitions and to one crucial theorem. There seems to me and to other commentators to be a serious lacuna in the proof of this theorem, which motivates the revision of the protophysics of time to be offered in the following section.

According to Janich, "An instrument on which a point-body (pointer) moves uniformly is called a *clock*."[16] This definition is somewhat elliptical, since in fact to say that a point-body moves uniformly requires reference to at least one other point-body, such that it and the clock have operations "invariable to one another,"[17] and this phrase in turn is defined in terms of an appropriate prescientific conception of comparisons of motions. The theorem which Janich asserts is the following:

let two pairs of clocks U_1, U_2 and U_3, U_4 be given. Without restricting general validity it shall then be shown that the pair U_1, U_3 mutually agrees and is equal.[18]

As I understand this theorem, U_1 and U_2 are two instruments that satisfy relative to one another the relations of invariable operation and additional relations which suffice to establish that both are clocks; and likewise regarding U_3, U_4. No such relations are assumed in the premises of the theorem between U_1 and U_3, between U_2 and U_3, between U_1 and U_4, and between U_2 and U_4. Those assertions are the conclusions of the theorem. And yet, at a crucial point in the offered proof it is asserted "by supposition, all four clocks are invariable relative to each other,"[19] a statement which either is illegitimate or, if legitimate because it was part of the intended meaning of the premises, trivializes the theorem. In other words, Janich has not established to my satisfaction that time t according to one class of clocks which satisfy all the norms of his protophysics

14. H. Tetens (1982).
15. P. Janich (1985); J. Pfarr (1976, 1984).
16. P. Janich (1985), p. 152.
17. *Ibid.*, p. 146.
18. *Ibid.*, p. 153.
19. *Ibid.*, p. 154.

of time is a linear function of the time t' according to another class which also satisfies all these norms. This criticism is given with greater elaboration and an example by Pfarr,[20] who attributes the same criticism to A. Kamlah.

It therefore seems to me that – even if one abstains from the problem of the comparison of clocks moving relative to one another, where the protophysicists part company with relativistic kinematics – there is a profound difficulty in protophysical chronometry.

4. THE NEWTONIAN ALTERNATIVE

I propose that the solution to Janich's difficulty is to be found in a famous passage in the Scholium of Book I of Newton's Principia:

Absolute time . . . is distinguished from relative by the equation or correction of the apparent time. For the natural days are truly unequal, though they are commonly considered as equal and used for a measure of time; astronomers correct this inequality that they may measure the celestial motions by a more accurate time. It may be that there is no such thing as an equable motion whereby time may be accurately measured. All motions may be accelerated and retarded, but the flowing of absolute time is not liable to any change. The duration or perseverance of the existence of things remains the same, whether the motions are swift or slow, or none at all; and therefore this duration ought to be distinguished from what are only sensible measures thereof and from which we deduce it, by means of the astronomical equation.[21]

Janich is more sympathetic in his treatment of this passage than most of the philosophical commentators on Newton, as one might expect in view of his commitment to Newtonian kinematics. His sympathy, however, rests upon his own reinterpretation:

Now the Newtonian propositions concerning absolute space or absolute time are neither empirical nor can they be understood as hypotheses for explaining any phenomena whatever. . . . A sympathetic reading of the scholium . . . can presume methodological principles behind these statements; these principles – loosely put – determine the relation of real objects to ideal objects. 'Absolute space' or 'absolute time' are thus nothing more than methodological means – in modern terminology – to secure the invariance of physical laws with respect to displacements in perceptible space or measurable time. The formulation of a principle, which . . . would be a dynamical requirement indeed encounters manifold difficulties. At least one of these difficulties is unresolvable, namely formulating a principle which would give a definition of inertial movement without already having spoken of space, time and mass.[22]

20. J. Pfarr (1976, 1979).
21. I. Newton, Bk. I, Scholium to Definitions.
22. P. Janich (1985), p. 213.

However, much more can be said in defense of Newton's thesis than Janich admits. The passage quoted from the Scholium and the detailed working out of celestial mechanics in the *Principia* suggest an interpretation of "the flowing of absolute time" that goes beyond methodology: that *absolute time is the time which is implicit in dynamics, as formulated in the three laws of motion.* (The modern expositor who has presented definitively the case for this interpretation is H. Stein, in his "Newtonian Space-Time".[23]) Dynamics makes it possible systematically to determine how a time-keeping device which is a "sensible measure" of time deviates from absolute time, and thereby to make corrections to its readings. The process of correction is by no means algorithmic, and yet it is more specific than a methodological program that can only promise ideal clocks in principle. The procedures of making corrections open the possibility that corrected clocks of radically different types will asymptotically approach agreement, thus providing a solution to the problem of the multiplicity of non-agreeing clocks, which Janich's chronometry was unable to handle. The solution is obtained by extending the resources of chronometry in a manner which the protophysicists regard as illegitimate – namely, by using the laws of physics.

The realization of a Newtonian alternative to the protophysics of time can be divided into two parts. (A) Upon supposition that the laws of motion are known, it is necessary to show how they can be used for the purpose of obtaining asymptotically accurate sensible measures of absolute time. (B) It is necessary to indicate how the connection was made between the life-world and the laws of motion. The advisability of this order of the two parts will be clear from the following discussion. In this discussion the Newtonian alternative will not be worked out in detail; the intention is only to indicate that it is a feasible program.

Three types of measures of time will be briefly considered from a Newtonian point of view: (i) the rotation of the earth, (ii) the positions of planets and natural satellites, and (iii) mechanical clocks. Atomic clocks are omitted from the list, since their principles of operation are not Newtonian even in first approximation.

(i) *The rotation of the earth.* If the earth were rigid, perfectly symmetrical about its axis of rotation, and subject to no external torque, and if the "fixed" stars were truly fixed, then it follows from Newtonian dynamics that its period of rotation, in absolute time, would be constant, and hence a period of rotation would be a natural unit of a perfect clock. Furthermore, it is a clock that could be read anywhere on earth, thus providing

---

23. H. Stein (1967), part I.

local clocks. Roughly, the procedure of reading is to observe the transit of specified stars past the meridian, which is the great circle of the celestial sphere passing through the pole of the celestial sphere and the zenith. Corrections to this idealized picture are required from a Newtonian point of view for several reasons. The "fixed" stars move, and indeed are accelerated relative to each other. The earth is not perfectly symmetrical about any axis, and the axis about which it is most nearly symmetrical, from the south to the north poles, does not coincide exactly with the axis of rotation, with the result that there is a wobble of the former axis about a fixed direction in the celestial sphere. The moment of inertia of the earth is variable, due to such factors as the melting of ice, the rising of the seas, and disturbances of the core. And finally, there is a torque on the earth due to the gravitational action of the sun, moon, and planets – an action which would not affect the angular momentum of a spherically symmetrical body, but which has a nonnegligible effect because of the equatorial bulge. Discrepancies between the period of the earth's rotation and the two other measures of time have been explained semiquantitatively from the laws of dynamics by taking into account the relevant geophysical data.[24] The failure to achieve a complete quantitative account of the discrepancies is what one expects from the uncertainties of the data and the fearful difficulties of the requisite calculations. The most decisive result was that of H. Spencer-Jones,[25] who showed that discrepancies between observations and calculated positions of the sun and Mercury could be consistently explained by a single correction for the rotation rate of the earth, and the same correction would also explain the discrepancy concerning the moon, provided that variation in the orbital velocity of the moon is taken into account. The additional complication concerning the lunar discrepancy is not ad hoc, since the moon's contribution to the tidal friction which slows the earth's rotation would affect its own orbital acceleration, whereas the corresponding effect on the orbital accelerations of Mercury and the sun are negligible.[26]

(ii) *Ephemeris time.* When the masses, positions, and velocities of a system of gravitationally interacting bodies are known at one moment, then by Newtonian dynamics and gravitation theory any coordinate x of one of them can be expressed as a function of absolute time t and the initial data. The solution obtained by appropriate approximations can be inverted, and the time can be expressed as a function of x, which in practice will be presented in tabular form. If, in particular, x is the longitude

24. W. Munk and G. MacDonald (1960); A. Grünbaum (1973), p. 67.
25. W. Munk and G. MacDonald (1960), pp. 181–83.
26. *Ibid.*, p. 201.

of the sun on the ecliptic, as measured from an initial point, then the tabulated time is called "ephemeris time." This measure of time is a good approximation to Newtonian absolute time, limited in accuracy by the uncertainties of the initial conditions, the neglect of small perturbing forces, and the inaccuracies of calculation. Each planet and satellite in the solar system in principle provides an analogous "sensible measure." There is good agreement among these variant ephemeris times, though in the case of Mercury a general relativistic correction to Newtonian gravitation theory is required.[27]

(iii) *Mechanical clocks.* According to Newtonian dynamics, an unperturbed pendulum swinging with a fixed amplitude, a fixed length, and in a fixed gravitational field executes its periods of motion in equal absolute time intervals. Variation in the amplitude due to damping, variations in the length of the pendulum due to temperature changes and stretching, and temporal variations of the local gravitational field cause variations in the length of the period. Corrections can be made for the first two sources of variation. In a sophisticated pendulum clock a force is occasionally supplied in order to maintain a nearly steady amplitude, and detailed considerations of dynamics are required in order to minimize the interruption of a period by the application of the force; also, compensations are made for thermal expansion and contraction in order to maintain a nearly fixed length of the pendulum.[28] The local variation of the gravitational field does not affect a quartz clock, since it operates by a piezo-electric process, with very slight imposition of an external current in order to maintain a steady amplitude of crystal vibration. There is quite good agreement among carefully constructed quartz clocks when they are relatively new, and the deviation between a new one and an old one (of the order of a year) can be understood dynamically in terms of a gradual change of the elastic coefficients of the crystal.[29]

To summarize, dynamics has been used to explain discrepancies among the three different types of clocks and in order to correct the discrepancies. When a fully quantitative account of the discrepancies is not possible – especially between the rate of rotation of the earth and the rates of the other two types of clocks – the shortcoming is readily understood in terms of the complexity of perturbations, without any indication of the lapsing of the absolute time implicit in the laws of dynamics. In pure mathematics the phrase "Newton's method" refers to a procedure for finding

27. G. Clemence (1957), p. 6.
28. Encyclopedia Britannica (1970), vol. 5, pp. 934–36.
29. Encyclopedia Britannica (1970), vol. 21, pp. 1160–61.

roots of equations by successive approximations. The phrase could also appropriately be applied in chronometry: to the process of using dynamical corrections to time-keeping devices in order to obtain sensible measures that are increasingly good approximations to the time which "flows equably without relation to anything external." The specificity of the dynamical means which have been used to obtain these increasingly good approximations suffices to show that – contrary to Janich – Newton's cryptic statements can and should be construed as more than methodological maxims.

The second part of the realization of a Newtonian alternative to the protophysics of time concerns the connection between the life-world and the laws of dynamics. On this matter Newton offers little in the way of historical or systematic reconstruction, saying merely, "Hitherto I have laid down such principles as have been received by mathematicians, and are confirmed by abundance of experiments."[30] How much Newton failed to tell us can be discerned from the mass of psychological and epistemological literature – extending from the Platonic dialogues and the classical inquiries concerning human understanding to modern works on phenomenology, genetic epistemology, and cognitive psychology – which has explored the propositional content and the modalities of our knowledge of the life-world, and the mass of historical studies of the evolution of mechanics. It is not wise to volunteer the reconstruction that Newton declined to supply. Three remarks, however, will indicate the feasibility of this reconstruction, along lines quite different from that of the protophysicists.

(a) Our primitive psychological time sense – which provides us with a sense of passage and a discrimination among past, present, and future – supplies a rough intuitive comparison of lengths of time intervals, provided that they are short enough and close enough to each other. The establishment of a rhythmical beat in music, which must be either a genetic capability of the species or a very primitive cultural achievement, is inseparable from the comparison of at least two adjacent intervals and probably more than two. It may be claimed that the capability of establishing a rhythmical beat depends upon regular breathing or the pulse, which share the generic characteristics of crude periodic clocks. There may indeed be some such causal connection, but it should be noted that the rhythmical beat can be established *ad libitum,* without regard to commensurability with the two obvious mechanical clocks of the body. There is another remarkable property of the time sense implicit in a rhythmical beat. It is sometimes said that the fundamental problem of chronometry

30. I. Newton, Bk. I, Scholium following Corollary VI to the "Axioms, or Laws of Motion."

is the impossibility of preserving a standard unit of time – for example, the interval from the stroke of midnight of New Year's Day, 1900, to the stroke of midnight of New Year's Day, 1901 – for making comparisons with other intervals of time.[31] If, however, one is content with time intervals of the order of a musical beat, then the freshness of memory effectively does preserve the unit for a while. A skeptic may well object to an uncritical reliance upon short-term memory, but the epistemological literature shows how difficult it is to give a rational reconstruction of human knowledge without granting some tentative credence to memory.[32]

(b) A primitive version of the concept of force plays just as important a role in the experience of the life-world – including its technology – as the concepts of space, time, and motion. The techniques of building, working upon clay and metals, and engaging in military combat instantiate the concept of force as much as the techniques of agriculture and cooking instantiate the concept of time, and there is no reason to believe that in early epochs there was less articulate consciousness of the former than of the latter. Obviously, these primitive concepts had to be refined and controlled, by investigators of genius, in order to effect the transition from a prescientific to a scientific world view. But there is no reason to grant that either historically or in rational reconstruction the transition had to be performed in two stages, one of which consisted of the refinement of the concepts of space, time, and motion, and the second of the refinement of other concepts, including that of force. The early recognition that a nearly uniformly moving ball is slowed by friction suffices to exhibit a primitive concept of force mingled with a primitive kinematic concept.

(c) The life-world provided a rich array of phenomena that could suggest the idea of a time which "flows equably without relation to anything external." Many different processes that are perceived to be roughly uniform by reference to the intuitive time sense turn out to be roughly synchronized with each other – for example, the intervals in which the sun rises by equal angles through its diurnal course, the intervals in which a clepsydra goes from full to empty, and the intervals in which a man steadily walks ten thousand paces. Furthermore, in the absence of strong impulsive forces, bodies are perceived to move quite smoothly; and since there is an intuitive sense of comparison of short time intervals, this perception cannot be construed away as the tautology that equal time intervals are those in which smoothly moving bodies traverse equal spatial intervals. So much rough agreement and rough synchronization among processes of great diversity conceivably could, and historically did, serve as heuristics for the idea of equable time flow which was not bound to a

31. For example, Reichenbach (1958), p. 116.
32. For example, C. I. Lewis (1946), chapter 11.

specific operational definition. And one asks no more of the experience of the life-world than heuristics. The critical assessment of what is heuristically suggested is the business of mature (or maturing) science.

5. TWO FURTHER IMPLICATIONS

The protophysicists maintain that classical kinematics, essentially that of Galileo and Newton, is methodologically so entrenched that it cannot be refuted by any physical discovery, including the nonclassical character of the laws of mechanics and electromagnetism. A complete critique of their thesis – that kinematics must remain classical whereas dynamics and other parts of physics are relativistic – is beyond the scope of this paper.[33] The revision of the protophysics of time proposed in Section 4 is nevertheless relevant to this critique, for it undercuts the following crucial argument:

The optic-independent transformation properties of a *kinematic reference system*, i.e., of a geometrical coordinate system marked on a rigid body by four points not lying in a plane and the clock connected with this system, depend solely on the ideatively required constancies of the body and of the clock. Where, then, indeed 'dynamical reference systems' are defined, as say by the condition that in a dynamical reference system a body uninfluenced by forces moves uniformly and in a straight line, the bounds of transformation properties controlled by kinematics have been overstepped."[34]

It was argued in Section 4 that the gap in protophysical chronometry could be filled in principle and to a large extent in practice by relying upon dynamics, and that there is no violation of the principle of methodical order in using dynamics to refine chronometry. This argument can be extended to kinematics. It is true that the concept of an inertial frame rests upon at least a tacit reference to force, and in this sense it is a dynamical concept. But there is no vicious circularity in developing in tandem a kinematics that singles out the family of inertial reference systems from all possible reference systems, and a dynamics that permits the investigation of whether a body is subject to forces.

It is important to note that not only relativistic but also classical kinematics makes fundamental use of the concept of an inertial reference system. Newton makes a separate assertion of the first law of motion, instead of treating it as a corollary of the second law in the force-free case. Kinematics is established by the separate assertion of the first law, but a dynamical concept is implicit in the assertion. By singling out the family of inertial reference systems, Newton drastically curtailed Leibniz's thesis of the "equipollence of hypotheses" (i.e., the invariance of

33. J. Pfarr (1976), pp. 149–77.
34. P. Janich (1985), p. 164.

the laws of physics under any coordinate transformations),[35] as he had to do in order to achieve a workable dynamics. The bifurcation between classical and relativistic kinematics does not hinge upon abstention from or recourse to the concept of an inertial frame of reference. It hinges upon what laws are found to hold in all inertial reference systems – specifically upon the non-isotropy or isotropy of light velocity.

Another implication of the proposal of Section 4 concerns the general enterprise of epistemology. There are epistemologists who draw upon the results of the natural sciences, particularly biology and psychology, in their assessments and reconstructions of human knowledge. They are too heterogeneous in their philosophical views to constitute a "school" of naturalistic epistemology, and they differ widely in their ways of combining empirical and analytical elements in their theories of knowledge.[36] Generically, however, the naturalistic epistemologists are open to the charge of circularity: their epistemology relies at least in part upon results of the empirical sciences, but the justification of the methods of the empirical sciences must be provided by epistemology. The principle of methodical order is manifestly violated, and the moral is that epistemology in general and scientific epistemology in particular must be developed in abstention from the results of the empirical sciences.

The Newtonian alternative proposed in Section 4 suggests a defense of naturalistic epistemology against the foregoing objection. Just as the principle of methodical order need not be construed to require that chronometry must precede dynamics, it also need not be construed to require that epistemology must be prior to the empirical sciences. Tentative epistemological starting points, derived from the life-world, are indeed essential in order to begin the enterprise of systematic inquiry, but these starting points are subject to refinement and correction in the light of subsequent discoveries. There is a dialectical interplay between epistemology and the substantive content of the empirical sciences. This dialectic does not commit a *petitio principi,* in the sense that what eventually is purported to be demonstrated has actually been presupposed. It must be acknowledged that the dialectic process is not algorithmic, and one has no guarantee that the succession of corrections and refinements will converge. To this extent, the naturalistic epistemology which I envisage may fail to satisfy all the desiderata of the protophysicists and of other constructive epistemologists. However, this shortcoming is more than compensated for by openness, freedom from rigid commitments, capability of self-criticism, and possible attunement to the natural world.[37]

35. H. Stein (1977), part I.
36. H. Kornblith (1985).
37. A more extensive presentation of this argument is given in A. Shimony (1981).

Acknowledgment. This work was supported in part by the National Science Foundation, grant no. SES-8908264.

REFERENCES

Clemence, G. M. (1957), "Astronomical Time." *Reviews of Modern Physics* 29: 2–8.
Encyclopedia Britannica (1970), vols. 5 and 21, "Clock" and "Time Measurement." Chicago: Encyclopedia Britannica, Inc.
Grünbaum, A. (1973), *Philosophical Problems of Space and Time,* 2nd ed. Dordrecht: Reidel.
Janich, P. (1984), "Commentary on 'Protophysics of Time and the Principle of Relativity'." In R. Cohen and M. Wartofsky (eds.), *Physical Sciences and History of Physics,* pp. 191–97. Dordrecht: Reidel.
Janich, P. (1985), *Protophysics of Time: Constructive Foundation and History of Time Measurement.* Dordrecht: Reidel.
Janich, P., "Time measurement and kinematics." Unpublished manuscript, Philosophy Department, University of Marburg.
Kant, I. (1950), *Critique of Pure Reason,* transl. N. K. Smith. New York: Humanities Press.
Kornblith, H., ed. (1985), *Naturalizing Epistemology.* Cambridge, MA: MIT Press.
Lewis, C. I. (1946), *An Analysis of Knowledge and Valuation.* La Salle, IL: Open Court.
Lorenzen, P. (1976), "Theorie des technischen Wissens." In G. Böhme (ed.), *Protophysik.* Frankfurt: Suhrkamp.
Lorenzen, P. (1982), *Elementargeometrie.* Mannheim: Bibliographisches Institut.
Lorenzen, P. (1987), *Constructive Philosophy,* transl. K. R. Pavlovic. Amherst: University of Massachusetts Press.
Munk, W., and MacDonald, G. (1960), *The Rotation of the Earth: A Geophysical Discussion.* Cambridge: Cambridge University Press.
Newton, I. (1947), *Principia,* ed. F. Cajori. Berkeley: University of California Press.
Pfarr, J. (1976), "Protophysik der Zeit und Spezielle Relativitätstheorie." *Zeitschrift für allgemeine Wissenschaftstheorie* 7: 298–326.
Pfarr, J. (1979), "Zur Eindeutigkeit der Zeit in der Protophysik." In W. Balzer and A. Kamlah (eds.), *Aspekte der physikalischen Begriffsbildung,* pp. 147–66. Braunschweig: Vieweg.
Pfarr, J. (1984), "Protophysics of Time and the Principle of Relativity." In R. Cohen and W. Wartofsky (eds.), *Physical Sciences and History of Physics,* pp. 159–89. Dordrecht: Reidel.
Reichenbach, H. (1938), *Experience and Prediction.* Chicago: University of Chicago Press.
Reichenbach, H. (1958), *The Philosophy of Space and Time.* New York: Dover.
Shimony, A. (1981), "Integral Epistemology." In M. Brewer and B. Collins (eds.), *Scientific Inquiry and the Social Sciences,* pp. 98–123. San Francisco: Jossey-Bass.

Stein, H. (1967), "Newtonian Space-Time." *Texas Quarterly* 10: 174–200.

Stein, H. (1977), "Some Philosophical Prehistory of General Relativity." In J. Earman, C. Glymour, and J. Stachel (eds.), *Foundations of Space-Time Theories*. Minneapolis: University of Minnesota Press.

Tetens, H. (1982), "Relativistische Dynamik ohne Relativitätsprinzip (über das Verhältnis der Protophysik zur Relativitätstheorie." *Philosophia Naturalis* 19: 319–29.

18
The transient now

This work was originally prepared for the Library of Living Philosophers volume, *The Philosophy of Paul Weiss* (Lewis E. Hahn, ed.), to be published by the Open Court Publishing Company, La Salle, Illinois.

I. INTRODUCTION

In Chapter VI ("Time") of Paul Weiss's first book, *Reality,* there are many incisive phenomenological reports concerning time, among them the following:

To specify the ordered moments of time as being past, present, and future is not to characterize time, but the picture of it, a time still undifferentiable from an eternal, unchanging structure. Passage is part of its essence.

Suppose that the series 1, 2, 3 represents the moments of time, and that 1 is past, while 2 is present and 3 is future. There will be no time unless when 3 becomes present, 2 becomes past. Being present cannot then be an irrevocable, intrinsic character of these moments, for then they would all be present together.[1]

Weiss never revoked these early philosophical comments, but he did elaborate them. For instance, in *First Considerations* one finds the following:

That time is more than a line is also evident from the fact that it is lived through. A line has all its distinguishable parts co-present, but the "line of time" is constantly being drawn and just as constantly erased.[2]

The present paper is essentially an exploration of the ideas expressed in these three passages. I agree with them at the commencement of my analysis and continue to agree with them when the analysis is completed. However, my mode of analysis – which combines considerations from phenomenology, analytic philosophy, and natural science – is quite different from Weiss's, and I shall proceed without further reference to his texts. In a critical comment on Weiss's philosophical explanations I wrote: "Whether or not you like the compliment, I feel that you are a remarkable phenomenologist, and that is where you have made your greatest contribution

1. Paul Weiss, *Reality* (Princeton, NJ: Princeton University Press, 1938), p. 226.
2. Paul Weiss, *First Considerations* (Carbondale, IL: Southern Illinois University Press, 1977), p. 154.

to philosophy."[3] The present paper may be taken as an illustration of this remark, or it may simply be read as my exploration of a deep problem that Paul Weiss has thought deeply about, which is my best way of offering homage to him.

The following three sections of this paper will be devoted to three aspects of the problem of transiency. Section 2 will examine the famous paradox of transiency posed by J. M. E. McTaggart and will endorse the very thoughtful but little-known solution proposed by David Zelicovici. Section 3 will examine the claim made by Adolf Grünbaum, Rudolf Carnap, Hermann Weyl, and others that *transiency, becoming, present, now,* and cognate concepts are subjective in character. Section 4 will argue that Zelicovici's solution is compatible with the space–time structure of special relativity theory.

2. MCTAGGART'S PARADOX AND ZELICOVICI'S SOLUTION

McTaggart[4] sharply distinguished two kinds of temporal discourse, and he has been followed with slight variation by many later writers on time (e.g., Broad, Gale, Schlesinger, and Zelicovici).[5] The "A-series" divides the linearly ordered set of temporal instants into past, present, and future. Notions which rely essentially upon this division – such as *passage from past to present, now, becoming,* and *transiency* – are referred to as "A-determinations" in the literature influenced by McTaggart. The "B-series" is the linear ordering of temporal instants by the transitive, asymmetric, and irreflexive relation *precedes,* without any reference to past, present, and future; and the notions which depend upon *precedes* are called "B-determinations." When a temporal metric is introduced, it may be used to formulate further A-determinations, such as "being ten seconds earlier than the present." Likewise, further B-determinations can be formulated in terms of a temporal metric, such as "t_1 is ten seconds earlier than t_2."

McTaggart also introduced the "C-series", which consists of events, each located in a set of space–time points. Because of problems of ordering events which have temporal duration, it seems to me best to avoid this locution of McTaggart. The temporal language concerning events can be borrowed from the two types of discourse concerning instants themselves

3. Abner Shimony, "Some Comments on Philosophic Method," in Weiss (1977), p. 188.

4. John McTaggart Ellis McTaggart, *The Nature of Existence* (Cambridge: Cambridge University Press, 1927), chapter 33.

5. Charles Dunbar Broad, *Scientific Thought* (New York: Harcourt, 1923); Richard M. Gale, *The Language of Time* (New York: The Humanities Press, 1968); George N. Schlesinger, *Aspects of Time* (Indianapolis, IN: Hackett, 1980); David Zelicovici, "A (Dis)Solution of McTaggart's Paradox," *Ratio* 28 (1986): 175–95; David Zelicovici, "Temporal Becoming Minus the Moving-Now," *Noûs* 23 (1989): 505–24.

(A-determinations and B-determinations), without any commitment regarding the metaphysical question of the dependence or independence of time upon events and the converse. For simplicity, I shall restrict my attention mainly to idealized events located at temporal instants. If an event E is located at the temporal instant t, and if t is described by the A-determination "present", then E also is described as present; if E and E' are located at the temporal instants t and t' respectively, and t precedes t', then E and E' can also be described by the B-determination "E precedes E'."

McTaggart implicitly assumes much of Newtonian space–time structure, specifically that every space–time point has a definite temporal location t, so that the linear ordering of instants determines a partial ordering of space–time points (also designated by "precedes" without any danger of confusion). Hence, for each pair of space–time points P and P', either P precedes P' or P' precedes P or P and P' have the same time (i.e., are simultaneous). Likewise, there is a partial ordering of the instantaneous events. These assumptions will be maintained in the present paper until Section 4.

McTaggart's paradox of transiency will now be formulated in a way that differs somewhat from his own formulation, in order to anticipate the analysis that follows.

(a) *A property of an event is not relative to the time of judgment or vantage point of discussion.* In this respect, an event differs from a persisting thing, which may have different attributes at different times. Indeed, the metaphysical category of event is introduced for the purpose of giving a changeless analysis of changing persisting things: that is, the career of a persisting thing is analyzed into instantaneous events, which may differ among themselves in various respects, but each of which is changeless, having exactly the properties that it has. The event as a particular entity may be identified by specifying its instant, but the properties of the entity thus identified are the same from any temporal vantage point of judgment. (It may be noted that even if we consider temporally extended events instead of restricting attention to instantaneous events, proposition (a) continues to hold, since the properties manifested at each instant of the extended event are independent of the vantage point of judgment, even though the properties at two different instants may differ.)

(b) *A-determinations of events are unusual properties* (differing, for example, from ordinary physical properties such as electromagnetic field strength and from ordinary mental properties such as sensations of tone), *and nonetheless, for all their peculiarity, they are properties.*

(c) *But A-determinations are relative to the time of judgment, for an event which is present when the time is t will not be present when the time is t'* (t' *being different from t*).

Clearly, the conjunction of (a), (b), and (c) is inconsistent, even though each of these three premisses is plausible. That is McTaggart's paradox.

Zelicovici gives two different formulations of his solution to McTaggart's paradox, which he clearly intends to be equivalent, though they are unequal in lucidity. The first formulation[6] consists in asserting that thesis (a) simply does not hold of A-determinations, because these are not "usual" properties. It was indeed noted in the statement of premiss (b) that A-determinations differ from ordinary properties. Nevertheless, it appears ad hoc to exempt them from the plausible premiss (a) without some deeper analysis.

The second formulation, which I much prefer, draws upon a famous argument that Kant uses in order to demonstrate the invalidity of the ontological argument for the existence of God.[7] Zelicovici states,

The attribution of pastness operates somewhat like the attribution of existence in the Kantian sense: it applies to the event complete with all its properties, and only defines its status (for Kant existence, for us pastness).[8]

Zelicovici's formulation is very illuminating. It utilizes the generally recognized profundity of Kant's treatment of "existence" or "being" in the following famous passage: "'*Being*' is obviously not a real predicate, that is, it is not a concept of something which could be added to the concept of a thing."[9]

Kant warns specifically against interpreting the subject-predicate form of the sentence "this thing exists" as a genuine instance of attribution of a property to the thing in question. He illustrates his thesis with some well-known homely and humorous remarks:

the real contains no more than the merely possible. A hundred real thalers do not contain the least coin more than a hundred possible thalers. For as the latter signify the concept, and the former the object and the positing of the object, should the former contain more than the latter, my concept would not, in that case, express the whole object, and would not therefore be an adequate concept of it. My financial position is, however, affected very differently by a hundred real thalers than it is by the mere concept of them (that is, of their possibility). For the object, as it actually exists, is not analytically contained in my concept, but is added to my concept (which is a determination of my state) synthetically; and yet the conceived hundred thalers are not themselves in the least increased through thus acquiring existence outside my concept.[10]

6. Zelicovici (1986), p. 188.
7. Immanuel Kant, *Critique of Pure Reason,* trans. N. Kemp Smith (New York: The Humanities Press, 1950), pp. 500–7.
8. Zelicovici (1986), p. 189.
9. Kant, p. 504.
10. *Ibid.*, p. 505.

Now if one examines some of the arguments against the objectivity of A-determinations, one finds striking similarities to the thesis that Kant analyzes. For instance, Grünbaum writes:

But I am totally at a loss to see that anything non-trivial can possibly be asserted by the claim that at 3 P.M. nowness (presentness) inheres in the events of 3 P.M. For all I am able to discern is that the events of 3 P.M. are indeed those of 3 P.M. on the day in question.[11]

Zelicovici's *tour de force* consists in recognizing the affinity of the conceptual poverty of *nowness* (a recognition somewhat disguised by his earlier locution of "non-ordinary property") to the conceptual poverty of "existence" that Kant insists upon. There is indeed a kind of richness in both *nowness* and *existence,* but it is existential in character, entirely different from conceptual determination.

Zelicovici's analysis can be carried one step further. He suggests only an analogy ("somewhat like") between an A-determination and the attribution of existence. In my opinion, however, there is nearly an identity. Suppose, for simplicity, that one takes a possible event rather than a thing as the subject of an existential statement (thereby departing from Kant's treatment of the category of substance in objective judgments). Specifically, let E denote a possible instantaneous event. Then to assert that E exists is to say that it is present at some time – in other words, to assert an existentially quantified A-determination. There is no other way for a possible instantaneous event to exist than to be *now* at some instant! The appropriate A-determination that is implicit in the assertion of existence of a possible temporally extended event can also be constructed, but it is somewhat more complicated than the one just given because of the complexity of structure of the subject.

One of the attractive features of Kant's analysis of existence is that his analysis does not depend upon the doctrinal framework of his critical philosophy – for example, of his doctrines of "transcendental idealism" and "empirical realism." A non-Kantian philosophy that speaks freely of things in themselves and of our knowledge of them, whether direct or indirect, can incorporate Kant's analysis of existence statements, construing them in agreement with Kant as not asserting conceptual determinations, and yet construing existence claims in a "realistic" manner that is alien to Kant. Consequently, if a non-Kantian philosophy accepts Zelicovici's proposal for assimilating the analysis of A-determinations to Kant's analysis of existence assertions, then it can regard the ascription of "now" or "past" to a moment of time or to an event as objectively true or objectively false, without reference to a knowing subject.

11. Adolf Grünbaum, "The Meaning of Time," in E. Freeman and W. Sellars (eds.), *Basic Issues in the Philosophy of Time* (LaSalle, IL: Open Court, 1971), p. 211.

3. THE THESIS THAT A-DETERMINATIONS ARE SUBJECTIVE

There have been attempts to do justice to the phenomenology of A-determinations by treating them as subjective. The most complete exposition of this point of view is by Adolf Grünbaum, who presents the following thesis: "presentness or nowness of an event requires conceptual awareness of the event or, if the event itself is unperceived, of the *experience* of another event simultaneous with it."[12] He makes several commentaries on this thesis, of which the most important but also the most ambiguous is the following:

> I am *not* offering any kind of *definition* of the adverbial attribute now, which belongs to the conceptual framework of tensed discourse, solely in terms of attributes and relations drawn from the tenseless (Minkowskian) framework of temporal discourse familiar from physics. In particular, I avowedly invoked the present tense when I made the nowness of an event E at time t dependent on someone's knowing at t that he is *experiencing E*. And this is tantamount to someone's judging at t: I am experiencing E *now*. But this formulation is *non*viciously circular. For it serves to articulate the mind-dependence of nowness, *not* to claim erroneously that nowness has been eliminated by explicit definition in favor of tenseless temporal attributes or relations. In fact, I am very much less concerned with the adequacy of the specifics of my characterization than with its thesis of mind-dependence.[13]

The ambiguity that I find in this passage stems from an apparently deliberate abstention from a commitment concerning the ontological status of awareness, experience, and in general of mental events. I shall consider two extreme positions, neither of which can I ascribe with confidence to Grünbaum, for he may espouse an intermediate position or he may simply keep his opinion in abeyance on this philosophical problem. Nevertheless, it is illuminating to examine the conjunction of the thesis of mind-dependence of A-determinations with each of these two extreme positions, for the philosophical implications of the two cases are quite different.

Position (i) asserts that mental entities (minds, mental events, occasions of awareness) are ontologically fundamental, neither derivative from nor constructed from physical entities in any way (e.g., in the way that macrophysical systems are constructed from microphysical components). Of course, this position can be subdivided, notably into (a) a fundamental mentalism, like that of Leibniz and Whitehead, according to which the systems ordinarily considered to be physical are constructed in some sense from mental entities; and (b) a dualism, like that of Descartes and

12. *Ibid.*, p. 207.
13. *Ibid.*, p. 209.

Locke, in which both physical and mental entities are fundamental, neither being reducible to or constructible from the other. This finer taxonomy is not important for the problem of the status of transiency. What is crucial is that if A-determinations are mind-dependent, and in addition mental entities are ontologically fundamental, then A-determinations have an uneliminable status in nature. To put the matter differently, if A-determinations are mind-dependent, they may indeed be subjective, but in position (i) the very concept of subjectivity is cleansed of the pejoratives of illusion, derivativeness, and superficiality. I can make this point more forcefully by adopting the terminology of Whitehead's philosophy of organism.[14]

According to Whitehead, the ultimate concrete entities in the cosmos are *actual occasions,* which are occasions of experience, though he construes "experience" so broadly that in some dim way even the minimal temporal events in the career of an electron have a kind of protomentality. Transiency is a fundamental notion for Whitehead. An actual occasion in its subjective immediacy is present, but when its moment of immediacy is completed (a process which has some temporal breadth according to Whitehead, but this point is not essential to the considerations of this paper) it remains objectively immortal; the occasion in its particularity can be *prehended* by subsequent occasions and serves as an ingredient in their immediate experience. As objectively immortal the occasion has all the content that characterized it in the last stage of its subjective immediacy. The difference between the occasion as present and the occasion as past does not lie in content, and in this sense there is no *property* of presentness. Whitehead's treatment of transiency can be subsumed (with one qualification) under Zelicovici's, though of course with more metaphysical detail. The qualification is that according to Whitehead the occasions of the future do not exist as actualities but only as real potentialities with incomplete specification, and therefore the thesis that the content of an occasion is independent of its A-determination holds only with regard to presentness and the various degrees of pastness, but not with regard to futurity.

Position (ii) holds that physical entities are ontologically fundamental and that mental entities are reducible to or constructible from physical entities. There is, of course, a vast literature about the exact sense in which theories of mind are supposed to be reducible to physical theories and psychological concepts are definable in terms of physical concepts, but these complications do not affect our main line of analysis. The point is that when a purely physicalist ontology is conjoined with the

14. Alfred North Whitehead, *Process and Reality* (New York: Macmillan, 1929).

thesis that A-determinations are mind-dependent, it then follows that A-determinations are as illusory and derivative as purely sensory *qualia*. Grünbaum says something similar:

assuming the causal dependence of mental on physical events, why is the mind-dependence of becoming more puzzling than the fact that the raw feel components of mental events, such as a particular event of seeing green, are not members of the *spatial* order of physical events?[15]

The view that A-determinations are illusory does not stand up under analysis, whether or not it is based upon a physicalistic ontology. To demonstrate this assertion, consider an event E in the career of a normal human being (ideally instantaneous, but the argument goes through even if E has temporal breadth τ centered about t). Suppose that the content of E is fully specified. If physicalism is correct, then the full specification of the physical properties of the organism specifies everything, and all mental properties will be automatically determined, whether or not there is an explicit linguistic definition of each psychological concept in terms of physical ones (for our concern is ontological rather than epistemological). If physicalism is not correct, then a complete specification requires mental and perhaps psychophysical properties in addition to physical ones. What concerns us at this point is the completeness of the specification, not the solution to the mind–body problem. By premiss (a) in McTaggart's paradox of transiency (not challenged by Zelicovici) the complete specification of the properties of E is intrinsic to E, and is not relative to the time of judgment. Now an illusion at time t in the career of the human being under consideration is one of the properties of E. Hence, if nowness is an illusion of the subject at (or within $\tau/2$ of) t, then it is part of the complete specification of E and is not relative to the time of judgment. Thus *nowness as an illusion* applies to the event E as well when time t is long past as it does at t itself – and the singling out of a particular instant as *now, even as an illusion,* evaporates.

An elaboration of the foregoing argument may be enlightening. There is a very important principle linking epistemology and ontology, one that is pervasive in the literature of empiricism from Berkeley to the sense-data theorists of the early twentieth century and implicit in other philosophical writings: that *even though the distinction between appearance and reality is maintained, a minimal condition on ontology is to recognize a sufficient set of realities to account for appearances* qua *appearances.* I cannot believe that no name has been given to this principle, but since I do not recall reading one, I propose "the Phenomenological

15. Grünbaum, pp. 216–17.

Principle."[16] Applied to the thesis that A-determinations are illusory, the Phenomenological Principle is devastating. Specifically, consider Grünbaum's query: "why is the mind-dependence of becoming more puzzling than the fact that the raw feel components of mental events . . . are not members of the *spatial* order of physical events?" The answer to this query is that the status of raw feel components *qua* appearances is not being challenged in the quotation, but only their status as attributes of spatially located physical things. By contrast, if an A-determination is nothing but an illusion, then it isn't even an illusion! An A-determination cannot function as an appearance without transiently singling out an instant of time (or more broadly, an interval of time). But if it is only an illusion, then it cannot function this way as an appearance, by the argument given in the preceding paragraph. Nowness must either be more than an illusion or less than an illusion. If it is less, then no justice has been done to the phenomenology of time. If it is more, then either Zelicovici's existentialist analysis is required, or something equally strong and explanatory, which I have yet to see.

Tangentially, I remark that another application of the Phenomenological Principle is to devastate pure physicalism. If awareness is nothing but an illusion, supervenient upon the physical state of an organism, then it cannot even be illusion. Consequently, the dialectical openness to pure physicalism that was granted in stating position (ii) must be short-lived.

4. A-DETERMINATIONS AND RELATIVISTIC SPACE–TIME STRUCTURE

In this section I shall investigate whether the solution proposed by Zelicovici to McTaggart's paradox of transiency can be maintained when Newtonian space–time is replaced by Einstein–Minkowski (or special relativistic) space–time. It is necessary first to characterize the structure of the latter.

According to special relativity theory,[17] space–time is an affine space associated with a vector space V (to be specified below) in the following sense: An ordered pair P, Q of space–time points uniquely determines a vector v belonging to V, which we may call *the directed vector from P to Q*; for any space–time point P and vector v there is a unique space–

16. The principle of "saving the appearances" has a quite different meaning: to order or explain or predict phenomena on the basis of a theoretical scheme.
17. More details about the space–time structure of special relativity can be found in many texts, an excellent one being Wolfgang Rindler, *Introduction to Special Relativity,* 2nd ed. (New York: Oxford University Press, 1991).

time point Q such that the directed vector from P to Q is v; if v is the directed vector from P to Q and w is the directed vector from Q to R, then the vector sum $v+w$ is the directed vector from P to R; and the directed vector from P to Q is the null vector 0 if and only if P is identical with Q. V is assumed to be a real four-dimensional vector space endowed with a nondegenerate inner product (u, v), where nondegeneracy means that (u, v) equals zero for all u only if v is the null vector. Furthermore, V has the further property that if four of its members v_0, v_1, v_2, v_3 are mutually orthogonal (i.e., $(v_i, v_j) = 0$ if $i \neq j$) and if each v_i has a nonzero inner product with itself, then exactly three of the inner products (v_i, v_i) have one sign and exactly one has the other. We shall assume, as a matter of convention, that only one of the (v_i, v_i) is positive (and this one is taken to be $i = 0$), whereas the other three, with $i = 1, 2, 3$, are negative.

All the familiar concepts of relativistic space–time theory can be defined straightforwardly in terms of this simple and austere structure. Here are the most important concepts. If $(v, v) > 0$, then v is *timelike;* if $(v, v) < 0$, then v is *spacelike;* and if $(v, v) = 0$, then v is *null* or *lightlike.* If P is an arbitrary space–time point, then the set of all space–time points Q such that the directed vector from P to Q is timelike constitutes the *interior of the light cone* of P. This set consists of two disconnected parts, the interiors of the forward and the backward light cones respectively. If Q belongs to the former and Q' to the latter, and if v and v' are the directed vectors from P to Q and Q' respectively, then (v, v') is negative; and also one says that v is *forward directed* and v' is *backward directed.* The boundary points of the interior of the light cone are those points Q such that if v is the directed vector from P to Q then $(v, v) = 0$; the boundary points are said to be *on* the light cone of P. The forward light cone of P consists of the points in the interior of the forward light cone or on it; and the backward light cone is similarly defined. Hence the forward and backward light cones intersect at the one point P. Finally, the *exterior* of the light cone of P consists of all those points Q such that the directed vector from P to Q is spacelike. Each of these geometrical concepts has a straightforward physical meaning. If Q belongs to the backward light cone of P, then a signal can be sent from Q to P; the signal can be carried by a means slower than light, such as the motion of a massive particle, if Q is in the interior of the backward light cone, but only by some entity with the velocity of light if Q is on the light-cone. If Q belongs to the forward light cone of P, then a signal can be sent from P to Q, by means slower than light if Q is in the interior but only with the velocity of light if Q is on the boundary. Finally, if Q belongs to the exterior of the light cone of P, then no signal can be sent from P to Q or from Q to P.

Space–time can be coordinatized by choosing a space–time point O as the origin and a quadruple of mutually orthogonal nonnull basis vectors v_0, v_1, v_2, v_3. As noted before, it is implicit in the structure of V that exactly one of these four is timelike, and this one is conventionally chosen to be v_0. It is convenient to choose the v_i such that $(v_0, v_0) = 1$, and $(v_1, v_1) = (v_2, v_2) = (v_3, v_3) = -1$. Every space–time point P is now assigned a quadruple of coordinates in the following way: Let v be the directed vector from the origin O to P, and express v as

$$(1) \qquad v = c_0 v_0 + c_1 v_1 + c_2 v_2 + c_3 v_3,$$

where c_0, c_1, c_2, c_3 are real numbers; it is always possible to find real numbers c_i which make Eq. (1) true, and this can be achieved in only one way. The quadruple (c_0, c_1, c_2, c_3) constitutes the coordinates of P relative to the specified origin and set of basis vectors. In particular, c_0 is the *time coordinate* of P. For aesthetic purposes the further condition is imposed upon v_0 that it be forward directed. This condition implies that if P is distinct from O, then it has a positive time coordinate if (but definitely *not* only if) it belongs to the forward light cone of the origin, and a negative time coordinate if (but definitely *not* only if) it belongs to the backward light cone of the origin.

We can now ask about the applicability of McTaggart's characterization of the B-series to relativistic space–time. The answer turns out to be complicated but clear. Once the timelike vector v_0 is chosen, the B-series is well-defined; but there is an infinity of allowable choices of v_0 with different consequences for the temporal ordering of space–time points, and nothing intrinsic to space–time structure singles out one choice from all the others. Consequently, there is no *natural* B-series. Some comments are needed.

(1) First, assume that v_0 has been chosen. It is easily shown that the specification of the three spacelike vectors v_1, v_2, v_3 do not affect the time coordinate of any space–time point P. Furthermore, even though the choice of the origin does affect the time coordinates, it does not affect the *temporal ordering* of space–time points. If $t(P)$ is the time coordinate of P with O as the origin, and $t'(P)$ is the time coordinate of P with O' as the origin, then

$$(2) \qquad t'(P) = t(P) - t(O'),$$

and hence the ordering of space–time points according to the magnitudes of the time coordinates is independent of the choice of the origin. Thus, when v_0 is specified, each space–time point P determines an *instantaneous space* $S(P)$ of space–time points simultaneous with itself, consisting of all points P' such that the directed vector from P to P' is orthogonal to v_0. When an origin is chosen, all members of $S(P)$ are assigned the same time coordinate as P itself. The instantaneous space $S(P)$ can thus be considered to be an *instant,* and it follows from what has been said that the set of these instants is a one-dimensional linearly ordered continuum. Thus a B-series is determined, independent of the origin and of the choice of the spacelike vectors once v_0 is specified, but dependent on v_0.

(2) Now consider the B-series associated with two distinct timelike vectors v_0 and v_0' such that v_0' does not equal cv_0 for any real number c, and again for convenience assume that both v_0 and v_0' are forward directed. Choose an origin O, and let $t(P)$ be the time coordinate of P relative to v_0 and $t'(P)$ be the time coordinate of P relative to v_0'. Then it is a simple mathematical exercise to find two space–time points P_1 and P_2 such that each is in the exterior of the light cone of the other and

(3a) $t(P_1) < t(P_2);$

(3b) $t'(P_1) > t'(P_2).$

The time of P_1 is earlier than the time of P_2 relative to v_0, and is later than the time of P_2 relative to v_0'. Hence, one cannot define an absolute or natural B-series that is independent of the choice of the timelike basis vector. Of course, another strategy for saving the idea of an absolute B-series is to try to find some intrinsic physical reason for one choice among the various possible timelike vectors; but to find such a reason is equivalent to giving a reason for a preferred inertial reference frame, which is contrary to the entire conception of relativity theory. It is beyond the scope of this paper to present the experimental reasons for rejecting a preferred inertial reference frame.

(3) The exterior of the light cone of a space–time point P is sometimes thought of as consisting of "generalized contemporaries" of P, and in this sense all of space–time can be divided into three parts: the past of P, the generalized contemporaries of P, and the future of P (with no reference to transiency). However, it is impossible to construct a natural or absolute B-series in this way. The reason is that the relation of generalized contemporaneity is intransitive. It is a straightforward exercise to construct three points P, Q, R such that P and Q are outside each other's light cones and hence are generalized contemporaries, and Q and R are outside each other's light cones and hence are generalized contemporaries, but this relation fails for P and R. Although this fact is well known to students of relativity theory, there have been many conceptual tricks for evading or disguising it. I am spared the labor of commenting on the

confusions engendered by these tricks because Howard Stein has already given penetrating critiques of them.[18]

(4) Instead of attempting to fit a B-series in an absolute manner to the entirety of space–time, a natural and physically significant construction of a B-series can be achieved by restricting attention to certain interesting subsets of space–time – namely, to *timelike world lines.* A timelike world line is a one-dimensional curve in space–time, any two distinct points P and P' of which have the property of being in the interior of the light cone of the other, and such that there exists no space–time point whose forward light cone contains the entire curve and no space–time point whose backward light cone contains the entire curve. Informally, a timelike world line is the track in space–time of a particle moving for all eternity at speeds permitted by the special theory of relativity – that is, less than that of light. Clearly the points of a timelike world line are linearly ordered, with P preceding P' if and only if the latter is in the future light cone of the former, and hence the timelike world line has the order of a B-series without resorting to any coordinatization. Of course, if there is a coordinatization, and if v_0 is a future-directed timelike vector, then for any P and P' belonging to the world line $t(P) < t(P')$ if and only if P precedes P' in the sense just stated, and $t(P) = t(P')$ if and only if $P = P'$. It is also possible to parametrize the timelike world line by choosing an origin O on it and computing the "proper time" $\tau(P)$ along the world line:

$$(4) \qquad \tau(P) = \int_O^P \frac{dt}{(1 - v^2/c^2)^{1/2}},$$

where the infinitesimal of time dt along the world line and the instantaneous velocity v both depend upon the coordinatization, but the integral $\tau(P)$ depends on O but not otherwise on the coordinatization. Then, for any pair of points P and P' on the timelike world line, $\tau(P) < \tau(P')$ if and only if P precedes P' in the sense stated previously, and $\tau(P) = \tau(P')$ if and only if $P = P'$.

After all of these preliminary considerations of B-determinations we can turn to the question of A-determinations. The most natural way to consider this question within the framework of Einstein–Minkowski space–time is to focus on a timelike world line, which may be idealized as the track of a particle endowed with a mind. Hermann Weyl formulated his thesis of the subjectivity of A-determinations as follows:

18. Howard Stein, "On Einstein-Minkowski Space-Time," *Journal of Philosophy* 65 (1968): 5–23; "A Note on Time and Relativity Theory," *Journal of Philosophy* 67 (1970): 289–94.

The objective world simply *is,* it does not *happen.* Only to the gaze of my consciousness, crawling upward along the life line of my body, does a section of the world come to life as a fleeting image in space which continuously changes in time.[19]

But this statement is inadequate in the same way as Grünbaum's thesis of the subjectivity of becoming. Whatever constitutes subjectivity is a set of properties of the organism – purely physical if the ontology is physicalistic, purely mental if it is mentalistic, and psychophysical if it is dualistic. At any point on the world line there is a complete set of properties, including those that constitute the subjectivity of the organism with which the world line is associated. Hence it is true from any vantage point, including any point along the world line, that at any other point on the world line the subjective experience is fully specified. Hence, no "fleeting image in space which continuously changes in time" can be ascribed to subjectivity. The phenomenology of a transient *now* can be accounted for only by something that is not encapsulated in a bundle of properties. The argument of Zelicovici, based upon Kant's analysis of "existence," maintains that this something is existential. Something fleeting does indeed traverse the world line, but that something is not subjective; it is the transient *now,* which as a matter of objective fact is momentarily present and thereafter is past. Without this minimal amount of objectivity there cannot even be an illusion of transiency.

Even if one accepts the existential character of nowness, one may claim that it is relative to a specific world line, thereby diminishing its objective status. In other words, if L_1 and L_2 are two distinct timelike world lines, each being the track of some idealized unextended mind-endowed particle, it may be conjectured that the transient *now* on L_1 is uncorrelated with the transient *now* on L_2. There is, however, one important piece of phenomenological evidence regarding temporal correlations on two world lines: If the conscious organisms associated with L_1 and L_2 possess the usual means of communication and if L_1 and L_2 intersect at a space–time point P, then if one of the organisms judges P to be *now* so does the other. The ordinary sharing of experience depends upon this fundamental consensus about the fleeting present. But one can go a step further, in view of the argument of the preceding paragraph that nowness cannot be attributed to subjectivity: the agreement between the two organisms associated with L_1 and L_2 rests on a fact independent of their consciousness. Stripping away the nonessentials leaves only the fact of intersection of L_1 and L_2 at the point P as the reason why the nowness of P on L_1 entails the nowness of P on L_2 and conversely. The conclusion is

19. Hermann Weyl, *Philosophy of Mathematics and Natural Science* (Princeton, NJ: Princeton University Press, 1949), p. 116.

that A-determinations must be ascribed to the point P itself, not to P as associated with one or another world line. I agree with Čapek's statement that "on each particular world-line any particular 'now moment' divides unambiguously the past from the future,"[20] but his excellent statement can be strengthened by the ascription of A-determinations to space–time points themselves.

Certain other A-determinations are corollaries of what has just been said. Suppose that P is *now* and Q belongs to the light cone of P. If Q belongs to the backward light cone then a signal can be sent from Q to P, and as a matter of physical fact an event at Q may have influenced an event at P, and therefore Q is past in the sense of an A-determination. Likewise, if Q is in the future light cone of P, then Q is future as an A-determination. In sum, the attribution of nowness to a space–time point P, which is the central A-determination, can be combined with structural features of Einstein–Minkowski space–time (with no further reference to transiency) in order to generate an infinity of A-determinations.

If P_1 and P_2 fall outside each other's light cone, then nothing can be inferred about an A-determination of one from an A-determination of the other. Suppose the contrary. Suppose, for example, that the nowness of P_1 entailed the pastness of P_2. Then there would be a derivative B-relation between them – that P_2 precedes P_1 in a natural sense. Of course, by assumption this would not be so in the sense of P_2 belonging to the backward light cone of P_1. It would then mean that somehow those time-like forward-directed vectors v_0 with the property of entailing a time co-ordinate of P_2 smaller than the time coordinate of P_1 would be distinguished from the other forward-directed timelike vectors, and this in turn is equivalent to selecting a privileged subset of the inertial frames of reference. This conclusion would not break the special relativistic thesis of the equivalence of all inertial frames as strongly as the singling out of a particular inertial frame, but nevertheless it definitely does break the equivalence. In other words, if an A-determination of a space–time point P implied an A-determination of a space–time point outside its light cone, then the structural properties of Einstein–Minkowski space–time would be violated, even though that structure is neutral with respect to the status of transiency.

It may be objected, however, that the wrong conclusion has been drawn from the conflict between the structure of Einstein–Minkowski space–time and the proposition that there are relations of A-determination between

20. Milič Čapek, "Einstein and Meyerson on the Status of Becoming in Relativity," *Actes du XIe Congrès International d'Histoire des Sciences* (Varsovie, 1967), p. 138. See also M. Čapek, *The New Aspects of Time. Its Continuities and Novelties* (Dordrecht: Kluwer Academic Publishers, 1991), *passim,* for a defense of the objectivity of becoming.

spacelike separated space-time points. Instead of holding on to the former and rejecting the latter, which was the conclusion of the preceding paragraph, one might claim that the phenomenology of temporal passage shows the inadequacy of Einstein–Minskowki space-time structure. The claim is that we have an experience of nowness that is not just here-now but is extended – an experience of a "unison of becoming"[21] or of extended contemporaneity. I am very skeptical of such arguments. Upon introspection, I certainly find no direct experience of contemporary objectively existing distant events. There is indeed a presented visual scene in my experience, with a spatial organization, but the relation between the contents of this perceptual space and objectively existing distant events may be (and I think *is*) causal and indirect. As an objective entity in nature, the entire spatially organized visual scene may be located in a very small region of space-time. A stronger case may be made for the immediate experience of the presentness of the body, but the evidence of the finitude of signal propagation along the nerves shows that this phenomenology also must be subjected to careful critical analysis. What is particularly needed is an examination of the specious present, as contrasted with the idealized instantaneous present, and this project is beyond the scope of this paper. Fortunately, Stein has given an important part of the requisite analysis, arguing that

the set of events contemporaneous with a specious present will always be a spatially extended one. . . . this spatial extent – though finite – is in fact *and in principle, as a matter of physics,* always, in a certain sense, immensely large.[22]

The fact that Stein's analysis is carried out entirely within the framework of special relativity theory greatly strengthens the thesis that the phenomenology of becoming does not overflow this framework.

It has also been maintained that quantum-mechanical nonlocality, as exhibited in the experimental violations of Bell's Inequality, entails causal relations between spacelike separated events, in violation of the structure of Einstein–Minkowski space-time. This question is also beyond the scope of the present paper, but I have argued elsewhere that quantum-mechanical nonlocality "peacefully coexists" with special relativity theory, because quantum-mechanical correlations between spatially separated systems cannot be exploited for sending messages faster than light.[23]

21. Whitehead, p. 189.
22. Howard Stein, "On Relativity Theory and Openness of the Future," *Philosophy of Science* 58 (1991): 159.
23. Abner Shimony, "Events and Processes in the Quantum World," in R. Penrose and C. Isham (eds.), *Quantum Concepts in Space and Time* (Oxford: Oxford University Press, 1986), pp. 191–93.

A final remark is mainly historical. It is often maintained that Einstein exorcized transiency from nature by introducing the space–time structure of special relativity theory. That this conclusion is conceptually false is the main point of the arguments of the present section. That it is also historically false is shown by a remarkable reminiscence by Carnap of a conversation with Einstein at the Institute for Advanced Study between 1952 and 1954:

Once Einstein said that the problem of the Now worried him seriously. He explained that the experience of the Now means something special for man, something essentially different from the past and the future, but that this important difference does not and cannot occur within physics. That this experience cannot be grasped by science seemed to him a matter of painful but inevitable resignation. I remarked that all that occurs objectively can be described in science; on the one hand the temporal sequence of events is described in physics; and, on the other hand, the peculiarities of man's experience with respect to time, including his different attitude towards past, present, and future, can be described and (in principle) explained in psychology. But Einstein thought that these scientific descriptions cannot possibly satisfy our human needs. . . . I definitely had the impression that Einstein's thinking on this point involved a lack of distinction between experience and knowledge. Since science in principle can say all that can be said, there is no unanswerable question left. But though there is no theoretical question left, there is still the common human emotional experience, which is sometimes disturbing for special psychological reasons.[24]

According to the present paper, there *is* a theoretical question about *now,* but it can be answered satisfactorily along the lines proposed by Kant and Zelicovici. I like to think that Einstein would be happy with this proposed solution, and I hope that Paul Weiss will be persuaded.

Acknowledgments. The research for this paper was supported in part by the National Science Foundation, grant no. DIR-8908264, and by a fellowship from the National Endowment for the Humanities in 1991.

24. Rudolf Carnap, "Intellectual Autobiography," in P. A. Schilpp (ed.), *The Philosophy of Rudolf Carnap* (LaSalle, IL: Open Court, 1963), pp. 37–38.

PART E

The mental and the physical

19

Quantum physics and the philosophy of Whitehead

I. INTRODUCTION

One of the virtues which Whitehead claims for his philosophy of organism is that it provides a conceptual framework for quantum theory (*SMW* chapter 8, *PR* 121–2 and 145).[1] The theory which he has in mind is the 'old' quantum theory, consisting of the hypotheses of Planck (1901) and of Einstein (1905) that electromagnetic energy is emitted and absorbed in quanta, together with Bohr's model of the atom (1913) in which discontinuous transitions were supposed to occur between discrete electronic orbits. The philosophy of organism was presented in a preliminary form in the Lowell Lectures of 1925 (published in the same year as *SMW*) and in its most systematic form in the Gifford Lectures of 1927–28 (published as *PR* in 1929). It was during the years 1924–28 that de Broglie, Schrödinger, Heisenberg, Born, Jordan, Bohr, Dirac, and others developed the 'new' quantum theory, which was more systematic than the old, much more successful in its predictions, and more revolutionary in its departures from classical physics. Whitehead never refers to the new quantum theory,[2] and it would be unreasonable to expect that even so imaginative a philosopher and scientist could have anticipated it except in the most general terms. Nevertheless, it is important in evaluating the philosophy of organism to determine how well its physical implications agree with quantum theory and with contemporary microphysical theory in general. To do this is the primary purpose of the present essay. It will appear that

This work originally appeared in Max Black (ed.), *Philosophy in America,* London: George Allen & Unwin, 1965. Reprinted by permission of the publisher.

1. The abbreviations of the titles of Whitehead's works and the editions to which the page numbers refer are as follows:

    SMW – Science and the Modern World (New York: Macmillan 1925).

    PR – Process and Reality (New York: Macmillan 1929).

    AI – Adventures of Ideas (New York: Macmillan 1933).

2. Hereafter the qualification 'new' will be omitted.

the agreement is only partial and that there are several crucial discrepancies both in detailed predictions and in general spirit.

The second purpose of this essay is to suggest the possibility of a modified philosophy of organism, which would preserve Whitehead's essential ideas while according with the discoveries of modern physics. This is a very ambitious programme, and only a few tentative speculations on how one might proceed will be presented. It is, furthermore, a somewhat hazardous programme, for one can cite many examples in the history of thought of philosophical schemes which attempted to conform to current science but succeeded only in amplifying scientific error. One can but hope that the comparison of quantum theory with a philosophical system of such great scope as Whitehead's will not only improve the philosophical system, but will throw light upon some of the conceptual difficulties in modern physics.

II. IMPLICATIONS OF THE PHILOSOPHY OF ORGANISM FOR MICROPHYSICS

In *SMW* Whitehead criticizes classical physics for supposing that there exist material entities without any intrinsic mental characteristics, and for supposing that these entities are simply located, i.e. '*here* in space and *here* in time, or *here* in space–time, in a perfectly definite sense which does not require for its explanation any reference to other regions of space–time' (*SMW* 72). Both of these suppositions, he claims, are instances of the 'fallacy of misplaced concreteness', which consists in regarding abstract characteristics of things as their complete and concrete natures. The philosophy of organism, which is an attempt to avoid misplaced concreteness, requires a radically different conception of physical reality from that of classical materialism. In this Section some propositions of 'Whiteheadian physics' (omitting relativity theory) will be listed, partly on the basis of Whitehead's explicit remarks and partly by inference. For this purpose, the following theses of the philosophy of organism are relevant.[3]

(1) The ultimate concrete objects in the universe[4] are the 'actual occasions', each of which has proto-mental characteristics and can be characterized as a unit of experience. The word 'experience' is obscure, and various of Whitehead's statements have the effect of postulating that the

3. Except for thesis (5), for which a special reference is given, these theses are selected from the 'categories of explanation' and 'categoreal obligations' (*PR* 33–41) with much condensation, rearrangement, and simplification.

4. The one exception to this statement is the actual entity God, who shares some but not all of the characteristics of the actual occasions.

word may be applied to entities usually considered to be inorganic, without entirely vitiating its ordinary intension.

(2) Every actual occasion is distinguishable from every other in virtue of its intrinsic character and not merely because of its external relations to the rest of the world.

(3) An actual occasion 'prehends' each occasion antecedent to it, i.e. it recognizes the experience of the antecedent occasions in qualitative detail, though with loss of immediacy and shift of emphasis. It is the relation of prehension which prevents actual occasions from being simply located (cf. *PR* 208).

(4) The temporal duration of an actual occasion is finite, and even though phases of the becoming of an occasion can be distinguished, each phase is only derivatively real and is incomplete without reference to the entire occasion.

(5) Each actual occasion occupies a definite spatial region and is indefinitely divisible, but the parts have only derivative reality relative to the whole occasion (*PR* 434–35).

(6) The total set of prehensions of antecedent occasions by an occasion in process of becoming does not suffice to determine that process in all its details. There is, thus, an element of freedom in the process, negligible in low-order occasions but permitting radical novelty in those of higher order.

The foregoing fragment of Whitehead's philosophy not only contains implicitly most of his views on physics, but also contains his explanation of the existence of the subject-matter for special sciences such as physics. Most occasions are almost entirely constituted by their prehensions of antecedent occasions, since spontaneity and originality are usually negligible. As a result, the world, or at least that part of it contained in our 'cosmic epoch', is populated largely by 'enduring objects'. An enduring object is a temporally ordered chain of actual occasions, all sharing a common defining characteristic and sharing it because it is the dominant element in the prehensions of each successive occasion in the chain. Thus the enduring object is, in a sense, self-sustaining. Even in a heterogeneous society of occasions there may be characteristics common to all or nearly all members, so that the prehensions of new occasions will be virtually uniform in certain respects and the common characteristics will tend to persist. Conformity to a basic law of physics is an outstanding example of a persistent set of characteristics in a heterogeneous society. Special laws, such as those of biology, may hold in sub-societies of the vast society governed by the basic physical laws. The special laws may not be derivable from the more basic laws, and yet they may presuppose the order established by the latter. Similarly, the basic laws governing a heterogeneous society determine its tolerance for various kinds of simple enduring objects, with the result that the number of species of elementary particles may be small, although the number of exemplars of each may be enormous. (Cf. *AI* 257 and 264, *PR* 138–40.)

It follows from Whitehead's view of the nature of physical laws that they cannot be derived from his philosophical first principles. In fact, since he supposes that the propagation of dominant characteristics in a society is subject to lapses, he infers that the type of order expressed by physical laws may decay or change. Nevertheless, Whitehead's philosophical principles do seem to have at least the following physical consequences, which would presumably be valid in any cosmic epoch.

(i) Even though physics abstracts from the detailed content of an actual occasion, it cannot overlook the spatial and temporal extendedness of occasions. *Hence, the most direct microphysical consequence of Whitehead's scheme is atomicity.* Thus, the physical fact that energy is transferred in quanta follows from the supposition that 'physical energy . . . must then be conceived as an abstraction from the complex energy, emotional and purposeful, inherent in the subjective form of the final synthesis in which each occasion completes itself' (*AI* 239), in conjunction with the thesis that the actual occasion is an extended atom of experience. Furthermore, there must be a temporal atomicity in physical processes, since the individual actual occasion is not divisible into concrete events, one of which is earlier than the other (*PR* 107). Whitehead seems to believe that from the fundamental atomicity of actual occasions there follow other types of physical atomicity – e.g. the integral character of elementary particles and the indivisibility of electric charge into units smaller than the charge of the electron – but he is vague on this point.[5] One would also expect him to claim that the elementary particles of physics exhibit spatial extendedness in spite of their integral character, since each occasion constituting a link in the career of a particle has an internal spatial structure. Whitehead does not make such a claim explicitly, but perhaps it is implicit in his discussion of 'vibratory organic deformations' of a proton (*SMW* 195).

(ii) *Elementary particles should be capable of creation and destruction.* According to Whitehead an elementary particle is an especially simple kind of enduring object, and the continuation of any enduring object depends upon the degree to which each new occasion in the appropriate neighbourhood will re-enact the experience of earlier occasions. Since the experience of a new occasion is partially colored by prehensions of other occasions than those constituting the enduring object, as well as by

5. He is also careless, for example in saying that 'Electrons and protons and *photons* are unit charges of electricity' (*AI* 238, italics not in the original text).

exercise of its intrinsic spontaneity, the elementary particle will almost certainly end after a finite number of links. An inverse argument explains how an elementary particle can be initiated.

(iii) *A consequence of atomicity, according to Whitehead, is the association of some sort of vibratory motion with all elementary particles.* An elementary particle is a chain of occasions, all having nearly the same internal development, so that the particle has a definite periodic structure with a definite frequency. It is clear, however, that Whitehead conceives of waves as more generic than particles, for particles always exhibit some characteristics of waves but not conversely. Thus:

the doctrine, here explained, conciliates Newton's corpuscular theory of light with the wave theory. For both a corpuscle, and an advancing element of wave front, are merely a permanent form propagated from atomic creature to atomic creature. A corpuscle is in fact an 'enduring object'. The notion of an 'enduring object' is, however, capable of more or less completeness of realization. Thus, in different stages of its career, a wave of light may be more or less corpuscular. A train of such waves at all stages of its career involves social order; but in the earlier stages this social order takes the more special form of loosely related strands of personal order. This dominant personal order gradually vanishes as the time advances. Its defining characteristics become less and less important, as their various features peter out. The waves then become a nexus with important social order, but with no strands of personal order. Thus the train of waves starts as a corpuscular society, and ends as a society which is not corpuscular. (*PR* 53–54)

(iv) *As a result of thesis* (6) *strict determinism cannot hold in physics.* However exhaustively the antecedents of a physical event are specified, the character of the event cannot, in principle, be predicted with certainty.

(v) *The specification of the state of a composite system containing several elementary particles is equivalent to the specification of the states of the individual particles.* This follows from the natural identification of the state of an elementary particle at a given time as an actual occasion, together with Whitehead's theses that actual occasions are pre-eminently real and that all groupings of occasions have derivative status.

Several further physical consequences of Whitehead's philosophy appear plausible, except for doubts regarding the extent to which physics abstracts from concreteness.

(vi) If concrete reality is considered, there is clear asymmetry between past and future, since the occasions of the past are fully determinate while those of the future are not. *It is reasonable, consequently, to expect the asymmetry between the past and the future to be exhibited in the laws of microphysics,* and not merely in macroscopic laws such as those of thermodynamics.

(vii) An occasion prehends, with suitable gradations of relevance, all previous actual occasions. *Consequently, the physical properties of an elementary particle (e.g. its charge or its magnetic moment) should be slightly modified by the inclusion of the particle in a highly organized society such as an animal body.*

(viii) If the occasions of an enduring object are considered in full concreteness, they will exhibit 'aging' from earlier to later parts of the chain, simply in virtue of the accumulation of prehensions. *Consequently, one expects a systematic development of physically observable characteristics of elementary particles* – perhaps a drift towards instability, indicative of primitive feelings of satiation, or perhaps a drift towards greater stability, indicative of the entrenchment of a habit.

III. COMPARISONS WITH CURRENT PHYSICS

Propositions (i)–(viii) will now be examined in the light of current physics. Whenever possible, these propositions will be confronted with direct experimental evidence. In most cases, however, one can do no more than compare them with their counterparts in quantum theory, so that experimental evidence can be invoked only to the extent that quantum theory as a whole is experimentally confirmed.[6]

(i) Part of Whitehead's conception of atomicity is in excellent agreement with current physics. Specifically, energy is transferred in quanta, and matter has a granular structure in the small which prevents indefinite divisibility. There is also evidence, although it is not decisive, that the integral character of an elementary particle is compatible with spatial extendedness and internal structure. Thus, experiments in which protons are scattered by protons indicate that these particles have a definite charge and current distribution, even though it is not possible to subdivide their charge into discrete parts.[7] Furthermore, it is difficult to envisage how

6. Individual references will not be given for each of the propositions of quantum theory mentioned in the following discussion. Although these propositions are explained in every standard exposition of quantum theory, several books are particularly worth noting for their treatment of topics that are philosophically significant: D. Bohm, *Quantum Theory* (New York 1951), J. von Neumann, *Mathematical Foundations of Quantum Mechanics* (Princeton 1955), F. London and E. Bauer, *La Théorie de l'Observation en Mécanique Quantique* (Paris 1939), P. A. M. Dirac, *The Principles of Quantum Mechanics,* 3rd edition (Oxford 1947).

7. Cf. R. Hofstadter, F. Bumiller, and M. R. Yearian, *Reviews of Modern Physics,* vol. 30 (1958), p. 482. The standard model for interpreting these experimental results pictures the observed proton as a 'cloud' of virtual particles, fluctuating in and out of existence. Although the cloud exhibits a statistically describable structure, the virtual particles are supposed to be 'bare' and perhaps without structure. Cf. E. M. Henley and W. Thirring, *Elementary Quantum Field Theory* (New York 1962), pp. 77–78, 219–31. This model is so different from Whitehead's conception of a spatially extended actual occasion that one must hesitate to say that the scattering experiments really support his conception.

a particle which is localized at a mathematical point could exhibit spin angular momentum; but one should be cautious in advancing this argument, since in actual calculations theoretical physicists are able to treat the spin simply as a 'quantum number' without any commitments to the extendedness or non-extendedness of the particles. With regard to temporal atomicity, which is the most radical of Whitehead's assertions concerning atomism, the testimony of current physics is unfavourable, but not decisively so. One of Heisenberg's uncertainty relations is $\Delta E \Delta t \sim h$, i.e. the uncertainty of the duration of a physical process can be reduced only at the price of increasing the uncertainty of the energy of the system during the process. The limitation which Heisenberg's relation places upon the sharpness of temporal specification is reminiscent of Whitehead's proposition, yet it is not the same. Whitehead is not attributing an *indeterminate* stretch of time to the actual occasion, but rather a *determinate finite* stretch; he denies temporal definiteness only to the phases within the occasion, for 'this genetic passage from phase to phase is not in physical time' (*PR* 434). Moreover, there is no hint in Whitehead's work that an occasion of short duration has a less definite energy than one of longer duration. On the contrary, he says that every actual occasion, when it has completed its process of becoming, is completely definite with respect to every family of attributes (*PR* 38, Category of Explanation XXV). It should be noted, incidentally, that although the duration of each actual occasion is indivisible, Whitehead does not assume a lower limit to the set of all durations (*SMW* 198). Nevertheless, his proposal of temporal atomicity would be supported if the postulation by March, Darling, *et al.* of a minimum length or minimum space–time region proved successful in removing the troublesome 'divergences' of quantum field theory.[8] So far their postulates have not led to outstanding successes, but in view of the great mathematical difficulties and the large number of variants to be examined their failure is not decisive.

(ii) The Whiteheadian proposition that elementary particles can be created and destroyed has been strikingly confirmed by experiments, with regard to the 'stable' as well as to the 'unstable' particles. For example, physicists have not only observed ordinary beta-decay, in which the unstable neutron decays into three stable particles, the proton, the electron, and the anti-neutrino, but they have also detected inverse beta-decay, in

---

8. B. T. Darling, *Physical Review,* vol. 80 (1950), p. 460, A. March, *Quantum Mechanics of Particles and Wave Fields* (New York 1951). Their postulates are motivated by the desire to eliminate the 'divergences' of quantum field theory. Because of the quantum-mechanical proportionality of energy to frequency and the fact that a minimum length would imply a maximum frequency, their postulates would cut off the high range of frequencies responsible for the theoretically computed infinite energies.

which a proton and an anti-neutrino are annihilated and a neutron and positron are created. Thus, the integral character of elementary particles is not associated with permanence, contrary to Democritean atomism but in accordance with Whiteheadian physics.

(iii) Whitehead's account of the association of waves with particles is only superficially in accord with quantum theory. The passage from pp. 53–4 of *PR*, quoted in (iii) of the preceding Section, shows that he conceives of a wave front as a set of actual occasions (a 'nexus') with each occasion occupying a small region of the front. The mutual relations among contiguous occasions, whereby they begin and end in unison, ensure that all parts of the wave front are in phase with each other. Whitehead's picture is reminiscent of Schrödinger's attempt to interpret the wave function of a particle as a description of an ordinary physical field, which manifests particle-like properties whenever the region of high field intensity is very small. This interpretation was abandoned because Schrödinger's own equation for the time dependence of the wave function implies that in the absence of confining forces an initially concentrated wave packet will disperse.[9] In order to account for the experimental fact that particles do not disperse, Born proposed that the physical content of the wave function was to determine the probabilities of experimental outcomes (e.g., $|\psi(x, y, z)|^2\, dx\, dy\, dz$ is the probability of finding the particle in a small volume $dx\, dy\, dz$). In the Born interpretation, the wave function characterizes the state of the particle in its entirety and does not describe the physically real parts of a field. There is a sharp discrepancy between this point of view, which at present is accepted by most physicists, and Whitehead's attribution of primary reality to the occasions of a wave front.

Deeper insight into this discrepancy can be achieved by considering the following fundamental principle of quantum theory, called 'the superposition principle': if $u_1, u_2, u_3, \ldots$ represent physically possible states, then the combination $\sum c_i u_i$, where the c_i are arbitrary complex numbers, represents a superposition of the original states which 'overlaps' each of them in a certain sense and which, moreover, is itself a physically possible state. The nature of the 'overlap' is mostly clearly exhibited if there is an observable property A of the system having definite values a_i in the states represented by the u_i, such that all the a_i are different. If the u_i are then specified as being of a standard length ('normalized' to 1), then according to quantum theory the probability of finding that $A = a_i$ when a measurement of A is performed is $|c_i|^2/|c_1|^2 + |c_2|^2 + \cdots$. It must be insisted that

9. Cf. M. Born, *Atomic Physics* (Fifth edition, New York 1951), p. 93 and pp. 142–44, for discussion of reasons for abandoning Schrödinger's interpretation in favour of Born's and for a general discussion of attempts to rationalize wave–particle dualism.

$\psi = \sum c_i u_i$ is a maximum specification of the system; it is not a statistical description of a system which really is in a state represented by a definite, but unknown, one of the u_i. The superposition therefore has the counter-intuitive characteristic of being a state in which the observable A is objectively indefinite and not merely unknown. It follows that the uncertainty principle of Heisenberg, which limits the simultaneous determination of complementary quantities such as position and momentum, refers to the objective properties of the particle and not simply to human knowledge about these properties. The wave function $\psi(x, y, z)$ can now be understood in terms of the superposition principle: roughly, $\psi(x, y, z)$ represents a superposition of states in each of which the particle has a position localized within a small region $dx\,dy\,dz$, with the numbers $\psi(x, y, z)\,dx\,dy\,dz$ serving as the expansion coefficients in the superposition (i.e., they are the c_i in $\sum c_i u_i$). It should now be obvious that the quantum-mechanical account of the waves associated with particles is entirely alien to the spirit of Whitehead's philosophy. In particular, the postulation of indefinite values of observable quantities, as required by the quantum-mechanical account, would be repugnant to Whitehead. He might admit that indefiniteness is characteristic of the spatial or temporal parts of actual occasions, but surely not of the complete occasions, for the twenty-fifth Category of Explanation, on p. 38 of *PR*, asserts:

The final phase in the process of concrescence, constituting an actual entity, is one complex, fully determinate feeling. . . . It is fully determinate (*a*) as to its genesis, (*b*) as to its objective character for the transcendent creativity, and (*c*) as to its prehension – positive or negative – of every item in its universe.

An important technical disagreement follows from the fundamental conceptual difference between Whitehead's analysis of wave–particle dualism and that of quantum theory. According to Whitehead, a sharp frequency may be characteristic of a well-localized particle, since the periodicity is essentially due to the reiteration of a basic pattern in successive actual occasions. In quantum theory, on the other hand, a sharp frequency is characteristic of a particle which has a definite momentum, and therefore, by the uncertainty principle, a completely indefinite position; and conversely, the wave function of a well-localized particle has a very broad spectrum of frequencies.

A final note on Whitehead's account of waves and particles is to point out a striking disconfirmation by experiment. He explicitly states, in the passage quoted above from p. 53 of *PR*, that the corpuscular character of light is gradually lost as the light is propagated – a reasonable remark given his general analysis of particles. However, the phenomenon which most vividly illustrates the particle aspect of electromagnetic radiation, the photo-electric effect, is entirely independent of the age of the radiation. In fact, photo-electric cells are attached to telescopes in order to study starlight which has travelled for millions of years.

(iv) Quantum theory and Whiteheadian physics are both indeterministic, but in quite different ways. According to quantum theory, the state of a physical system evolves in a continuous and fully deterministic manner as long as the system is isolated. Probability enters only when a measurement is made of an observable which does not have a sharp value in the state of the system prior to the measurement. The interruption of the deterministic evolution of the state does not contradict the Schrödinger equation, since a measurement requires the interaction of the system with another system – a macroscopic piece of apparatus and perhaps a conscious observer. According to Whitehead, the evolution of an isolated system (the concept of which is an idealization, since an actual occasion prehends all occasions in its past) cannot be entirely deterministic, because of the element of freedom in each occasion. If one attempts to adjudicate between these two different accounts of indeterminism by considering the success of statistical predictions, one must certainly prefer quantum theory, for its statistical predictions are remarkably good, while Whitehead's proposals are too vague to yield any quantitative statistics. Nevertheless, there may be strong reasons for exploring the hypothesis that chance events occur in isolated systems, since, as will be discussed in Section IV, the quantum-theoretical account of indeterministic transitions leads to conceptual difficulties.

(v) The Whiteheadian treatment of the state of a composite system is at odds with a quantum-mechanical principle which has attracted little attention in spite of its revolutionary philosophical implications: *that a several-particle system may be in a definite state, i.e. may have as definite properties as quantum theory permits, without the individual particles being in definite states.* To illustrate this principle consider two systems I and II and let $\phi_1(I)$ and $\phi_2(I)$ represent states of I in which observable A has values a_1 and a_2 respectively $(a_1 \neq a_2)$, and $\psi_1(II)$ and $\psi_2(II)$ represent states of II in which observable B has values b_1 and b_2 respectively $(b_1 \neq b_2)$. Let $\phi_1(I) \otimes \psi_1(II)$ represent that state of the composite system I plus II in which system I is described by $\phi_1(I)$ and system II by $\psi_1(II)$; and let $\phi_2(I) \otimes \psi_2(II)$ be similar. The superposition principle can now be invoked to affirm the physical possibility of a remarkable state, namely, the one represented by

$$\Psi = \frac{1}{\sqrt{2}}(\phi_1(I) \otimes \psi_1(II)) + \frac{1}{\sqrt{2}}(\phi_2(I) \otimes \psi_2(II)).$$

Neither I nor II is in a definite state when Ψ represents the state of I plus II. A rigorous proof will be omitted, but the following rough argument indicates the essential reason: if one tries to claim that I is to some extent described by ϕ_1, and to some extent by ϕ_2, the claim is vague unless the expansion coefficients of ϕ_1 and ϕ_2 are specified, but in the above expression for Ψ it is clear that the expansion coefficients refer to states of II. One might try to reconcile the existence of such states of several-particle systems with Whitehead's consequence (v) by identifying the state of a several-particle system with a single actual occasion, and identifying the individual particles at a given moment with subdivisions of the occasion. Divisions of an occasion could be reasonably expected to lack the specificity implied in saying that an entity is in a definite state, for 'in dividing the region we are ignoring the subjective unity which is inconsistent with such divisions' (*PR* 435). This attempt at reconciliation fails, however, because quantum mechanics permits the parts of a system described by Ψ to be indefinitely far apart spatially.[10] An actual occasion which not only is macroscopically extended, but even broken into non-contiguous parts, is evidently contrary to Whitehead. To be sure, he is reticent about the exact extent of an occasion, but he seems to fear that the identification of a large scale process as a single occasion will remove the barriers to Spinozistic monism (cf. *PR* 10–11).

It is appropriate at this point to interpolate a discussion of another quantum-mechanical principle concerning several-particle systems, even though it has no clear counterpart in Whiteheadian physics: *that if a system contains several elementary particles of the same species, they must all play the same role in the system as a whole.* Consider, for example, a system composed of two pi-mesons, and suppose ϕ_1 and ϕ_2 each represents a possible state of a single pi-meson. Then the principle forbids the composite system to be in the state represented by $\phi_1 \otimes \phi_2$, for in this state pi-mesons I and II have different roles; but the principle permits the state represented by

$$\frac{1}{\sqrt{2}}(\phi_1(\mathrm{I}) \otimes \phi_2(\mathrm{II})) + \frac{1}{\sqrt{2}}(\phi_2(\mathrm{I}) \otimes \phi_1(\mathrm{II})),$$

for this state is 'symmetrized' with respect to the two particles.[11]

To fit symmetrization into the framework of Whiteheadian physics is a delicate matter, since it implies a kind of loss of identity of the individual particles. In particular, the simple model of particle I as a chain of

---

10. Cf. D. Bohm and Y. Aharonov, *Physical Review*, vol. 108 (1957), p. 1070.
11. Symmetrization is actually a property of states of systems composed of particles with 'integral spin'. When the particles have 'half-integer spin', as do electrons, neutrinos, neutrons, and protons, the state of the composite system must be 'anti-symmetrized'. However, the difference between symmetrization and anti-symmetrization is irrelevant for the present purpose, which is to insist upon the identity of roles of all particles of the same species.

occasions $O_1, O_1', O_1'', \ldots$ and of particle II as a chain $O_2, O_2', O_2'', \ldots$ will not work, because individual occasions are always distinguishable in virtue of intrinsic characteristics (thesis 2 of Sect. II) and hence the two chains are distinguishable. A possible reply is that physics does not treat actual occasions *in concreto* and hence can fail to take cognizance of the respect in which the chains $O_1, O_1', O_1'', \ldots$ and $O_2, O_2', O_2'', \ldots$ are different; but the symmetrization obtained in this manner would appear to be a coincidence rather than a general law.

A more promising explanation is that the chains $O_1, O_1', O_1'', \ldots$ and $O_2, O_2', O_2'', \ldots$ intersect and the occasions which are shared by several chains have a quite different character – the physical manifestation of which is symmetrization – from the character of the occasions prior to the merger of the chains or after their separation. This explanation conforms to Whitehead's general view that the corpuscular nature of a variety of occasions is highly special and easily dissipated. However, the explanation encounters the same difficulty that was noticed in the preceding paragraph. Composite systems containing particles of the same species can have indefinitely large spatial extent, as in the case of the conduction electrons in a bar of metal. Since all the particles play the same role, it becomes necessary to identify the state of a macroscopically extended system with a single actual occasion, contrary to Whitehead's conception of the actual occasion as a microscopic entity.

(vi) In contemporary physics the only laws which involve a definite direction of time are macroscopic, the outstanding one being the thermodynamic law of entropy increase. No microphysical law has yet been discovered which is not invariant under reversal of the direction of time. If this peculiarity remains a permanent feature of physics, it would constitute evidence detrimental to Whitehead's scheme, in which the asymmetry of past and future is essential. The only defence of a Whiteheadian physics would then be the desperate one that the asymmetry of past and future is one of the features of concrete reality neglected by physics. However, since the discovery of parity non-conservation by Yang and Lee *et al.*, the confidence of physicists in some of the physical symmetry principles has been shaken, and many suspect that a violation of time-reversal invariance will also be detected.[12]

(vii) The Whiteheadian expectation that the physical properties of an elementary particle are slightly modified when the particle enters a structural society is counter to the reductionist spirit of physics, chemistry, and biophysics. Many of the predictions of these sciences rest upon the assumption that such properties of elementary particles as charge, mass,

12. Cf. P. Morrison, *American Journal of Physics,* vol. 26 (1958), p. 358.

and magnetic moment are unchanged by the incorporation of the particles into highly structured macroscopic objects. Of course, the same predictions would be made if the changes due to incorporation are extremely small. Since Whitehead gives no indication of the amount of modification to be expected there can be no crucial experiment.

(viii) The 'aging' of elementary particles implied by Whitehead's philosophy is contrary to current physical theory and is not confirmed by experiment. The intrinsic properties of newly created elementary particles seem to be no different, at least statistically, from those of particles which have endured a long while. Particularly significant is the decay rate in a population of unstable particles. For example, the number of neutrons decaying per unit time is proportional to the number of neutrons in the population (with allowance, of course, for statistical fluctuations), and is independent of the 'age' of the population. It is reasonable to infer that the probability that a given neutron will decay during an interval of time is independent of the age of the neutron, thus suggesting that no physically significant changes occur in the neutron due to aging. Again, however, Whitehead's statements permit no quantitative estimate of the change to be anticipated in the decay rate, and therefore no crucial experiment is possible.

IV. PROPOSALS FOR RECONSTRUCTION

The discrepancies noted in Sect. III between Whiteheadian physics and current microphysics constitute strong disconfirmation of Whitehead's philosophy as a whole. The possibility remains, however, of constructing a philosophical system, Whiteheadian in its general conceptions though not in details, and according with the fundamental discoveries of science. A few tentative suggestions will be given here concerning the initiation of such a large philosophical undertaking.

A useful first step is to distinguish both in the philosophy of organism and in quantum theory those elements which are radical by the standards of classical physics from those which are conservative. Most radical in Whitehead's philosophy are the attribution of proto-mental properties to entities normally considered to be physical, and the postulation of prehension as the fundamental relation between occasions. The assumption of the complete definiteness of the occasion in its final stage is conservative, although the correlative assumption of indefiniteness in the early stages of concrescence is not. Also conservative is his reductionist assumption that the characteristics of a nexus are entirely determined by the characteristics of its constituent occasions. Quantum theory is conservative in supposing that certain quantities initially introduced in the study of macroscopic physical objects – especially spatio-temporal position, energy, momentum, angular momentum, charge, and magnetic moment – can be used meaningfully in characterizing microscopic entities. On the other hand, the superposition principle is radical, for it has the consequence that a physical quantity can have an indefinite value in a maximally specific state of a microscopic system and has a sharp value only in exceptional states (the 'eigenstates' of that quantity). Quantum theory is also radical in its treatment of the relation between the state of a several-particle system and the states of its constituent particles.

The foregoing juxtaposition suggests a programme of reconstruction: *to graft the radical elements of quantum theory onto the radical elements of the philosophy of organism, by assuming that elementary entities have proto-mental characteristics while treating the states of these entities in accordance with the combinatory principles of quantum theory.* The synthesis contemplated here does not seem forced from a Whiteheadian point of view. Whitehead often engages in a dialectical analysis which is reminiscent of the quantum-mechanical treatment of complementary quantities,[13] but he never achieves what is most remarkable in quantum theory – a set of systematic rules for predicting statistically what will appear when a shift is made from one description to a complementary one. A modification of Whitehead's philosophy in accordance with the combinatory principles of quantum theory would perhaps make explicit and precise certain tendencies that are implicit and haphazard in his work.

Such a modification of Whitehead's system would surely change the conception of an actual occasion. For example, it would be impossible to maintain that an actual occasion, in its final phase, is definite with respect to every family of attributes (*PR* 38, Category of Explanation XXV). Instead, in accordance with the uncertainty principle, the specificity of any attribute is always attained at the price of indefiniteness of other attributes. In particular, an occasion may have a quite sharp location in time and an arbitrarily short duration, provided that properties complementary to duration are sufficiently indefinite.[14] It is also possible that the actual occasions may lose their status of being (along with God) the only

13. This is most striking in Whitehead's theology: 'It is as true to say that God is permanent and the World fluent, as that the World is permanent and God is fluent. . . .' (*PR* 528). Whitehead explains that 'In each antithesis there is a shift of meaning which converts the opposition into a contrast' (*ibid.*). Dialectical analysis is also exhibited in the more mundane parts of his philosophy, for example in his statements that an actual occasion is prehended in its concreteness and yet with loss of immediacy.
14. According to the Heisenberg relation $\Delta E \Delta t \sim h$, the property complementary to duration is energy. But if Whitehead's thesis is maintained that physical energy is an abstraction from emotional and purposive energy (*AI* 239), then Heisenberg's relation may require supplementation.

ultimate real entities, and they may appear instead only as special cases of ultimate reality. Elementary particle theory and quantum electrodynamics may provide a hint as to the more general form of ultimate reality: i.e., it might be some kind of 'field' of diffused primitive feeling, of which the actual occasions are 'quanta' existing whenever there are individual loci of feeling. The hypothesis of diffused feelings is no more of an extrapolation from psychological data than is Whitehead's attribution of proto-mental characteristics to elementary particles, and indeed our everyday experience of sensitivity pervading the whole human body may possibly be construed as confirmation of the hypothesis.

The physical evidence concerning composite systems, discussed in (v) of Sect. III, suggests a quantum-theoretical refinement of Whitehead's treatment of the relation between the nexus and its constituent occasions. Whitehead conceives of the nexus in a reductionist manner, as the totality of its constituent occasions, and he supposes the internal relations exhibited in a nexus to be completely explicable in terms of the prehensions of earlier occasions of the nexus by each new occasion. Quantum theory, on the other hand, treats a composite system in a subtle manner, which at first seems paradoxical: it allows the state of the composite system to be described, in a certain sense, in terms of its components, and yet it permits the composite system to be in a definite state even when its components are not. This treatment, of course, is intimately bound up with the superposition principle. Thus, in the example cited in the previous Section, the composite system I plus II is in the state represented by

$$\Psi = \frac{1}{\sqrt{2}}(\phi_1(\text{I}) \otimes \psi_1(\text{II})) + \frac{1}{\sqrt{2}}(\phi_2(\text{I}) \otimes \psi_2(\text{II})),$$

which is clearly describable in terms of the states ϕ_1 and ϕ_2 of component I and ψ_1 and ψ_2 of component II; yet, because of the character of the superposition, neither I nor II is in a definite state. If this quantum-theoretical treatment of the whole–part relationship is introduced into the philosophy of organism, it opens a number of possible lines of exploration. For example, the 'field of feeling', the existence of which was hypothesized in the preceding paragraph, might be characterized holistically as being in a definite state. This state could always be described as a superposition of field states, in each of which there is a definite set of actual occasions, just as in quantum electrodynamics the state of the entire electromagnetic field can be described as a superposition of field states in each of which there is a definite set of photons. Moreover, special states of the 'field of feeling' can exist in which the superposition is in part reduced, so that definite actual occasions exist. Consequently, the existence of many independent loci of feeling, which is an essential aspect of the experienced plurality of the world, is permitted – though not required – in this modified Whiteheadian scheme. A nexus of the type described by Whitehead is also permitted but is a rather special case: it occurs when there is a network of occasions sharing some common characteristic, each in a definite single-quantum state. What the modified scheme permits which Whitehead's does not is the existence of a composite system more complex than a nexus: an '*n*-quanta system' in which each of the *n* occasions is so correlated with the others, *via* the superposition principle, that none is in a definite single-quantum state.

An evident advantage of the quantum-theoretical modification of the philosophy of organism is that it removes some of the discrepancies with modern physics noted in Sect. III – a virtue which is not surprising, since the modification was inspired by these discrepancies. A further advantage is a possible improvement in treating the question of 'simple location'. Whitehead's rejection of simple location in *SMW* is a dramatic criticism of classical physics, but his sketch of an alternative in *PR* is somewhat disappointing. His alternative is essentially to postulate the relation of prehension, whereby an actual occasion is felt in complete detail in the initial phase of each later occasion. Even if the ambiguities inherent in the conception of prehension can be dispelled, the relation of prehension can at best provide a kind of multiple location in time, i.e. the occasion as 'subject' and the same occasion as 'ingredient' in later subjects. By contrast, the quantum-theoretical modification exhibits a breakdown in simple location in space in two respects: first, the quantum state of an individual occasion may be such that its position is indefinite; and, secondly, a composite system can have spatially separated components which are not in definite states, but which are so correlated with each other that the composite system is in a definite state. Finally, there is an advantage which was briefly mentioned earlier but which deserves amplification: the possibility that the modification of Whitehead's philosophy will greatly improve the account of high-order mental phenomena – which, after all, are the only mental phenomena we know about without resorting to radical hypotheses and extrapolations. Because Whitehead conceives actual occasions to be microscopic in size, and because the human personality at any moment has a unity which entitles it to the status of an actual occasion, he is led to the strange doctrine of a microscopic locus of high-order experience wandering through the society of occasions that compose the brain:

Thus in an animal body the presiding occasion, if there be one, is the final node, or intersection, of a complex structure of many enduring objects. . . . There is also an enduring object formed by the inheritance from presiding occasion to presiding occasion. This endurance of the mind is only one more example of the general principle on which the body is constructed. This route of presiding occasions probably wanders from part to part of the brain, dissociated from the physical material atoms. (*PR* 166–67)

The question of the location of mentality is extremely complicated, but introspection seems to indicate that it is diffused throughout the body, and neuro-physiology has not yielded evidence of extreme localization. Various of the concepts of the modified Whiteheadian scheme seem relevant in describing high-order mentality: the concept of an indefinitely located actual occasion, the concept of a field of feeling which is generally diffused but occasionally quantized, and the concept of quantum correlations among the components of a composite system. Which one of these, or which combination of them, will be most fruitful is a matter of speculation, but all of them seem preferable to Whitehead's microscopic localization of high-order mentality.[15]

This list of advantages must be weighed against some strong reservations. The first concerns a particular proposal made above rather than the general programme of a quantum-theoretical modification of the philosophy of organism. The proposal was to consider the entire 'field of feeling' as being in a definite state and to consider actual occasions as quanta of this field. One may wonder, in view of the notorious difficulties of quantum field theories, whether they are suitable models for a metaphysical scheme describing all of reality, and one may wonder whether a single field suffices for this purpose. Setting aside these doubts, however, one may still be sceptical about the adequacy of this proposal to account for the experienced plurality of the world. It was noted that the existence of a definite state of the entire field is permissive of individual loci of feeling, but is permissiveness sufficient? Is it a mere contingency, characteristic perhaps of our cosmic epoch, that there are definite actual occasions, or are there deep-lying reasons why this should always be so? Merely raising these questions exhibits the vagueness of the proposal and its need for supplementation.

A second reservation arises from the apparent absence of any clear psychological manifestations of the superposition principle, which one would expect if the combinatory principles of quantum theory apply to actual occasions. There are, to be sure, psychological phenomena which at first sight could be construed as evidence of the superposing of mental states – e.g., perceptual vagueness, emotional ambiguity, conflict of loyalties, and the symbolism of dreams. Yet in all these cases the quanti-

---

15. Some speculations on the application of quantum theoretical concepts to mentality are given in D. Bohm, *Quantum Theory* (Englewood Cliffs, New Jersey 1951), pp. 168–72; and in his *Wholeness and the Implicate Order* (London 1980), pp. 192–213.

tative characteristics of the superposition principle, as it is exhibited in physics, are missing.[16] This reservation, however, is not decisive, for one can optimistically reply that the absence of confirming evidence is not equivalent to the presence of disconfirming evidence. It is possible that the psychological manifestations of the superposition principle are too delicate to be detected by the introspective, behaviouristic, and physiological techniques in current use.[17] It is also possible that these techniques are sufficient, but that no one has yet been sufficiently serious about using the superposition principle in psychology to design a good experiment.

A third reservation is that quantum theory, in spite of its striking successes in physics, is beset by a serious conceptual difficulty, which should perhaps be resolved before the theory is incorporated into a philosophical system. This conceptual difficulty concerns the 'reduction of a superposition', which occurs if an observable A, having sharp and distinct values in the states represented by u_i, is measured when the state is represented by $\sum c_i u_i$. As a result of the measurement there is a non-deterministic transition from the initial state to a final state represented by a definite one of the u_i. Although many textbooks and popular accounts say that the transition occurs when a microscopic system interacts with the macroscopic measuring apparatus, this explanation is inadequate, since the Schrödinger equation, which is the equation governing the evolution of the state, implies that the final state of the apparatus plus microscopic system will also be a superposition with respect to the observable A. Some theoreticians have concluded that the reduction of the superposition does not take place until the result of the measurement is registered in the consciousness of an observer. This desperate conclusion is unsatisfactory for several reasons.[18] It suffices to say here that if observation is a natural process, it is difficult to understand why a non-deterministic transition should occur when an observation is made, while all other natural processes are deterministic. One possible solution to the problem of the reduction of superpositions is to suppose that the evolution of the quantum state is to some extent stochastic and hence only approximately governed by the Schrödinger equation. The non-deterministic reduction of a superposition could then occur in a system remote from anything ordinarily called 'an observer'. In this way the superposition principle could

---

16. This point is discussed in A. Shimony, *American Journal of Physics*, vol. 31 (1963), p. 755.
17. Dr Karl Kornacker pointed out in a private communication that the interference effects characteristic of superpositions are no more to be expected in gross emotional and perceptual phenomena than in macroscopic physical phenomena.
18. E. Wigner, article in *The Scientist Speculates*, ed. I. J. Good (London 1962); also the paper by A. Shimony cited in note 16.

be maintained in microphysics, but at the price of changing the dynamics of quantum theory. The success of a solution along this line would remove the third reservation about a quantum-theoretical modification of Whitehead's philosophy, for the modification depends primarily upon extending the application of the superposition principle. The proposal of a chance element in the evolution of the state does not disrupt the programme envisaged here, for in fact this proposal is closer to Whitehead's version of indeterminism than is the indeterminism of current quantum theory. One could even speculate that Whitehead's account of the concrescence of an actual occasion provides some insight into the way in which the reduction of the superposition occurs. Such speculations, however, are rather empty until stochastic generalizations of the Schrödinger equation are proposed and their physical consequences are studied.

A methodological remark is appropriate in conclusion. It has been tacitly assumed throughout this paper that the hypothetico-deductive method is an appropriate instrument in philosophical inquiry. This assumption is in the spirit of Whitehead's philosophy, for he deliberately formulated a categoreal scheme which could be confirmed or disconfirmed only by examining its remote consequences (*PR* 7–8). Regrettably, the difficulties of employing the hypothetico-deductive method in philosophy are illustrated only too clearly in this paper. The conclusions which can be drawn from philosophical first principles are generally qualitative, and therefore their confrontation with experience lacks sharpness. When they are confirmed, it is gross rather than fine confirmation, and when they are disconfirmed there is often a plausible way of saving the appearances. The moral, however, is not that the hypothetico-deductive method should be abandoned. Rather, it is to seek refinements of philosophical first principles and liaisons of these principles with scientific hypotheses, in such a way that sharp predictions and fine confirmations may result.[19]

19. I am deeply indebted to Dr Howard Stein for his criticism of an early draft of this paper and, more important, for stimulating and suggestive conversations over many years concerning the topics which it treats. I am grateful for a careful reading of the manuscript by Prof. J. M. Burgers.

Added in proof: An experiment by J. W. Christenson *et al.*, described in *Physical Review Letters*, vol. 13 (1964), p. 138 (and reported in *The New York Times,* Aug. 5, 1964), indicates a violation of 'time-reversal invariance' (see paragraph (vi) in Sect. III).

20
Reflections on the philosophy of
*Bohr, Heisenberg, and Schrödinger**

Many of the pioneers of quantum mechanics – notably Planck, Einstein, Bohr, de Broglie, Heisenberg, Schrödinger, Born, Jordan, Landé, Wigner, and London – were seriously concerned with philosophical problems. In each case one can ask a question of psychological and historical interest: was it a philosophical penchant which drew the investigator towards a kind of physics research which is linked to philosophy, or was it rather that the conceptual difficulties of fundamental physics pulled him willy-nilly into the labyrinth of philosophy? I shall not undertake to discuss this question, but shall cite an opinion of Peter Bergmann, which I find congenial: he learned from Einstein that "the theoretical physicist is . . . a philosopher in workingman's clothes" ([1], q. v.).

The problems with which I am preoccupied concern the philosophical implications of quantum mechanics – either epistemological, bearing on the extent, validity, and character of human knowledge; or metaphysical, bearing on the character of reality. Although quantum mechanics is not a system of philosophy, one can wonder whether it is susceptible to coherent incorporation in a philosophical system. I propose to examine the thought of three masters of quantum mechanics – Bohr, Heisenberg, and Schrödinger – not with a critical or historical intention, but in hope of finding some enlightenment concerning the problems posed by contemporary physics. I can say in advance that enlightenment will continue to elude us; nevertheless, the ideas of Bohr, Heisenberg and Schrödinger are rich and evocative for new studies.

Certain general principles of Bohr's philosophy can be sketched without any reference to quantum mechanics, even though it was his efforts

This paper is a translation of 'Réflexions sur la philosophie de Bohr, Heisenberg et Schrödinger,' which was part of a symposium entitled *Les Implications Conceptuelles de la Physique Quantique,* published in *Journal de Physique* 42, Colloque C-2, supplément au no. 3 (1981), pp. 81–95. Permission for publishing this translation was kindly granted by Les Editions de Physique. This work appeared in R. Cohen and L. Laudan (eds.), *Physics, Philosophy, and Psychoanalysis,* Dordrecht. Copyright © 1983 by D. Reidel Publishing Company. Reprinted by permission of Kluwer Academic Publishers.
* This paper is dedicated to Adolf Grünbaum in honor of his lifetime of explorations of the interdependence of philosophy and the natural sciences.

to interpret the discoveries of the new physics that gave definitive form to his principles. Bohr always insists that scientific knowledge requires unequivocal description, a necessary condition for which is a distinction between the subject and the object ([2], p. 101). The success of our communication in everyday life concerning the positions and motions of macroscopic objects shows *a posteriori* that we can use these descriptions unequivocally, but one finds no assertion in the essays of Bohr that such concepts are *a priori*, like the categories of the understanding of Kant. In Bohr's opinion, the clarity of this macroscopic description does not at all imply that atomic objects are less existent than macroscopic objects ([2], p. 16). Rather, because of the indirectness of our knowledge of atomic objects and, even more, because of the quantum of action, an unequivocal description of an atomic phenomenon must "include a description of all the relevant elements of the experimental apparatus" ([2], p. 4).

From time to time Bohr indicates that his epistemological theses do not commit him to a metaphysics: for example, "the notion of an ultimate subject as well as conceptions like realism and idealism find no place in objective description as we have defined it" ([2], p. 79). In place of a metaphysics Bohr proposes a purely epistemological strategy – the mobility of the separation between the subject and the object ([2], pp. 91–92). I am very grateful to Aage Peterson, who was Bohr's assistant for seven years, for his testimony [3] – supplementary to the writings of Bohr but agreeing with them – concerning his renunciation of metaphysics. Bohr believed that even psychology must recognize this renunciation, because "in every communication containing a reference to ourselves we, so-to-speak, introduce a new subject which does not appear as part of the content of the communication" ([2], p. 101). The general project of elaborating an epistemology which rejects in principle the support of a metaphysics is reminiscent of the epistemological system of Kant. Although Bohr disagrees with some Kantian ideas concerning the structure of human knowledge, like the possibility of synthetic *a priori* judgments, he shares with Kant the renunciation of all knowledge of the 'thing-in-itself'.

The well-known proposals of Bohr concerning quantum mechanics follow, for the most part, from his epistemological theses in conjunction with the physical discovery of the quantum of action. The latter prevents the observation of all the properties of a physical object by a single experimental arrangement, or even the combination of all these properties in a single picture. But because of the mobility of the separation between the object and the subject, one can give complementary descriptions to a physical system. The range of possible descriptions is so rich that no experimental predictions can in principle exceed the means of the quantum formalism, and in this sense the formalism is complete. As for the analysis of Einstein, Podolsky, and Rosen, Bohr says:

> Of course there is in a case like that just considered no question of a mechanical disturbance of the system under investigation during the last critical stage of the measuring procedure. But even at this stage there is essentially the question of an influence on the very conditions which define the possible types of predictions regarding the future behavior of the system. Since these conditions constitute an inherent element of the description of any phenomenon to which the term 'physical reality' can be properly attached, we see that the argumentation of the mentioned authors does not justify their conclusion that quantum-mechanical description is essentially incomplete ([2], pp. 60–61).

Bohr is saying essentially that the argument of Einstein, Podolsky, and Rosen is fallacious, because it is founded upon the supposition that we can speak intelligibly of the state of a physical system without reference to an experimental arrangement, which is equivalent to speaking of the 'thing-in-itself'.

There is some good sense in these proposals. In my opinion, Bohr is one of the great phenomenologists of science, showing a rare subtlety concerning the connections between theoretical concepts and experimental procedures. His 'thought experiments', which disentangle phenomena from inessential complications, clearly exhibit this subtlety. Nevertheless, something is missing in his overall interpretation of quantum mechanics. Perhaps he has renounced prematurely and without definitive reasons one of the great projects of Western thought, which is to establish the mutual support between epistemology and metaphysics. Bohr advises us to renounce the explanation of conscious activity, because introspection modifies the mental content which one wishes to examine ([4], pp. 13–14). But we can object that "the explanation of conscious activity" consists in a theory which sets forth principles governing the mind, rather than in a chronicle of mental content. One can see that there is a *threat* of paradox in the acquisition of knowledge of the principles governing the acquisition of knowledge; but all the reasoning that I have seen along this line lacks the force of the well-known set-theoretical and semantical paradoxes which are based upon self-reference.

There is another reason for my skepticism concerning Bohr's renunciation of metaphysics. If this renunciation is presented as a matter of principle, how does it differ from a kind of positivism, according to which the content of an assertion is completely exhausted by its implications for experience? To be sure, in disavowing idealism ([2], pp. 78–79) Bohr probably rejects all kinds of positivism; and moreover, Bohr shows a very strong attachment to the presence of ordinary things, which he does not wish to interpret as packets of sense impressions. But one arrives at a point where Bohr's renunciation of metaphysics begins to appear like an artifice: he wants to avoid the assault launched by positivists on our realistic preconceptions, and at the same time the obligation to examine questions of ontology. One can wonder whether such an artifice will not lead to more obscurity than illumination.

Heisenberg often identifies himself with the Copenhagen interpretation of quantum mechanics ([5], pp. 3 and 8), and he shares a large part of the philosophical theses of Bohr. There are, however, at least some differences of emphasis between them, and perhaps also some more profound differences, which deserve to be pointed out.

One sees an affinity to Bohr, and also to Kant, in the following passage: "what we observe is not nature in itself but nature exposed to our method of questioning" ([5], p. 58). Nevertheless, Heisenberg does not accept as completely as Bohr the Kantian idea of the renunciation of knowledge of the 'thing-in-itself'. Circumspectly and yet significantly he says, "The 'thing-in-itself' is for the atomic physicist, if he uses this concept at all, finally a mathematical structure; but this structure is – contrary to Kant – indirectly deduced from experience" ([5], p. 91). To the extent that he accepts the attribution of the quantum state to the atomic particle in itself ([5], p. 195), he weakens the renunciation of metaphysics, which is one of the hallmarks of Bohr's philosophy.

I shall risk expressing an even stronger opinion: that Heisenberg enunciates a metaphysical implication of quantum mechanics more explicitly than the other pioneers of this science. Quantum mechanics requires, according to Heisenberg, a modality which is situated between logical possibility and actuality, which he calls "potentia" ([5], p. 53). (It should be noted that Margenau ([6], p. 300) used the similar concept of "latency" to characterize the quantum state prior to Heisenberg.) This modality is relevant in considering the question of what happens between two observations, a question to which Heisenberg's answer is "the term 'happens' is restricted to the observation" ([5], p. 52). In spite of this response, Heisenberg does not wish to present himself as a positivist, because according to quantum mechanics the system is characterized between two observations by a quantum state, in other words by a wave function. This state evolves continuously in time in a manner determined by the initial conditions, and because of the independence of this state from the knowledge of any observer it deserves the characterization 'objective'. The quantum state describes nothing actual, but "It contains statements about possibilities or better tendencies ('potentia' in Aristotelian philosophy)" ([5], p. 53). The historical reference should perhaps be dismissed, since quantum mechanical potentiality is completely devoid of teleological significance, which is central to Aristotle's conception. What it has in common with Aristotle's conception is the indefinite character of certain properties of the system. One does not find Aristotle saying, however, that a property becomes definite because of observation and that the probabilities of all possible results are well determined, whereby the quantum mechanical potentialities acquire a mathematical structure. These probabilities, which Heisenberg characterizes as "objective" (*ibid.*), do not result from the ignorance of the observer, as is the case in classical statistical mechanics. The following is a remarkable passage, in which Heisenberg allows himself the use of the metaphysical term "ontology" and indicates the structural complexity of the set of potentialities:

> This concept of 'state' would then form a first definition concerning the ontology of quantum theory. One sees at once that this use of the word 'state', especially the term 'coexistent state', is so different from the usual materialistic ontology that one may doubt whether one is using a convenient terminology. On the other hand, if one considers the word 'state' as describing some potentiality rather than a reality – one may even replace the term 'state' by the term 'potentiality' – then the concept of 'coexistent potentialities' is quite plausible, since one potentiality may involve or overlap with other potentialities ([5], p. 185).

Heisenberg's interpretation of the wave function as a collection of potentialities is based in large part upon a consideration of the interference of amplitudes in the two-slit experiment. He acknowledges the formal success of hidden variables models of de Broglie and Bohm, but he objects to the reality of waves in a configuration space of more than three dimensions ([5], pp. 131–132). Heisenberg's objections are lacking in rigor, but his intuition was correct and was justified by the profound theorems of Gleason [7] and Bell [8, 9, 10]. As a result of their careful work one now knows that for a hidden variables theory to be both free from mathematical contradictions and in agreement with experiment it must have two properties: (1) it must be 'contextualist', that is, the values of quantities must be determined in part by the measuring apparatus, and (2) it must be 'non-local' in the sense of Bell. If one does not find these properties to one's taste (particularly the non-locality, which violates relativistic conceptions of space–time), one is obliged to admit that the wave function gives a complete description to a physical system. Then, if one does not want to renounce metaphysics, there is no other reasonable ontological conception of the wave function than that of Heisenberg.

  Justifying the conception of the wave function as a collection of potentialities leads, however, to another metaphysical problem: how does the transition from potentiality to actuality take place? In other words, how does the reduction of the wave packet occur? It seems to me that Heisenberg offers two solutions, although he does not clearly distinguish them.

The first is essentially that of Bohr. Knowledge requires a separation between the subject and the object, even though the location of this separation is movable. Since the measuring apparatus is situated on the side of the subject, it is described in classical terms, in which one does not find quantum mechanical potentialities ([5], pp. 57–58). To the extent that this solution belongs to the general philosophy of Bohr, it has already been discussed above. It is difficult to see, however, how a strict adherence to the philosophy of Bohr would be compatible with Heisenberg's metaphysical doctrine concerning potentialities.

The second solution is in better agreement with this doctrine. It differs from the first solution in that it applies certain conceptions of quantum mechanics to the measuring process. Heisenberg suggests that the microscopic state of the measuring apparatus is indeterminate as a result of its interaction with the rest of the world ([5], p. 53). At this point he seems to say that the appropriate description of the apparatus ought to make use of a statistical operator (which is equivalent to the density matrix). He suggests further that the final statistical operator of the composite system consisting of atom plus apparatus is diagonal in a certain basis of vectors, each one of which is an eigenvector of a designated observable of the apparatus ([5], pp. 54–55). He hopes to capitalize upon the fact that the initial microscopic state of the apparatus is indefinite in order to arrive at the end of the measuring process at this diagonal statistical operator. If this were so, he could regard the designated apparatus observable as having a definite but unknown value, as in classical physics, and in that case the consciousness of the observer would not be the agent of the reduction of the wave packet: "we may say that the transition from the 'possible' to the 'actual' takes place as soon as the interaction of the object with the measuring device, and thereby with the rest of the world, has come into play; it is not connected with the registration of the result by the mind of the observer" ([5], pp. 54–55). This proposed explanation of the transition from potentiality to actuality is so clear that it is susceptible to being evaluated mathematically. Here is one of those rare cases in which a metaphysical question admits of a mathematical answer, as Leibniz hoped. Unfortunately from Heisenberg's point of view the result of the evaluation is negative. The dynamical law of quantum theory does not permit the statistical operator to evolve in the manner required by Heisenberg's proposed solution – a result first established by Wigner under special conditions [11], and then generalized by d'Espagnat [12], Shimony [13], and others.

To conclude, Heisenberg has drawn from quantum mechanics a profound and radical metaphysical thesis: that the state of a physical object is a collection of potentialities. But his discovery is incomplete, in that the transition from potentiality to actuality remains mysterious. T. S. Eliot has said (à propos of other things) that

> Between the potency
> And the existence
>
> . . . .
>
> Falls the Shadow.   ('The Hollow Men', [14], p. 104.)

I turn now to Schrödinger, who I believe was the most remarkable philosopher among the physicists of our century. I propose to extract from his works three very different groups of remarks. Although there are no contradictions among them, there are some tensions which deserve close study.

(1) The first group of remarks concerns the implications of the quantum mechanical formalism, when it is considered to be an objective description of nature and not just a means for making predictions. The most celebrated remark concerns the cat which is prepared in a superposition of a state of being alive and a state of being dead [15]. Schrödinger accepts – at least provisionally – the interpretation of the wave function as a collection of potentialities, but he insists upon the fact that the dynamical law of quantum mechanics prohibits a transition from potentiality to actuality. It is clear from context that Schrödinger is giving a *reductio ad absurdum* argument. He wishes to signal that the quantum mechanical formalism needs to be changed in some way.

In his comment upon the experiment of Einstein, Podolsky, and Rosen, Schrödinger emphasizes the non-separability of the state of two particles: "they can no longer be described in the same way as before, viz., by endowing each of them with a representative of its own. I would not call that *one* but rather *the* characteristic trait of quantum mechanics" [16]. In this comment he is not attempting a *reductio ad absurdum*. He recognizes that one is concerned with a radical metaphysical thesis, the experimental evidence for which was incomplete at the time of his writing. He asks whether a non-separable state of particles spatially distant from one another is realizable in nature and leaves the answer to this question to experiment. So far as I know he never commented upon the positive answer which Bohm and Aharonov [17] derived from the experiment of Wu and Shaknov [18], while he was still alive.

(2) The second group of Schrödinger's philosophical remarks is his polemic defending realism against a positivist interpretation of science. He grants that postulating material bodies, governed by the laws of physics, achieves order in our experience and produces an "economy of thought"; but he insists that the success of this postulate reveals something important

which goes beyond our conventions: "The fact alone that economy and a successful mental supplementation of experience, in particular extrapolation to the future, are at all possible, presupposes a definite quality of experience: *it can be ordered.* This is a fact that itself demands an explanation" ([20], p. 183).

Schrödinger considers the Copenhagen interpretation of quantum mechanics to be a positivist exercise ([21], pp. 202–205). In his opinion, the principle of complementarity evades the ontological problems posed by quantum mechanics, by insisting primarily upon the mutual exclusiveness of the conditions of different types of observations. By contrast, Schrödinger himself speaks of the physical reality of the quantum mechanical waves: "Something that influences the physical behavior of something else must not in any respect be called less real than the something it influences – whatever meaning we may give to the dangerous epithet 'real'" ([21], p. 198). He grants that up till now no one has constructed a faithful picture of physical reality (*ibid.*, p. 204). Nevertheless, he hopes that the renunciation of the conception of an individual particle endowed with individuality and the recognition of the primacy of waves will guide us towards the desired picture (*ibid.*, pp. 206ff). In any case, Schrödinger is unwilling to abandon his grand vision in favor of the Copenhagen interpretation: an 'either–or' which seems to him too facile ([20], p. 160).

(3) The third group of philosophical remarks which I shall cite is chosen from Schrödinger's speculative writings, which are concerned with appearance and reality, the self, God, and above all the relation between matter and mind. It is noteworthy that he almost never makes use of physics or of his philosophical analysis of physics in dealing with these questions. An explanation for his abstention can be found in the thesis that science is founded upon 'objectivation' – that is, "simplification of the problem of nature by preliminary exclusion of the cognizing subject from the complex of what is to be understood" ([20], p. 183); but the most profound philosophical problems are precisely those concerning the subject which has been excluded from the scientific picture of the world.

Briefly, his principal philosophical theses are the following.

(i) The dichotomy between mind and matter is ultimately artificial, even though it is useful for the conduct of our lives. "The 'real world around us' and 'we ourselves', i.e., our minds, are made up of the same building material, the two consist of the same bricks, as it were, only arranged in a different order – sense perception, memory images, imaginations, thought" ([22], pp. 91–92).

(ii) The difficulty of finding the place of mind in the scientific picture of the world is precisely that mind and matter are composed of the same elements: "To get from the mind-aspect to the matter-aspect or vice versa, we have, as it were, to take the elements asunder and to put them together again in an entirely different order" (*ibid.*, p. 92).

(iii) In spite of the illusion of a multiplicity of subjects in the world, each with its own feelings and thoughts, there is in fact only one Mind. In *What Is Life?* ([19], [20]) – a book which argues powerfully for the reduction of biology to physics and which advocated the idea of the chemical character of the genetic code a decade before Watson and Crick – the epilogue contains only one equation: "ATHMAN = BRAHMAN (the personal self equals the omnipresent, all-comprehending eternal self)."

The tension among the elements of Schrödinger's philosophy which I mentioned earlier ought now to be evident. On the one hand he defends physical realism against a positivist interpretation of science; on the other hand he proposes an idealist metaphysics which recalls that of *The Analysis of Sensations* [23] of the great positivist Ernst Mach, as well as Indian idealism. I have mixed reactions to the cornucopia of philosophical ideas which Schrödinger offers us.

First of all, his criticism of a positivist interpretation of science is excellent. It may not have been entirely fair to classify the Copenhagen interpretation as positivist, especially since both Bohr and Heisenberg reject this epithet for their views. However, the analysis made in the first part of this paper showed that the philosophical positions of Bohr and Heisenberg are, so to speak, 'metastable', and positivism is one of the stable states into which they could fall. Consequently, Schrödinger's criticisms are relevant, even if one has some doubts about his exegesis of their texts.

I share his hope for a completely physical solution to the problems of quantum mechanics, notably the problem of the reduction of the wave packet and the problem of non-locality. It is necessary to recognize a level of description on which physical discourse is appropriate, even if the fundamental ontology of the universe is idealist. In the language of Schrödinger, the world is susceptible to the operation of 'objectivation'. For the most part, the processes described by quantum mechanics do not go beyond this physical level. The apparatus by which the typically quantum phenomena are exhibited – interferometers, spectrometers, coincidence counters, etc. – belong as much to the realm of matter as any ordinary object. The essential question is whether the 'interphenomena' (Reichenbach's term) have a character as physical as that of the apparatus. Schrödinger agrees with Bohr and Heisenberg that no variant of classical physics can describe the 'interphenomena', but he insists, in strong opposition to them, that the purely physical means for achieving this description are far from being exhausted.

A little-explored line of research is to replace the usual linear law governing the evolution of the wave function (that is, the time-dependent

Schrödinger equation) by a non-linear law. This replacement would be consistent with the ontological primacy of waves, which Schrödinger never abandoned. There is even some textual indication ([24], p. 451) that Schrödinger considered the possibility of a non-linear modification of quantum dynamics, although he seems never to have made a specific suggestion. I find this line of research attractive, but must make two negative comments. The first is that this line of research at best promises to resolve the problem of the reduction of the wave packet and offers nothing concerning the problem of non-locality. The other is that the experimental results recently obtained [25] do not support the conjecture of non-linear dynamics.

Another line of research which does not go beyond the physical level is the study of relations between quantum mechanics and space–time structure. So far as I know, Schrödinger never made a suggestion along these lines, even though he was a profound investigator of space–time structure. There are, however, two reasons for taking seriously this line of research. One is the difficulty of applying the procedures of quantization to space–time itself. The other is the non-locality (in the sense of Bell) of certain experimental predictions corroborating quantum mechanics. I spoke elsewhere [26] of the possibility of "peaceful coexistence" of the non-locality of quantum mechanics and the relativistic structure of space–time, a possibility which is suggested by the fact that one cannot make use of quantum non-locality for the purpose of sending a message instantaneously. If, however, this peaceful coexistence does not succeed, then it would be necessary to take the radical step of postulating a modification of relativistic space–time structure.

It seems to me possible that all attempts to explain the reduction of the wave packet in a purely physical way will fail. There would then remain only one type of explanation of the transition from quantum mechanical potentiality to actuality: the intervention of the mind. I wish to emphasize that in my opinion it is very improbable that we shall be pushed to this extremity. Nevertheless, I think that Schrödinger was wrong in excluding this possibility *a priori*. Perhaps physical evidence will exhibit to us new restrictions upon the operation of objectivation (to use Schrödinger's own terminology) and will reveal some imperfections in the physical level – some fissures, so to speak, through which the essentially mental character of the world reveals itself. There are many people who embrace with enthusiasm the thesis of the indispensability of the mind for the reduction of the wave packet, a thesis to which I circumspectly allow only the possibility that it may be true. I allude particularly to the authors of a collection of parapsychological articles entitled *The Iceland Papers* [27]. Before the thesis in question could attain a status above pure speculation, it is essential to have careful experiments which are capable of repetition. I doubt that such experiments have already been carried out. One might say that my doubts indicate conservatism or even conformity, but to this accusation I offer the following response. With the help of three students I attempted to transmit a message by means of the reduction of the wave packet, an attempt which should have succeeded had the thesis in question together with an auxiliary hypothesis been true. Our result [28] was negative, and it presents an obstacle to the thesis of the indispensability of the mind for the reduction of the wave packet, although it is far from being definitive.

Returning to Schrödinger, I evidently cannot do justice to his metaphysics in a few pages. I wish to indicate, however, the possibility of formulating an idealist metaphysics which remedies some of the imperfections of his own. I find his thesis that the mind cannot be included in a scientific picture of the world quite unconvincing. The science of psychology – by which I mean the study of thought, sensations, and feelings, and not merely the study of behavior – has made enough progress to cast doubt upon this thesis. And Schrödinger's doctrine of a single Mind is difficult to reconcile with the immense body of evidence concerning private sorrows, hidden hopes, and secret conspiracies. He discusses briefly ([29], pp. 94–95) only one example of a pluralistic idealism, namely the monadology of Leibniz, which he dismisses because the monads are windowless and therefore cannot account for language and other communication. So far as I know, Schrödinger never mentions the pluralistic idealism of Whitehead, according to which the monads are endowed with windows, so to speak, since one of them can contribute to the sensations of another. Whitehead supposes that the instantaneous state of an elementary particle must be characterized in mental terms, like 'feeling', even though the sense of these terms must be extrapolated far beyond their normal usage. His great design is to integrate physics into a generalized psychology, as Maxwell integrated optics into electromagnetic theory. Whitehead rejects the above-cited thesis of Schrödinger that it is necessary "to take the elements asunder and to put them together again in an entirely different order" in order to relate the mind-aspect of the world to the matter-aspect. Whitehead rather regards the matter-aspect as an abridged version of the mind-aspect. "The notion of physical energy, which is at the base of physics, must then be conceived as an abstraction from the complex energy, emotional and purposeful, inherent in the subjective form of the final synthesis in which each occasion completes itself" ([30], p. 188).

I do not wish to deny the obscurity of Whitehead's exposition. Despite its obscurity, however, it offers a possibility which Schrödinger has denied: the possibility of integrating the mind into a scientific picture of the world.

In conclusion, I would like to make one additional speculation. Perhaps the great metaphysical implications of quantum mechanics – namely, nonseparability and the role of potentiality – have made the unification of physics and psychology somewhat less remote. Perhaps we are confronted with structural principles, which are applicable as much to psychological as to physical phenomena. If this should turn out to be the case, then the physical discoveries of Schrödinger would be more closely connected with his metaphysical preoccupations than he himself recognized.**

REFERENCES

[1] Bergmann, P. 1949. *Basic Theories of Physics* 1. New York: Prentice-Hall.
[2] Bohr, N. 1958. *Atomic Physics and Human Knowledge*. New York: Wiley.
[3] Petersen, A. 1968. *Quantum Physics and the Philosophical Tradition*. Cambridge, Mass.: M.I.T. Press.
[4] Bohr, N. 1966. *Essays 1958–1962 on Atomic Physics and Human Knowledge*. New York: Vintage.
[5] Heisenberg, W. 1962. *Physics and Philosophy*. New York: Harper.
[6] Margenau, H. 1949. *Philosophy of Science* 16, 287.
[7] Gleason, A. 1957. *Journal of Mathematics and Mechanics* 6, 885.
[8] Bell, J. S. 1964. *Physics* 1, 195.
[9] Bell, J. S. 1966. *Reviews of Modern Physics* 38, 447.
[10] Bell, J. S. 1971. In *Foundations of Quantum Mechanics,* ed. B. d'Espagnat. New York: Academic Press.
[11] Wigner, E. 1963. *American Journal of Physics* 31, 6.
[12] d'Espagnat, B. 1966. *Supplemento al Nuovo Cimento* 4, 828.
[13] Shimony, A. 1974. *Physical Review* D9, 2321.
[14] Eliot, T. S. 1936. *Collected Poems 1909–1935*. New York: Harcourt, Brace and Co.
[15] Schrödinger, E. 1935. *Naturwissenschaften* 23, 807, 823, 844.
[16] Schrödinger, E. 1935. *Proceedings of the Cambridge Philosophical Society* 31, 555.
[17] Bohm, D. and Aharonov, Y. 1957. *Physical Review* 108, 1070.
[18] Wu, C. S. and Shaknov, I. 1950. *Physical Review* 77, 136.
[19] Schrödinger, E. 1967. *What Is Life?* and *Mind and Matter*. Cambridge: Cambridge University Press.
[20] Schrödinger, E. 1956. *What Is Life? and Other Scientific Essays*. Garden City, N.Y.: Doubleday.
[21] Schrödinger, E. 1957. *Science, Theory, and Man*. New York: Dover.
[22] Schrödinger, E. 1954. *Nature and the Greeks*. Cambridge: Cambridge University Press.

** The research on which the paper is based was supported in part by the National Science Foundation, Grant no. SOC-7908987. I wish to thank Dr. Andrew Frenkel for his helpful suggestions.

[23] Mach, E. 1959. *The Analysis of Sensations.* New York: Dover.
[24] Schrödinger, E. 1936. *Proceedings of the Cambridge Philosophical Society* **32**, 446.
[25] Shull, C. G., Atwood, D. K., Arthur, J., and Horne, M. A. 1980. *Physical Review Letters* **44**, 765.
[26] Shimony, A. 1978. *International Philosophical Quarterly* **18**, 3.
[27] Puharich, A. (ed.). 1979. *The Iceland Papers.* Amherst, WI.: Essentia Research.
[28] Hall, J., Kim, C., McElroy, B., and Shimony, A. 1977. *Foundations of Physics* **7**, 759.
[29] Schrödinger, E. 1964. *My View of the World.* Cambridge: Cambridge University Press.
[30] Whitehead, A. N. 1965. *Adventures of Ideas.* New York: New American Library of World Literature. Originally published by the Macmillan Company, New York, 1933.

21

Wave-packet reduction as a medium of communication[1]

Joseph Hall,[2] Christopher Kim,[2] Brien McElroy,[2]
and Abner Shimony

> – I close these expositions . . . concerning the interpretation of quan-
> tum theory with the reproduction of a brief conversation which I had
> with an important theoretical physicist. He: "I am inclined to believe
> in telepathy." I: "This has probably more to do with physics than
> with psychology." He: "Yes" –
>
> A. Einstein[1]

Using an apparatus in which two scalers register decays from a radioactive source,
an observer located near one of the scalers attempted to convey a message to an
observer located near the other one by choosing to look or to refrain from looking at his scaler. The results indicate that no message was conveyed. Doubt is
thereby thrown upon the hypothesis that the reduction of the wave packet is due
to the interaction of the physical apparatus with the psyche of an observer.

I. INTRODUCTION

According to many of the writers on the foundations of quantum mechanics,[3] there is a genuine and unsolved problem of the "reduction of
the wave packet." Briefly, and with some simplifications, the problem can
be posed as follows. Consider a physical system S_I, such as an electron,
the states of which are represented in the standard quantum mechanical
manner[4] by vectors in an appropriate space, and suppose that in states
represented by vectors $u_1, u_2, \ldots$ the observable property $\mathcal{F}$ of S_I has the
distinct values $f_1, f_2, \ldots$. Suppose also that a physical system S_{II} is a
suitable apparatus for measuring $\mathcal{F}$ of S_I: i.e., if S_{II} is prepared in the
right kind of initial state and allowed to interact with S_I, which has been

This work originally appeared in *Foundations of Physics* 7 (1977), pp. 759–67. Reprinted
by permission of Plenum Publishing Corp.
1. Aided in part by National Science Foundation support to one of the authors (A.S.).
2. Physics Department, Boston University, Boston, Massachusetts (undergraduates).
3. See, for example, Schrödinger,[2] Wigner,[3] Furry,[4] and d'Espagnat.[5]
4. See, for example, von Neumann[6] or any standard textbook of quantum mechanics.

prepared in u_i, then after a certain time the composite system S_I-plus-S_{II} will enter a final state (represented by the vector η_i in an appropriate space), in which the apparatus observable $\mathcal{G}$ has the definite value g_i (where $g_i \neq g_j$ for $i \neq j$). If S_I was initially in a definite but unknown u_i, and therefore had at that time a definite value f_i of the observable $\mathcal{F}$, then a reading of $\mathcal{G}$ at the final time permits an inference of the initial value f_i. Thus, one sees schematically how information about properties of a microphysical system can be extracted by examining a macroscopic apparatus. A difficulty arises, however, if initially the state of S_I is a superposition of states with different values of the observable $\mathcal{F}$. Such a superposed state is represented by the "wave packet" $\sum c_i u_i$, where more than one of the coefficients c_i are nonzero. It then follows from the linearity of the quantum mechanical time-evolution operator that the final state of S_I-plus-S_{II} is represented by the wave packet $\sum c_i \eta_i$, with the same coefficients c_i. In this final state the apparatus observable $\mathcal{G}$ does not have a definite value, and no objective outcome of the measuring process has occurred. Apparently, then, no reduction of the wave packet has occurred during the physical process of measurement, and there is no agreement among physicists as to how or when or even whether the reduction occurs.

Although we concur that there is a genuine problem of the reduction of the wave packet, we do not intend in our paper to defend this opinion against those who maintain that it is a pseudoproblem.[5] Rather, we wish to focus attention upon one of the most radical proposals made by those who take the problem seriously: that the reduction of the wave packet *is a physical event which occurs only when there is an interaction between the physical measuring apparatus and the psyche of some observer.*[6] As

5. Many of the writings of Bohr present the view that there is no problem of the reduction of wave packets, but it is hard to point to one locus where his opinion is expressed with exemplary explicitness. Perhaps the best single reference is Ref. 7. See also von Weizsäcker[8] and Groenewold.[9] Criticisms of this view can be found in d'Espagnat,[5] Shimony,[10] Stein,[11] and Hooker.[12]

6. The most explicit statement which we have been able to find of this radical proposal is the following by Costa de Beauregard[13]: "Nous postulons que *l'événement aléatoire individuel,* appelé collapse du ψ, implique un *acte de conscience de l'observateur.* Ceci entraîne que la *probabilité essentielle* inhérente aux Quanta n'est ni objective, ni subjective, mais *indissolublement objective et subjective,* étant le nœud qui lie psychisme et réalité." See also Ref. 14, especially pp. 549, 551, 556, and 557. The same point of view has been attributed (in Refs. 10 and 14 and elsewhere) to London and Bauer,[15] because of the well-known passage: "Ce n'est donc pas une interaction mystérieuse entre l'appareil et l'objet qui produit pendant la mesure un nouveau ψ du système. C'est seulement la conscience d'un 'Moi' qui peut se séparer de la fonction $\Psi(x, y, z)$ ancienne et constituer en vertu de son observation une nouvelle objectivité en attribuant dorénavant à l'objet une nouvelle fonction $\psi(x) = u_k(x)$." In view of London's philosophical training as a student of Husserl, however, we now are inclined to believe that the

we understand this proposal, it presupposes a dualistic ontology, according to which there are both physical and mental entities in nature, and these not only coexist but interact. According to the proposal, the normal temporal evolution of the state of a physical system is governed by the time-dependent Schrödinger equation, which makes the time-evolution operator linear and thereby precludes the reduction of the wave packet. Interaction of the measuring apparatus with the psyche of the observer, however, requires an entirely different dynamical principle. As in the dualistic ontology of Descartes, the temporal evolution of a material system can be modified by interaction with a mind, and the reduction of the mind packet, according to the proposal, is a modern analog of Descartes' action of the soul upon the animal spirits of the body at the pineal gland.

Since the proposal is very radical, an explanation of the propriety of giving serious attention to it is in order. One reason is that if the reduction of the wave packet does pose a genuine problem, then it is a very hard one, and some physicists and philosophers have come to believe that no easy, nonradical solution will succeed. Since the mind–body problem is a perennial unsolved problem (which classical physics somehow managed to by-pass without solving), one could conjecture that the two problems are intermeshed. A second reason is methodological. We feel that there is value in exhibiting concretely how a radical hypothesis can be subjected to careful experimental scrutiny. The thesis that the scientific method combines openness to theoretical innovation with a critical insistence upon experimental test is generally accepted, but exemplifications of the thesis are uncommon and instructive.

The experiment which we report below is based upon taking seriously the proposal that the reduction of the wave packet is due to a mind–body interaction, in which both of the interacting systems are changed. It follows that if two observers interact with the same apparatus, then it may be possible in principle for one observer to send a message to the other by means of the wave packet reduction. Successful communication in this manner would constitute a spectacular confirmation of the proposal under consideration. On the other hand, the failure to communicate in a situation designed to be favorable to communication would be prima facie evidence against the proposal, and attempts to explain away the failure would have to be examined very critically.

attribution is incorrect and that the passage quoted should be given a phenomenological interpretation. We also believe that it would not be correct to attribute a dualistic ontology to Wigner, since in his most explicit statement [16] he has asserted that the content of the consciousness of the ultimate subject is the only "absolute" reality.

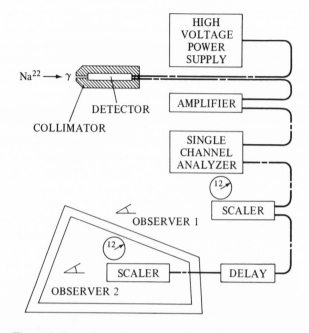

Figure 1. Experimental arrangement (not drawn to scale).

2. THE CONCEPTION AND DESIGN OF THE EXPERIMENT

A sample of sodium-22 atoms, which are gamma emitters, is taken to be the system S_I. The system S_{II} is a complex apparatus, consisting essentially of a sodium iodide scintillation detector, a series of amplifiers for amplifying pulses from the detector, and two scalers in separate rooms on each of which all detection events are registered (see Figure 1). In the quantum mechanical theory of gamma emission from a single atom a nucleus is regarded as being in a superposition of an "undecayed" state and a "decayed" state, and the absolute square of the coefficient of the latter determines the probability of the decay. Our S_I is a many-particle system, but it also can be regarded as being in a superposition of an "undecayed" and a "decayed" state. According to the scheme of the measuring process sketched in Section 1, after an appropriate time interval from the initial interaction of S_I with S_{II}, the composite system is in a superposition of two states, one in which a detection of an emitted photon has occurred and one in which it has not. The occurrence and nonoccurrence of a detection event are here taken to be the values of the macroscopic observable $\mathcal{G}$. Now, according to the radical proposal under consideration, a reduction of this superposed state of S_I-plus-S_{II} occurs only when an observer looks at one of the two scalers of the apparatus S_{II}. Either of the two scalers can be used for this purpose, and if there are two observers A and B, one near each scaler, then presumably the first one to look is responsible for the reduction of the wave packet. (We abstain from considering complications, such as the possibility that A looks first, but in an absent-minded manner, so that B is the first to become fully conscious of a reading.)

Now suppose that A exercises an option to look or to refrain from looking at his scaler, but if he looks he does so before B. *If A's looking actually changes the physical state of S_I-plus-S_{II}, then possibly B's experience when he looks is different according to whether A has or has not looked.* In other words, B's experience if he is presented with a wave packet which is already reduced (either to the "decayed" or to the "undecayed" state) would be different from his experience if he is responsible for the reduction of the packet. It must be emphasized that the proposition that B's experience differs in the two cases is not simply a consequence of the proposal that the psyche of an observer is responsible for the reduction of the wave packet, but is an additional hypothesis. If this additional hypothesis is true, then it is possible for A to send a message to B by choosing to look or not to look at his scaler. One "bit" of information can be communicated in this way if B can determine with certainty whether he is or is not responsible for the reduction, but less than a "bit" can be communicated if B cannot do so with certainty (i.e., "the channel is noisy"). If A keeps a record of when he looks and when he refrains from looking on successive occasions, and B keeps a record of when he feels that he is responsible for the reduction of the wave packet and when not, then a comparison of the two records should show whether A has communicated with B.

3. EXPERIMENTAL PROCEDURE

A collimator sleeve and over 2 in. of lead shielding were used to cut down background radiation and to slow the detection rate. The amplifiers and multichannel analyzers were adjusted so as to accept only well-separated high-energy pulses. As a result, the counting rate was slowed to approximately one event per 30 sec. In order to have A make his observation before B, in case both are looking at their scalers during the same time interval, a time delay of 1 μsec was inserted into the circuit to B's scaler. Observer A was provided with a random number list, and every $\frac{1}{2}$ min he decided by consulting it whether to initiate a 15-sec period of observing his scaler. He recorded whether he looked, and in the event that he looked

Table I. *Summary of experimental data:*
Numbers of decisions and of correct
decisions in individual runs and in sum

	Total decisions	Correct decisions	Percent correct
1	78	42	53.8
2	89	40	44.9
3	84	39	46.4
4	38	21	55.3
5	52	23	44.2
6	54	31	57.4
7	47	24	51.1
8	61	29	47.5
9	51	28	54.9
Composite	554	277	50

he also recorded the numbers indicated on his scaler. *B* observed every event of registration on his scaler, and he decided case by case, by introspection upon the quality of his experience, whether the wave packet had already been reduced (noted on his record by "O") or not (noted by "F"). He could also record ignoring a detection (noted by "X"), if he did not feel confident about his decision or if pulses arrived so closely together that there would be low probability that *A* could observe one without observing the other. To eliminate unwanted transmission of information, the scalers employed were soundless. Before actually taking data *B* was "trained" by observing several events in a row which *A* had already observed, and then several more which *A* refrained from observing. Three different experimenters (J.H., C.K., and B.M.) took turns at serving as observers *A* and *B*r

4. RESULTS

Table I presents the results of nine runs. The column headed "Total decisions" gives the number of "O" and "F" decisions made in each run, neglecting the events noted by "X." The column headed "Correct decisions" gives the number of agreements between *B*'s decision and *A*'s choice of observing or refraining from observing. The data are in excellent agreement with the hypothesis that there is probability one-half of a correct decision in each event, with stochastic independence of correct decisions on distinct events. According to this hypothesis, which implies a binomial distribution of correct decisions, the expected value of the total number of correct decisions is 277, and the standard deviation is $[\frac{1}{2}(1-\frac{1}{2})554]^{1/2} =$ 11.8. The total number of correct decisions is exactly 277, despite moderate deviations in individual runs. We conclude that almost certainly there was no communication between A and B.

5. DISCUSSION

As indicated above, one would expect communication from A to B in our experiment if both of two propositions are true: (i) That interaction of the psyche of an observer with the physical apparatus is responsible for the reduction of the wave packet, and (ii) that there is a phenomenological difference between making an observation which is responsible for the reduction of a wave packet and making one that is not. A defender of proposition (i) may explain the negative result of our experiment by denying proposition (ii). This explanation is not convincing, however, without some account of the mind–body interaction which would make it plausible that the psyche can be causally efficacious upon the wave function of a physical system and yet be insensitive to certain gross differences among wave functions. Someone may contend that both (i) and (ii) are true, but that observer B was not properly trained to be sensitive to the differences between the reduced and the unreduced wave packet. Or, he might contend that the μsec delay of the pulse to B's scaler does not suffice for A to be unequivocally responsible for the reduction of the wave packet in case both of them make observations.[7] Any one who tries to explain our negative result in either of these two ways, or by other criticisms of the details of our experimental procedure, is urged to devise a variant experiment which will be free from the flaws attributed to ours.

We wish to note that a version of our experiment was performed once in the past. Ron Smith, a student of one of us (A.S.), performed an experiment at MIT in 1968 using a cobalt-57 source.[17] His observer B made 67 decisions, of which 40 were correct. This result disagrees sharply with ours. His sample of data, however, is much smaller than ours, and he

---

7. Even if A observed prior to B only in half of the cases in which A decides to look at his scaler, we still should expect that B's record would agree with A's in more than half the cases. Indeed, if B could tell with certainty whether he is responsible for reducing the wave packet, and if A chooses to look in one half of the cases and looks before B in half of these, then one would expect agreement between the records of B and A in three-quarters of the cases. A further experiment is needed in which A's observations will unequivocally precede B's whenever the former chooses to observe. In the new experiment Costa de Beauregard's suggestion in Refs. 13 and 14 that a factor of willing is involved in communication can be taken into account, by providing observers A and B with positive motivation to transmit and receive a message.

did not, to our knowledge, make additional observations to check his remarkable initial run.

Finally, we wish to mention a comment made by C. F. von Weizsäcker to a talk in which the present experiment was proposed: "It is a very interesting experiment in telepathy, whose possibility I do not object to. But even if positive it only shows that A knows what B had done, not that A has reduced the wave packet." In reply let us suppose that results like those of Smith rather than like ours are obtained in subsequent refinements of the experiment, but suppose also that attempts at telepathic communication between A and B without using the systems S_I and S_{II} as auxiliaries consistently fail. Then it would be most unreasonable to claim that the instances of successful communication between A and B have nothing to do with reduction of wave packets.

Acknowledgments. We wish to thank Prof. Bernard Chasan, Dr. Bohdan Balko, and Mr. Andre Mirabelli for their valuable assistance.

REFERENCES

[1] A. Einstein, in *Albert Einstein: Philosopher–Scientist,* P. A. Schilpp, ed. (The Library of Living Philosophers, Evanston, Illinois, 1949), p. 683.

[2] E. Schrödinger, *Naturwiss.* **23**, 807, 823, 844 (1935).

[3] E. P. Wigner, *Am. J. Phys.* **31**, 6 (1963).

[4] W. H. Furry, Some Aspects of the Quantum Theory of Measurement, in *Lectures in Theoretical Physics,* Vol. VIII-A, W. Britten, ed. (Univ. of Colorado Press, Boulder, Colorado, 1965).

[5] B. d'Espagnat, *Conceptual Foundations of Quantum Mechanics* (Benjamin, Menlo Park, 1971).

[6] J. von Neumann, *Mathematical Foundations of Quantum Mechanics* (Princeton Univ. Press, Princeton, New Jersey, 1955).

[7] N. Bohr, On the Notions of Complementarity and Causality, *Dialectica* **2**, 312 (1948); reprinted in N. Bohr, *Essays, 1958–1962, on Atomic Physics and Human Knowledge* (Wiley, London, 1963).

[8] C. F. von Weizsäcker, The Copenhagen Interpretation, in *Quantum Theory and Beyond,* T. Bastin, ed. (University Press, Cambridge, 1971), p. 25.

[9] H. J. Groenewold, *Proc. Amsterdam Academy* (*B*) **55**, 219 (1952).

[10] A. Shimony, *Am. J. Phys.* **31**, 755 (1963).

[11] H. Stein, On the Conceptual Structure of Quantum Mechanics, in *Paradigms and Paradoxes: The Philosophical Challenge of the Quantum Domain,* Vol. 5 in the University of Pittsburgh Series in the Philosophy of Science, R. Colodny, ed. (Univ. of Pittsburgh Press, Pittsburgh, Pennsylvania, 1972).

[12] C. A. Hooker, The Nature of Quantum Mechanical Reality: Einstein versus Bohr, in *Paradigms and Paradoxes,* R. Colodny, ed. (Univ. of Pittsburgh Press, Pittsburgh, Pennsylvania, 1972).

[13] O. Costa de Beauregard, Du Paradoxe au Paradigme, *Epistemological Letters of the Institut de la Méthode,* 2nd issue (May 1974).

[14] O. Costa de Beauregard, *Found. Phys.* **6**, 539 (1976).

[15] F. London and E. Bauer, *La Théorie de l'Observation en Mécanique Quantique* (Hermann, Paris, 1939).

[16] E. P. Wigner, Two Kinds of Reality, in *Symmetries and Reflections* (Indiana Univ. Press, Bloomington, 1967).

[17] Ron Smith, unpublished, MIT (1968).

COMMENT

The experiment reported in this paper is flawed in two respects. First, the delay of 1 microsecond inserted into the circuit to *B*'s scaler is much too short and was accepted only because a longer delay would have required considerably more elaborate electronics. If the experiment is performed again, the delay time should be of the order of psychologically discriminable time intervals, in order to ensure that *B*'s observation is definitely later than *A*'s. Second, there was insufficient training of the observers to discriminate introspectively between the case of a reduced wave packet and one which has not been reduced. Olivier Costa de Beauregard was particularly critical of this feature of the experiment and suggested much longer training periods and possibly selection of sensitive observers (upon the assumption that there is a difference to be sensitive to!). The basic design of the experiment nevertheless appears to be sound, and both of these flaws can be eliminated by any group that is sufficiently motivated to aim at a decisive experiment.

Index